OUR WORLD
TRANSFORMED

READER'S DIGEST

OUR WORLD TRANSFORMED

ANSWERS TO 1,001 QUESTIONS ON WHY AND HOW TODAY'S WORLD IS CHANGING AT A PACE THAT LEAVES US BREATHLESS.

Published by The Reader's Digest Association Limited
LONDON • NEW YORK • SYDNEY • MONTREAL

About this book

As soon as children learn to speak, the why, what and how questions start: Why do stars shine? Why is the sky blue? Why do some people have freckles? Why can't I remember being born? Who invented writing? How do snakes climb trees? As adults, we often don't know the answers to these obvious, but classic, questions about the mysteries of nature, the human body, the universe and the science of everyday life.

Furthermore, researchers are constantly discovering new phenomena, engineers are developing new gadgets and our knowledge of the world keeps on expanding. What is a nano-taxi? Why do eggs burst in a microwave oven? What is a knockout mouse? What are the global effects of climate change?

Understanding the world around us can be great fun, and this book will bring you up to date with the latest findings as it guides you through the intriguing answers to 1,001 questions about science.

The Editors

OUR MYSTERIOUS UNIVERSE

THE MIRACLE THAT IS A HUMAN BEING

MYSTERIOUS NATURE

OUR BLUE PLANET

EXPLORING THE EVERYDAY

MARVELS OF SCIENCE AND TECHNOLOGY

Our mysterious
universe

What mysteries will be revealed as humans turn their attention to the vastness of space? Will our robot space probes find traces of life on other planets and moons? Will ever more powerful telescopes provide answers to the origins of the cosmos? Will the huge radio aerials now scanning the skies find evidence of extraterrestrial civilisations? Our planet Earth is just one tiny speck in the infinity of space. It seems inconceivable that there should not be life anywhere else.

Is the universe finite?

So far, astronomers still cannot say for certain. Most believe it is infinite, but even if it were finite, the universe would not have any boundary. It would curve in on itself so that – to put it simply – it would be shaped like a ball. And just like a ball the universe would have no spatial limits, but its volume would be finite.

The Andromeda Galaxy is one of our galaxy's larger neighbours. At 220 000 light-years in diameter, this island of stars is twice as big as our Milky Way galaxy and is thought to contain around 1 trillion stars.

How far can we see into space?

The most distant object visible to the naked eye is the Andromeda Galaxy, some 2.5 million light-years away from Earth. It is a spiral galaxy similar to the Milky Way – the galaxy in which we live. While the Andromeda Galaxy may seem very remote, using modern telescopes astronomers can see galaxies that are an incredible 13 billion light-years away. At this distance, observers come up against an insurmountable barrier, because every time we look into space we are simultaneously looking back into the history of the cosmos. When we look at a galaxy 13 billion light-years away, we are seeing it as it looked when light from it left on its journey, 13 billion years ago. At that time – close to the origin of the universe – the first galaxies had only just been formed. Before that, there was nothing to see – which means that we cannot see any deeper into space.

How long is a light-year?
Although statements like 'it took light-years' are frequently bandied about, the unit of measurement 'light-year' does not measure a period of time, but instead denotes a very great distance. A light-year is how far light travels in one year. Since light moves at a speed of around 300 000 km/sec., this is about 10 trillion kilometres. Given the huge scale of the universe, a light-year is an ideal unit for measuring the distances between the stars and galaxies.

What qualifies as a 'large' telescope?
When scientists talk about large telescopes, they are not referring to the height or width of the instrument, but to the diameter of the lens or mirror the telescope uses to collect light. Today, all the world's largest optical

telescopes are equipped with curved mirrors as light collectors. When a telescope is referred to as a 5m instrument, that means that its primary mirror is 5m in diameter. The world's largest optical telescopes are the twin 10m Keck Telescopes in Hawaii. The dishes used in radio astronomy are considerably larger. At 300m, the Arecibo radio telescope in Puerto Rico is the current world record holder.

How is distance measured in space?

Many different methods are used. Apollo 11 astronauts left a laser reflector on the Moon, making it possible to measure to the centimetre the precise distance between the Moon and the Earth.

The distances of other bodies in our solar system, such as the planets and asteroids, can be measured by radar – that is, by reflecting radio waves off them.

This method will not work with stars, which are much further away. To calculate these distances, astronomers use a technique that measures tiny differences in the position of a star when it is observed from opposite sides of the Earth's orbit, that is at

intervals six months apart. Since the diameter of the Earth's orbit is known, the change in angular position of a star at either side of the orbit can be used to calculate its distance. Known as the 'parallax method', this works for objects at distances of up to about 1000 light-years.

For still more distant objects, such as galaxies, astronomers must use indirect methods. One of these involves Cepheid Variables, a type of star that has the convenient property of varying in brightness at regular intervals. We know that the length of the intervals depends on a star's luminosity, so that by measuring a Cepheid Variable's brightness it is possible to calculate how far away it – and the galaxy it belongs to – is from us.

The Keck Telescopes stand at 4200m on the extinct Mauna Kea volcano in Hawaii. Here in the clean air, away from the light pollution of the big cities, astronomers can get a clear view of the cosmos.

How big is the universe?

According to the latest findings, the universe came into being about 13.7 billion years ago. In theory, we cannot receive any radiation more than 13.7 billion years old. Or, to put it another way, no light that reaches us from deep space can have travelled from more than 13.7 billion light-years away. Of course, this does not mean that the universe comes to an end at this distance. The universe could be very much larger than the area we are able to observe through our telescopes, and it could even extend to infinity.

Ever since the universe was created by a mighty explosion – the Big Bang – some 13.7 billion years ago, it has been expanding. The big question astronomers have long asked is whether the expansion will continue for ever, or whether at some time in the future the universe will contract back on itself.

Does the universe have a centre?

When it was discovered that nearly all galaxies are moving away from us, scientists wondered whether the Milky Way might be situated at the centre of the universe. Plausible as the idea may seem, it is incorrect. In reality, galaxies scarcely move, it is the space between them that stretches – just like the surface of a balloon when it is inflated. Every point moves away from every other point, but no one point is the central point on the balloon's surface. This would also be true if the universe were not curved like the balloon's surface but infinite. Although we may find this hard to imagine, even a universe of this kind would have no centre.

Are there also galaxies racing towards us?

Although the universe as a whole is expanding, there are still galaxies moving in our direction – for example, the nearby Andromeda Galaxy, which is travelling straight towards us at a rate of about 300km/sec.

To understand this apparent contradiction we have to differentiate between two phenomena: on the one hand the motion of galaxies through space, on the other hand the expansion of the cosmos itself. Random motions of galaxies in space are typically in the order of a few hundred kilometres per second. For galaxies that are very close to our Milky Way galaxy, the expansion is negligible and random motions dominate. For more distant galaxies, random motion is negligible compared to that of expansion. So all distant galaxies seem to be rushing away from us as a result of expansion, but nearby galaxies can be moving either towards us or away from us.

Will the universe keep on expanding for ever?

In theory, there are three possibilities as to what may happen to the universe in the future. The expansion of the cosmos could be held back by gravity, come to a complete standstill and then go into reverse. The

Edwin Hubble
(1899–1953)
proved in the 1920s
that the universe
contained other
galaxies besides our
Milky Way. His later
observations were
decisive in proving
that it was space
itself that
was expanding.

...that the universe is expanding?

In 1923, Vesto Slipher of the Lowell Observatory in Arizona discovered that, of the 41 nearest spiral galaxies, 36 were moving away from us. When other astronomers learnt of Slipher's measurements, they found it hard to believe that our Milky Way just happened to be located at the centre of the universe. For this reason, the German astronomer Carl Wirtz suggested that it wasn't galaxies that moved, but rather that it was space itself that was expanding. Soon afterwards, the cosmologist Howard Robertson showed, on the basis of Slipher's calculations and those of others as well as precise distance data produced by astronomer Edwin P Hubble, that more distant galaxies moved away from us at a faster rate than closer ones – just what we might expect to see if space was expanding.

Hubble backed up these findings a year later with additional measurements – and, in the end, took all the credit. Robertson had made the mistake of publishing his results in an obscure scientific journal for physicists and not, as Hubble did, in an astronomical publication.

universe would then collapse in upon itself causing a Big Crunch. The second possibility is that expansion could slow down but continue, in which case it would take an infinite amount of time to come to a standstill. The third possibility is that expansion will accelerate over time. Until 15 years ago astronomers were convinced that one of the first two scenarios was most likely. Then they made the unexpected discovery that stellar explosions in distant galaxies proved that expansion was increasing. Researchers don't yet know why this should be the case, but it seems that expansion is driven by a mysterious 'dark energy' that fills our cosmos.

As is happening with these spiral galaxies, the Milky Way and the Andromeda Galaxy will collide some time in the distant future.

Will the universe eventually become uninhabitable?

How long the universe remains habitable depends on whether it shrinks or expands. Should it eventually shrink, ending in a Big Crunch, this would end all forms of life.

But even in a cosmos that goes on expanding for all eternity – which is what seems likely if recent findings showing that expansion is accelerating prove to be correct – it seems certain that a time will come when life ceases. In an ever-expanding universe, energy and matter will increasingly thin out so that there will eventually be barely enough energy to sustain life of any kind. Life processes could slow down to counter this but, at some point life reaches a limit beyond which it cannot go on. All forms of life must be able to radiate energy, and that is only possible when the environment is cooler than the life form itself. There is a lower limit for the temperature of the universe. Although it is only an unimaginably tiny fraction of a degree above the lowest possible temperature (absolute zero, or -273.15°C), it is sufficient to deal the fatal blow to any remaining life.

In fractions of a second the Big Bang created energy, matter, space and time. The universe was born.

How loud was the Big Bang?

The Big Bang was not a 'normal' explosion – it did not even take place in existing space. Because it created not only matter and energy but also space and time there was no place from which the Big Bang could have been seen or heard from the outside. Observers (or listeners) would have found themselves right in the middle of the Big Bang – and would have been instantly vaporised due to its extremely high temperatures. Despite this, the American astrophysicist John Cramer once took the trouble to produce a computer simulation of the sound of the Big Bang. He compared the result to 'a large jet plane 100 feet [300m] off the ground, flying over your house'. But his description is somewhat misleading. In order to make the sound of the Big Bang audible, Cramer had to scale up the low-frequency sound waves by 100 000 billion billion times. The actual sound of the Big Bang is imperceptible to the human ear.

What was there before the Big Bang?

The Big Bang created not only matter and energy but also space and time. So does it make sense to ask about the time before the Big Bang? In the past, this was considered more of a theological question, but now increasing numbers of cosmologists and physicists are trying to find the answer. Perhaps, as one theory states, our universe was born out of the collapse of a former universe. According to another theory, new universes evolve from black holes. If so, our cosmos could be a kind of baby universe produced by another cosmos. It is doubtful whether such theories can ever be proved experimentally.

True or False?

You can still hear the Big Bang on any radio

The 'echo of the Big Bang' is not audible on a normal radio. Cosmic microwave background radiation peaks at a frequency of 160 GHz, which corresponds to a wavelength of 1mm. By comparison, the VHF band has a frequency of 100 MHz and a wavelength of 3 m. Ordinary radio aerials are not suitable for receiving the Big Bang's residual radiation.

What is cosmic microwave background radiation?

Cosmic microwave background radiation reaches the Earth from all directions and is like an echo of the Big Bang. The background pattern we see today emerged when the cosmos was only 400 000 years old. By that time its temperature had dropped to less than 2700°C, and what had been hot, opaque plasma became transparent gas through which electromagnetic radiation could travel freely. Since then, the universe has expanded and cooled so that the temperature of the cosmic background radiation is now only 2.7°C above absolute zero (-273.15°C). Thanks to tiny variations in the cosmic background radiation, we can hope to learn what happened at the time of the Big Bang.

Is the universe controlled by dark forces?

It seems that the matter from which the stars, planets and human beings are formed makes up only a tiny part of the universe. Research into cosmic background radiation has shown that our cosmos consists of 74 per cent dark energy, 22 per cent dark matter and only 4 per cent so called 'normal' matter. Dark matter holds galaxies together; without the gravity it provides, their rotation would cause them to fly apart. So far no one has been able to find out what dark matter consists of.

Dark energy is even more mysterious. It is thought to cause the expansion of the universe to accelerate. It may be that dark energy is a kind of inner tension within space. Dark energy may one day cause the expansion of space to become so rapid that all matter will be torn apart. If true, such a catastrophe is only likely to occur more than 100 billion years from now – so we don't need to start worrying about it just yet.

How old is the universe?

The most accurate measurement of cosmic microwave background radiation so far was made by the US satellite Wilkinson Microwave Anisotropy Probe (WMAP), which found our universe to be 13.7 billion years old. By comparison, Earth came into being some 4.5 billion years ago.

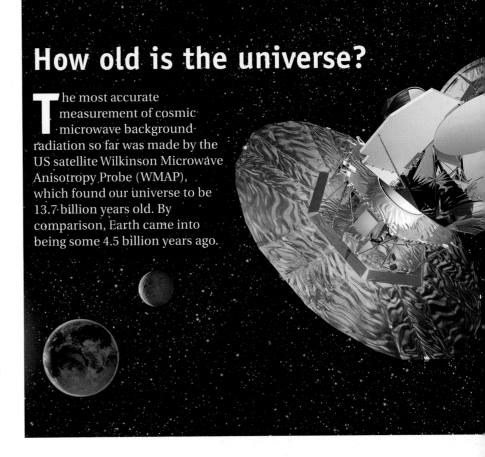

Are there other universes out there?

According to string theory – a relatively new theory that attempts to combine all the forces known to physics into a single unified whole – there may be any number of universes in existence. Because each of them might be governed by different natural laws, most of these universes could not sustain life: to find a combination of natural laws that would allow life to exist would be exceptional.

However rare such a combination may be, any universe containing an intelligent observer would inevitably be an exceptional universe of this kind. Some scientists believe this is one way of explaining why it is that the natural laws in our universe seem designed to allow life to exist – because we are here to observe it. Or it may be merely a matter of chance that this is how it is.

The false colours in this view of the sky show temperature variations in the cosmic microwave background radiation.

Looking through a telescope into space we can see vast numbers of stars and galaxies. Among the nearest is the Large Magellanic Cloud (bottom), a companion galaxy to our Milky Way.

What are galaxies?

Stars are not evenly scattered across the universe. Instead, they form large systems called galaxies, which can contain several hundred billion stars each. Galaxies exist in a variety of different shapes and sizes, from gigantic elliptical galaxies, through to spiral galaxies like our own Milky Way, and small, irregular galaxies such as the Large and Small Magellanic Clouds.

What are clusters of galaxies?

Galaxies themselves are not distributed randomly or evenly through the cosmos. They collect in groups and clusters.Our Milky Way belongs to the Local Group of galaxies that comprises about 30 star systems. Other galaxies are located in clusters that have several thousand members. One example is a cluster within the constellation Virgo, which is made up of more than 2000 galaxies

How many galaxies can we see?

Between September 2003 and January 2004, the Hubble Space Telescope took the longest ever exposure of the sky in the history of astronomy – more than 270 hours. Based on the number of galaxies visible in this spectacular image – dubbed the Hubble Ultra Deep Field – scientists estimate that with the technology available today some 50 billion galaxies can be observed.

How far is it to the nearest galaxy?

It is thought that the Canis Major dwarf galaxy in the constellation of the Great Dog is the galaxy closest to us. It is only 25 000 light-years away from our solar system and 42 000 light-years away from the centre of the Milky Way. This small system appears to be disintegrating as the gravitational pull of the much larger Milky Way is tearing it apart and absorbing its stars.

Are light rays always straight?

Light rays are subject to the effects of gravity. If a ray passes close to a star or galaxy its path is curved by an effect known as 'gravitational lensing'. This can produce some interesting effects. For example, scientists can observe quasars (see right), which would otherwise be hidden from us behind a much closer galaxy. The light curves around the intervening object, allowing us to see it. Depending on how objects are situated in relation to one another, the light from distant bodies can even be broken up into spectacular arcs or rings. The diversion of light also causes variations in brightness. If a star visible from Earth passes in front of another star, the brightness of the more distant star becomes momentarily more intense. If the star in front is also being orbited by a planet this can lead to a further increase in brightness. Thanks to this phenomenon, astronomers have discovered several planets orbiting other stars.

A quasar 1.4 billion light-years away from Earth can be observed through the Hubble Space Telescope.

What are quasars?

Quasars are among the brightest objects in the universe, radiating more energy than entire galaxies. After speculating for years about the mysterious nature of quasars, astronomers now know that they are located at the heart of some galaxies. They draw their energy from enormous black holes, which have masses billions of times greater than that of our Sun. Like cosmic vacuum cleaners, these gravitational monsters suck up matter from their surroundings, releasing unimaginable quantities of energy in the process.

How many stars are there in the sky?

With the naked eye and under favourable conditions – a moonless night well away from city lights – some 3000 stars are visible. But one look through even a modest telescope reveals that many more exist, especially in the shimmering strip of the Milky Way. Our galaxy alone contains about 200 billion stars.

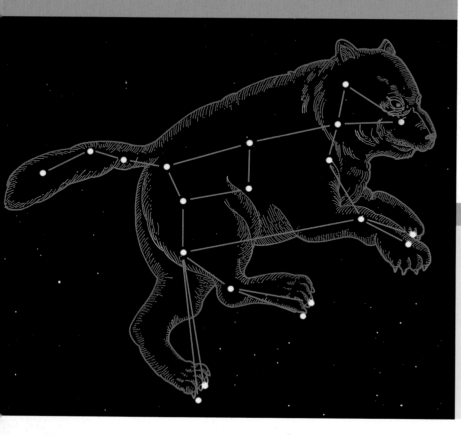

Who named the constellations?

From the earliest times, people arranged conspicuous stars in the night sky into patterns in which they believed they could see figures of living beings, gods or objects. The 12 well-known constellations that make up the zodiac date back to the Babylonians. Most of the other constellations in the northern sky have their origins in Ancient Greece. Two thousand years ago, Eratosthenes and Ptolemy wrote of 48 constellations, most of the names of which were taken from Greek mythology. More recently, more constellations previously unknown to Europeans have been introduced, especially in the Southern Hemisphere. The heavens are now home to creatures such as the Bird of Paradise and the Chameleon, as well as technical equipment, such as the Air Pump and the Chemical Furnace. In 1930, the International Astronomical Union drew up the list of 88 constellations and their boundaries, which is still used today.

How far away are the constellations?

Constellations are not true associations of stars; they merely help us find our way around the sky. Our ancestors arranged groups of bright stars more or less arbitrarily to form images. For this reason there is no answer to the question of how far away a constellation is. The stars in a constellation are situated at varying distances from Earth. The seven brightest stars in the Great Bear are between 55 and 93 light-years away, and the star with the designation Kappa in the Great Bear's front paw is 200 light-years away.

The name of the Great Bear constellation comes from Greek mythology – Zeus transformed the nymph Callisto into a bear and placed her in the sky.

Did you know...

... that constellations change?

If we were able to travel back in time to Ancient Greece and look at constellations from there, we would not notice any difference – and yet they do change. Stars don't stand still in space, and because the speed at which they move and their direction varies widely from star to star, they shift relative to one another when seen from our planet. But because of the huge distances involved, only in the course of tens of thousands of years does the difference become obvious.

Why do we see different constellations in summer and winter?

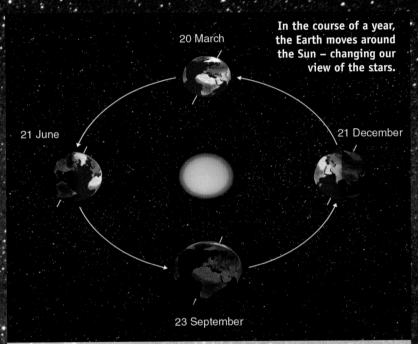

In the course of a year, the Earth moves around the Sun – changing our view of the stars.

20 March

21 June

21 December

23 September

The atmosphere scatters the light of the Sun so the sky looks bright by day and dark by night. We can see stars only at night, and therefore only one half of all the constellations at any time. But the position of the Sun in the sky – which is what prevents us from seeing some stars – varies over the course of the year. As the Earth follows its orbit around the Sun, its winter position is exactly opposite to the summer, with the result that the Sun is positioned against a background of the opposite side of the sky relative to the Earth. In winter we see constellations at night that were not visible six months earlier. For example, in the Northern Hemisphere summer the triangle of stars made up of Vega, Altair and Deneb is high in the sky, while the magnificent figure of Orion, with his prominent belt of stars, is visible in the northern winter.

Although it is huge, the constellation Ophiuchus, the serpent bearer, is difficult to make out in the sky.

What makes up the zodiac?

When we talk about the zodiac, we are referring to constellations through which the Sun moves in the course of the year. These constellations are mostly named after animals and cover an area about 14° wide running above and below the Sun's apparent annual path across the heavens. Even though we cannot see these stars by day, they are still there, hidden by the brightness of the Sun. Originally there were 12 signs of the zodiac but, since the adoption of constellation boundaries defined by the International Astronomical Union in 1930, the Sun now passes through a 13th constellation – Ophiuchus, the serpent bearer. The Sun crosses through Ophiuchus in moving from Scorpius to Sagittarius, but Ophiuchus is not included in the traditional signs of the zodiac.

Do the stars influence our destiny?

Although astrologers claim that our destiny is influenced by the position of the Sun, Moon and the planets in relation to the stars. Numerous research projects – some carried out according to criteria laid down jointly by astrologers and astronomers – have shown that such connections do not exist. When a daily horoscope in the newspaper seems unusually accurate it is because it is so cleverly composed that almost any reader can identify with it. It would be unfair to accuse astrologers of cheating the public. Many are simply trying to offer sensible advice about dealing with everyday problems and situations. All the same, their work is not based on scientific principles.

What are stars?

Stars are nothing other than suns. Just like our own Sun they, too, are huge balls of incandescent gas, inside which nuclear fusion transforms hydrogen into helium. The process releases enormous amounts of energy, which is why stars shine and radiate heat. It is only because stars are so far away that they appear in the night sky as no more than tiny points of light. In reality, many stars are considerably larger than our Sun.

Why do stars twinkle?

As starlight travels towards an observer on Earth it has to pass through the Earth's atmosphere, where the light is distorted by layers of air of differing temperature and thickness. Because these layers of air are also in constant and turbulent motion, from the observer's point of view the stars appear to flicker. Seen with the naked eye they seem to twinkle, while through a telescope they appear blurred. The effect is the same as when the air above a hot road surface in summer seems to shimmer. When stars are observed from space or from the Moon, which has no atmosphere, they do not twinkle. The brightness of some stars really does vary, although these variations will occur very slowly, over periods of days, weeks or even months.

As light passes through the different layers of moving air above a road surface in summer it appears to shimmer – this is the same process that makes stars appear to twinkle.

Why do stars appear pointed in photographs?

Some photographs, taken by large telescopes, show the brightest stars with points radiating from them. In reality, stars do not have points. They are huge spherical balls of gas, just like the Sun, and because they are so far away they appear as pinpoints of light in even the most powerful telescopes. On Earth, atmospheric turbulence blurs the images of stars, but this does not affect photographs taken from space. Instead, the size of the star's image is governed by light being bent at the edge of a lens, which scatters it into a small disc. At the same time, supports for the secondary mirror inside a telescope also diffract the light – and this is what produces the radiating points.

Why is the Sun yellow while the stars are white?

The light receptor cells – known as cones – in the human eye are considerably less sensitive than the cells responsible for distinguishing light and darkness – known as rods. As a result, in dim light we are barely able to discern colour – as the saying goes: at night all cats are grey. So most stars appear to us to be white. Only in the brightest stars can we see traces of colour, such as the reddish star Antares in the constellation Scorpius. The colour of a star enables astronomers to gauge its temperature. Red stars are cooler than the Sun, while blue stars are very hot.

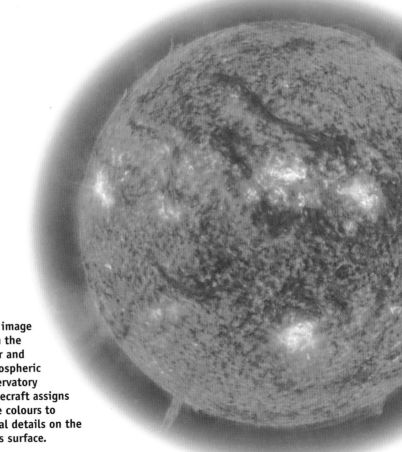

This image from the Solar and Heliospheric Observatory spacecraft assigns false colours to reveal details on the Sun's surface.

Does the Milky Way look like a Catherine wheel?

The Milky Way is the galaxy that contains our solar system. It is a spiral galaxy and contains about 300 billion stars. Viewed face-on, our galaxy looks like a gigantic Catherine wheel, while side-on it looks like a flat disc. Since we, the observers, are within this disc, we see most stars as part of a bright strip that extends across the sky, which is where the name Milky Way originally came from. All the stars visible with the naked eye are part of the Milky Way.

How old is the Milky Way?

The oldest objects in the Milky Way are the so-called globular star clusters, spherical clusters of hundreds of thousands of stars. The most recent calculations reveal globular star clusters to be about 13 billion years old, although determining their age is difficult and the figures are uncertain. The Milky Way itself could be as much as 800 million years younger. According to the latest findings, the universe came into being 13.7 billion years ago and the first stars lit up 400 million years after the Big Bang which suggests that the Milky Way cannot be older than 13.3 billion years.

The Milky Way is a fairly large spiral galaxy with three main components: a *disc*, in which the solar system resides; a central *bulge* at the core; and, an all-encompassing *halo*.

Can stars be bought?

There are many suppliers offering stars for sale, especially on the Internet. The sale is usually linked to the right to name the star. Purchasers even receive a certificate that appears to place an official stamp on the whole transaction. In reality, it is no more than a money-making scheme. The stars belong to no one and neither the purchase nor the name given to the star will be recognised by any official organisation.

How do newly discovered celestial bodies get their names?

Only the astronomers' professional organisation, the International Astronomical Union (IAU), can decide on names, and there are strict rules surrounding the naming of celestial bodies. Comets are always named after their discoverer. Wih asteroids, the discoverer is entitled to suggest a name but it must be approved by a committee of the IAU. These days, most stars are simply given a catalogue number that usually contains a set of celestial coordinates indicating the body's position in the sky.

On 30 January 1996 this comet was discovered by Yuji Hyakutake, for whom it was named. Officially, the comet is known both as Hyakutake and C/1996 B2.

The Star of Bethlehem is depicted in this 17th-century nativity scene.

What celestial body shone over Bethlehem?

There have been many attempts to explain the Star of Bethlehem – said to have guided the three wise men to the birthplace of Jesus – as a natural celestial phenomenon. It has been suggested that it might have been a comet or even a supernova, an exploding star. Many researchers believe that the most likely explanation is the meeting of two or more bright planets – a celestial event known as a conjunction. A number of these events occurred during the relevant time period. Among the possibilities are the rare triple conjunction of Jupiter and Saturn – in March, October and December – of 7 BC or the conjunctions of Venus and Jupiter in 3 BC and again in 2 BC. None of these hypotheses has so far been proven and it is possible that the Star of Bethlehem will never be explained.

This is how a planet the size of Jupiter might look as it orbits a distant sun.

Can we see planets orbiting other stars?

Astronomers have discovered more than 200 planets orbiting other stars, although they have only actually seen four and even then they are not sure that the new discoveries are really planets and not just very small stars.

In such cases, astronomers are faced with several problems. Normally, stars are millions of times brighter than their planets and planets, as seen from Earth, are very close to their stars.

So most extrasolar planets – or exoplanets – are very hard to see and can only be detected by indirect means. A large planet's gravity may be strong enough to pull its star from side to side as it orbits around it.

Alternatively, a planet may pass in front of its star at regular intervals, causing the star's light to fluctuate enough for the differences to be detected by sensitive instruments on Earth.

Who discovered...

... the first planet orbiting another star?

In 1995 Swiss astronomers Michel Mayor and Didier Queloz discovered a planet about half the size of Jupiter orbiting a Sun-like star. The planet takes just 4.2 days to complete its orbit of star number 51 in the constellation Pegasus. Four years earlier, the Polish astronomer Aleksander Wolszczan had already detected three planets outside our solar system. The planets are not circling normal stars – like our Sun – but neutron stars (see p.27). They are presumed to have been formed from the remnants of a stellar explosion some time in the past.

Where in the universe is life possible?

Earth is the only known example of an inhabited world. Life on Earth is based largely on the existence of liquid water. Scientists used to believe that there could only be liquid water on planets that orbited within a very narrow band around their star. This was because heat from the star would need to be great enough to melt ice, but not hot enough to evaporate all surface water. Within our solar system, only Earth fulfils these conditions, although in the distant past Mars might have met them as well.

We now believe that on one of Jupiter's moons there is an ocean lying beneath a thick layer of ice. The ice is melted by the tidal energy of the huge planet close by, and might be able to harbour life. The discovery means that the area around a star in which a populated world could exist has expanded considerably. But we must not exclude the possibility that quite different life forms exist that can survive without water.

Could we discover life on another planet?

Future developments in telescope technology could enable us to capture images of planets similar to Earth, and precise investigation of the light from such planets could show us whether they harbour life or not. The presence of life changes the atmosphere of a planet by affecting its chemical balance. A high oxygen level, like that found in the Earth's atmosphere, can be a sign of an inhabited world. Perhaps, within a few decades, we will be able to detect intelligent life on distant worlds if, like us, the extraterrestrials have polluted their atmosphere with artificially produced chemicals.

How can we detect another civilisation elsewhere in the galaxy?

Since 1960, researchers have been scanning the skies with huge radio antennae in search of signs of extraterrestrial life, so far without success. Isolated signals that could not be explained have been picked up, but these signals have never been repeated and so could not be explored further.

A huge telescope complex is now under construction in California, which can eavesdrop on a million stars through billions of radio channels. Also part of the project is the SETI Institute, dedicated to the Search for Extraterrestrial Intelligence. Frank Drake, the American radio astronomer and current president of the SETI Institute – who in 1960 led Project Ozma, a pioneering attempt at tracing extraterrestrial signals – reckons that within the next 20 years we will discover another civilisation in the Milky Way.

How great are the chances of there being extraterrestrial civilisations?

Scientists have speculated for years about how common developed extraterrestrial civilisations might be. In 1960, in an effort to put the guessing game on a firmer footing, Frank Drake devised an equation that could be used to calculate the number of technologically advanced civilisations in our galaxy. The snag is that many elements in Drake's equation are highly debatable. So it is not surprising that the results obtained with the equation vary a lot – from a single civilisation to as many as a million.

The surface of Jupiter's Moon Europa is covered with a thick layer of furrowed ice. Under this layer there is thought to be an ocean in which there could possibly be life.

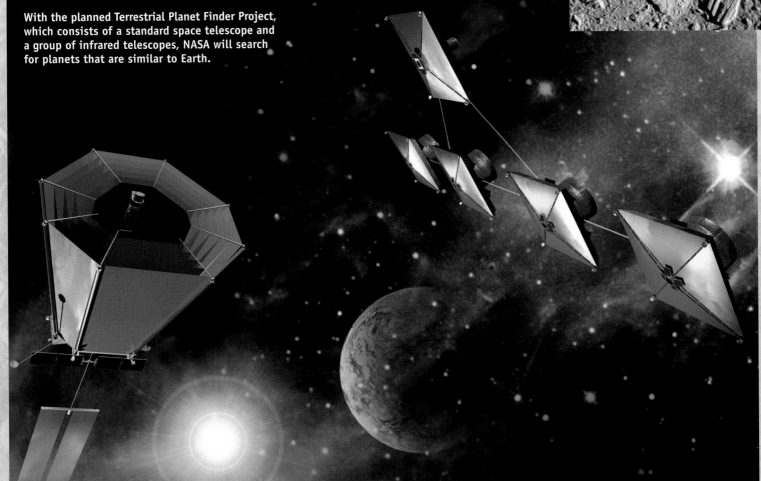

With the planned Terrestrial Planet Finder Project, which consists of a standard space telescope and a group of infrared telescopes, NASA will search for planets that are similar to Earth.

The life and death of stars

The Eagle Nebula is about 7000 light-years from Earth. New stars are constantly being formed from the material that makes up the nebula, so it could be viewed as a nursery for stars. The youngest suns are probably no more than 50 000 years old.

There is a relentless cycle of life and death going on in space. Astronomers can identify regions where new stars are being born as well as the remains of stars that have exploded. The stars appear unchanged to us because they take an almost inconceivably long time – millions and billions of years – to develop. Spectacular and rapid events like a supernova – a star exploding – are rare. The course of a star's life and its eventual death depend on the amount of mass it accrued during its formation.

How are the stars and planets formed?

The life of a star begins with a gigantic cloud of cool hydrogen gas with an initial temperature no more than a few degrees above absolute zero (-273.15°C). The cloud gradually collapses under its own gravity and clumps of material begin to form inside it, causing it to collapse at ever increasing speeds while the temperature of the gas rises steadily. The cloud eventually starts to glow, giving off thermal radiation that can be detected by infrared telescopes. The internal density and temperature increase to a point where nuclear fusion takes place, in which hydrogen is converted into helium and a new star is born.

Many stars of different sizes emerge from one large gas cloud. Star clusters made up of thousands or even millions of stars are created all at once. A rapidly rotating cloud of gas and dust (the remnants of the original cloud) floats around many of the young stars, and inside this cloud there are further clumps of matter that have the potential to develop into planets and other, smaller celestial bodies. It is thought that planets may be formed around most stars.

How old are the oldest stars?

The first stars in the cosmos were formed 400 million years after the Big Bang. Some stars are frugal in using their supply of fuel, so we are still able to see stars belonging to the first generations. This is especially the case with globular clusters, where we find stars that are more than 13 billion years old, although new stars are being formed all the time. One of the best-known starbirth regions is the Great Orion Nebula, which is 1600 light-years away and visible with the naked eye as part of the constellation Orion. The Hubble and Spitzer telescopes have revealed that many young stars in the Orion region have discs of matter rotating around them in which planets may be forming.

What determines how long stars live and how they die?

A star's mass determines its lifespan and manner of death.

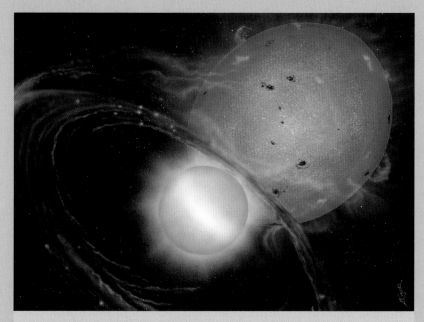

A red giant and a white dwarf orbit each other in a binary star system. The white dwarf acts like a gigantic vacuum cleaner, sucking up and absorbing matter from its larger companion star.

Large, high-mass stars use up their energy reserves more rapidly than smaller, low-mass stars. This is the why large stars shine brightly for no more than a few million years, while smaller stars can live for billions of years. Stars produce their energy as a result of nuclear fusion, during which hydrogen is converted into helium. Eventually all the hydrogen at the core of a star will be used up.

What are red giants and white dwarfs?

Once a star's hydrogen reserves are used up, the star enters the final phase of its life. It swells up to become what is known as a red giant. Our own Sun will one day turn into a red giant, and in so doing may even devour the Earth, along with several more of the inner planets of the solar system. Further fusion processes occur in the core of the enlarged star over millions of years, during which helium is converted into carbon and oxygen. When a star's stock of potential nuclear fuel has been exhausted, it collapses and turns into a white dwarf. At this stage the star is only about the size of Earth, and it takes billions more years for it to cool down.

What is a supernova?

When a star has a mass of more than eight times that of our Sun, its end is dramatic. The dying star hurls its outer layers into space with a mighty explosion, known as a supernova. At the same time, the star's interior collapses to form a small, extremely dense body, which can take the form either of a neutron star or a black hole.

What is a neutron star?

The matter in a neutron star is as densely packed as the matter in an atomic nucleus. A piece of neutron star the size of a pinhead would weigh more than one million tonnes. The immense pressure in the interior pushes the electrons into the atomic nuclei's protons – leaving nothing but neutrons behind. Neutron stars have a diameter of only about 20km but contain about the same mass as our Sun.

Can we see black holes?

If the mass of a star's collapsing interior is more than about three times that of our Sun, its gravity becomes so strong that even the neutrons cannot halt the process. The star continues to collapse until it turns into a black hole – the name given to a region of space where gravity is so powerful that nothing, not even light, is able to escape. This is why these gravity holes are 'black'. We are able to detect black holes because their colossal gravity draws in surrounding matter that, as it descends into the black hole, becomes very hot and begins to glow – thereby revealing the black hole's position.

Black holes don't only occur when a star dies. There are also monster black holes at the centres of galaxies that have a mass of millions if not billions of times that of our Sun. These are thought to have originated in the early stages of the universe, and there is even one of these super-massive black holes in the centre of our Milky Way.

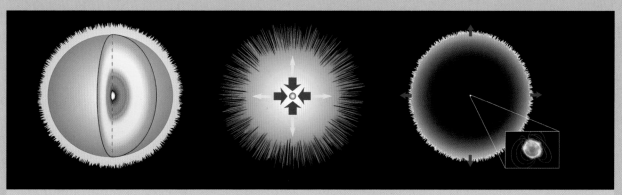

Stars (left) turn into supernovas when they have a mass that is at least eight times that of our Sun. With a supernova (centre), the star's external layer is expelled and the core collapses into itself. What remains after the explosion is a neutron star (right).

What sort of object is our Sun?

The Sun is a gigantic, glowing ball of gas made up mainly of hydrogen and helium. Its diameter is just over 100 times that of the Earth and its mass is 330 times greater. The Sun contains 98 per cent of all of the mass contained in our solar system, which it dominates with its immense gravity.

True or False?

The Sun gets lighter as it burns matter

The Sun converts about 700 million tonnes of hydrogen into helium every second, and it actually does get lighter as it does so – or, to use the correct scientific terminology, its mass is reduced. During nuclear fusion (see right), four hydrogen atoms are converted into a single helium atom, and one helium atom is lighter than four hydrogen atoms. The absent mass has not disappeared, but has been turned into energy – which can be calculated using Albert Einstein's famous formula, $E = mc^2$. The energy generated equals the mass times the square of the speed of light. So every second the Sun slims down by 5 million tonnes. This may sound like a huge amount, but during the life of the Sun it amounts to 0.01 per cent of its mass.

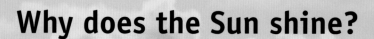

100 JAHRE RELATIVITÄT – ATOME – QUANTEN   55

$E = mc^2$

ALBERT EINSTEIN   DEUTSCHLAND   2005

A. Einstein

Why does the Sun shine?

Unlike the planets and our Moon, which merely reflect the Sun's light, the Sun emits its own light. With a surface temperature of around 6000°C it glows extremely brightly. Most solar radiation has a wavelength near the centre of the visible region of the electromagnetic spectrum (think of a rainbow) – which explains why we see it as being yellow. This is obviously not a coincidence, since most creatures on Earth have evolved to perceive the region of the spectrum in which the Sun gives its greatest amount of light. In addition, the Sun emits radiation in other areas of the electromagnetic spectrum, such as radio signals and X-rays as well as infrared and ultraviolet rays, which we can't see.

How does our Sun generate its energy?

Towards the centre of the Sun the temperature rises rapidly, reaching 15 million degrees Celsius at the core. In addition, the Sun's core is under immense pressure – 250 billion times normal atmospheric pressure on Earth. In these extreme conditions, the atomic nuclei of hydrogen atoms collide with each other at great speed and fuse to become helium nuclei. During this nuclear fusion, energy is released. This travels outwards from the interior of the Sun and is radiated from the Sun's surface. The released energy's journey from the Sun's centre to its surface takes around 10 million years. If the energy supply at the centre of the Sun were to be used up today (and somehow the Sun didn't collapse) it would take us 10 million years to notice.

What does the interior of the Sun look like?

Energy-producing nuclear fusion only occurs in the Sun's inner core. Although this contains about half the Sun's matter, it takes up only 1.6 per cent of its volume – so the inner core's radius is about a quarter of the size of the Sun's total radius. The energy initially travels outward in the form of radiation through a section of the Sun that is known as the radiative zone.

About 100 000km beneath the surface, the temperature drops sufficiently to cause the gas to absorb the radiation. This is why the Sun's external layer – the convective zone – bubbles like a boiling cauldron. The matter, heated by radiation, rises to the surface where it cools and then sinks back down to the edge of the radiative zone.

A section through the Sun. Above the core lies the radiative zone, and above that the convective zone (red). The thickness of the solar surface (photosphere) has been exaggerated in this illustration; in reality it is only a few hundred kilometres deep.

True or False?

The core of the Sun is dark

The temperature at the centre of the Sun is about 15 million°C. At this temperature X-rays are produced, which is why it is sometimes said that the centre of the Sun is too dark for human eyes since they cannot see X-rays. This is not true. If a gas is continuously heated, most of its radiation does shift to ever shorter wavelengths – from visible light to ultraviolet and then X-rays – but at the same time the brightness of the glowing gas increases at every wavelength. So even the centre of the Sun would be dazzling to human eyes.

Why are scientists hunting for neutrinos?

The process of nuclear fusion within the Sun creates a huge number of neutrinos as well as electromagnetic radiation. Neutrinos are mysterious elementary particles. They have minuscule mass and are able to pass through matter almost undisturbed, which is why, in contrast to radiation, the neutrinos that originate in the core of the Sun arrive at the Earth in only eight minutes. About 70 billion neutrinos strike each square centimetre of our planet every second. Physicists are deploying gigantic detectors in their efforts to catch at least a few of these elusive particles to help them understand a few more of the mysteries of the Sun's interior.

The Super Kamiokande Detector is 59m high and fitted with more than 11 000 photomultiplier tubes, which are used to detect neutrinos.

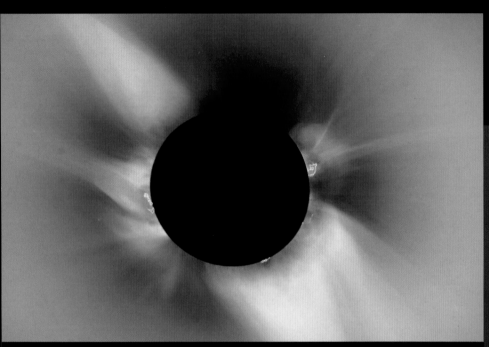

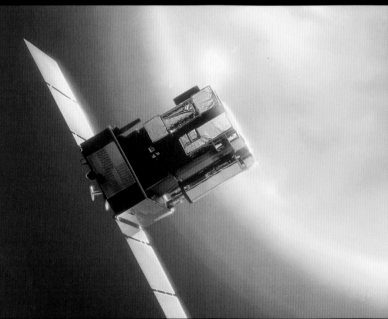

Views like this of the solar corona (left) are only possible with a solar telescope – unless there happens to be a solar eclipse.

The SOHO spacecraft (below) is a collaboration between the European Space Agency (ESA) and NASA. One of the satellite's many functions is to observe the solar wind.

What is the corona?

The corona is the Sun's thin external atmosphere. Sound waves and magnetic fields heat the gas in the corona to temperatures over 1 million°C. This high temperature causes the corona to emit X-rays. It also emits a lot of light, which is why the corona can be seen with the naked eye when the Sun's bright disc is covered, as is the case during a total eclipse of the Sun. In order to observe the corona, solar scientists use a special telescope, like the one aboard the Solar and Heliospheric Observatory (SOHO) spacecraft, which has been fitted with a device to cover the disc of the Sun. The corona stretches several million kilometres into space – two or three times the diameter of the Sun.

What is solar wind?

A constant stream of charged particles, mainly electrons, protons and the nuclei of helium atoms, is ejected from the hot corona of the Sun into space at a speed of several hundred km/sec. The solar wind causes our star to lose about 1 million tonnes of matter per second. The solar wind is of particular interest to scientists because by examining it they hope to determine the chemical composition of the original nebula from which the solar system was formed. In 2004, the Genesis mission was due to bring back to Earth particles captured from the solar wind. The project received a setback when the capsule containing the science experiments was badly damaged due to its parachute failing to open during its return to Earth.

How turbulent is the surface of the Sun?

E ven if we were to leave aside the fact that the Sun's surface is in constant motion because of the energy emitted from its core (see p. 29), it certainly couldn't be described as calm. Solar prominences are particularly impressive phenomena that extend outwards from the Sun's surface. They are low luminance arcs that extend from the Sun's surface into the corona, inside which gas from the corona is gathered together as a result of local magnetic fields. Solar prominences can be hundreds of thousands of kilometres long and extend several tens of thousands of kilometres above the Sun's surface. They can also break apart explosively, ejecting the matter they contain into space at speeds of up to 1000km/sec. Like the corona itself, solar prominences can only be observed by astronomers with the aid of special telescopes that obscure the Sun's bright disc.

This huge solar prominence was observed on the Sun in September 1999. The Earth (above left) is shown at the same scale.

How old is the Sun?

The most ancient rocks that scientists have discovered so far, both on the Earth and on the Moon, are about 4.5 billion years old. Since the Sun and its planetary system – which includes the Earth and its Moon – were formed at much the same time, the Sun must also be 4.5 billion years old.

This means our Sun is less than halfway through its life span. We know this, because astronomers have been able to calculate that the Sun's energy store will last for about 11–12 billion years, based on the Sun's mass and the amount of energy it emits. So we can expect the Sun to shine for another 7 billion years or so.

Has the Sun always been as bright as it is now?

At the time of its birth, the Sun was smaller, cooler and less bright than it is now – its brightness being approximately 70 per cent of what we experience today and the Earth also received less heat from the Sun. From a purely mathematical point of view, young Earth's temperature should have been about 25°C lower than it is now. The composition of the atmosphere at that time was different, and the larger percentage of carbon dioxide it contained generated a greenhouse effect. This prevented the young Earth from becoming entirely covered in ice.

Will the Earth be swallowed up by the Sun one day?

In about 5 billion years, when the hydrogen in the Sun's core has been exhausted, the Sun will swell up to become a red giant (see p. 27). Astronomers are not entirely sure just how large the Sun will eventually become, but it is likely that in this phase of its development the Sun will only consume the inner planets Mercury and Venus. Even if Earth is spared total annihilation – life on our planet will no longer be possible. A gigantic and hostile Sun will shine with 2300 times its current intensity, and will cause the Earth's crust to melt into an ocean of lava.

Islands in the cosmos. This photomontage shows the Sun, Earth and Moon to scale. At some point in the future the Sun will grow even larger.

Why does the Sun have spots?

Unlike Earth, the Sun has an extremely complicated magnetic field. A large part of this magnetic field runs beneath the surface of the Sun in the form of so-called magnetic tubes. Sometimes these tubes can break through the Sun's surface and, where they do so, the strong magnetic fields prevent hot matter from rising out of the Sun's interior, cooling the affected areas of the surface to a temperature of about 4000°C. Because of their lower temperature, these regions are not as bright as the rest of the Sun – which is about 1500°C hotter – and they appear as dark spots on the Sun's surface. Sunspots are between 2500km and 50 000km wide, which means they can be several times larger than Earth. The life span of individual sunspots varies from days to months.

This image of the Sun (left) was captured by the SOHO satellite (Solar and Heliospheric Observatory) in October 2003. Several large sunspot groups are clearly visible. The enlargement of a group of sunspots, taken in 2002 (below), shows their complex structure in detail.

Does the Sun always have the same number of sunspots?

Sunspot numbers rise and fall in a cycle of approximately 11 years. The last sunspot maximum was in 2000, so the next one is expected in 2011. The sunspot cycle is controlled by the Sun's magnetic field, which changes direction every 11 years.

Do other stars also have spots?

Our Sun is definitely not exceptional, and other stars also have dark spots in numbers that fluctuate in accordance with a regular cycle. The magnetic processes at work appear to be very similar. These 'starspots' can be even larger than the spots on our Sun, with many stars having between 10 per cent and 15 per cent of their surface covered in spots. In some cases, as much as half the entire surface is covered. In comparison, only about 1 per cent of the Sun's surface is covered during maximum sunspot periods.

Are sunspots visible to the naked eye?

Some sunspots are large enough to be seen with the naked eye, and there are historic records of sunspots that precede the invention of the telescope by a long time. The oldest written record of sunspot observations dates back to 47 BC in China.

To avoid permanent eye damage, never look directly at the Sun, unless you have a telescope fitted with special filters (dark glasses and exposed film are not adequate). Perhaps the best and safest way to view the Sun's disc is to project it onto a piece of paper with the help of a telescope.

When the Moon passes in front of the Sun

Since the earliest times, human beings have raised their eyes to the sky full of awe – and sometimes fear – when, in the middle of the day, darkness gradually descends, making the brightest stars visible in the sky. A total eclipse of the Sun is one of the most impressive dramas that the heavens have to offer.

One of the most important duties of anyone who claimed knowledge of the workings of heavenly bodies in times past was to predict eclipses accurately and in good time. (The execution of the royal Chinese astronomers Hi and Ho for failing to announce an eclipse in 2137 BC is probably the stuff of myth and legend).

The phenomenon of a solar eclipse has a simple explanation – the orbiting Moon passes in front of the Sun as seen from Earth. The fact that the Moon precisely covers the Sun's disc is simply a cosmic coincidence. The Sun is 400 times larger than the Moon but it is also 400 times further away from Earth. This makes both celestial bodies appear to be almost exactly the same size when viewed from Earth.

What is an annular solar eclipse?

The Sun and the Moon do not always appear the same size when viewed from Earth. This is because the Moon's orbit around Earth is elliptical and not circular, so the distance between Earth and the Moon fluctuates between 384 000km and 410 000km. As a result the Moon's apparent size in the sky

Central Europe saw its last total solar eclipse in 1999; there won't be another one for more than 130 years.

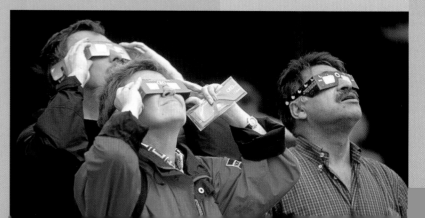

to the Earth's orbit around the Sun, the Moon's orbit around the Earth is inclined at an angle. The Moon usually passes either above or below the Sun as seen from Earth. A solar eclipse can only occur twice a year and a total eclipse of the Sun is even rarer.

There was a total solar eclipse in 2006 (visible from northern Africa and central Asia), none in 2007 and one in 2008 (visible from Russia, China, Canada and Greenland). Anyone wanting to witness a total solar eclipse will need to be fairly precise about their travel plans. The Moon's central shadow – known as the umbra – has a maximum diameter on the surface of the Earth of only about 200km. It is only within this umbra that the eclipse of the Sun is total, and then only for a few minutes.

Outside of this core shadow, in the Moon's half shadow, the Sun is only partially covered. A partial eclipse, even if the Moon covers 99 per cent of the Sun, still falls far short of the spectacle of a total eclipse.

What is the safest way to observe the Sun?

Looking at the Sun before it has been totally eclipsed can cause permanent damage to unprotected eyes. The only safe way to look at the Sun is through special eclipse spectacles, which are available from opticians. Ordinary spectacles that have been blackened with soot and exposed negative film do not provide adequate protection, because both allow heat

This image – digitally assembled from seven individual photographs – shows the progress of an annular solar eclipse that was visible in the skies over Spain in 2005.

also varies. If the Earth's satellite is at the furthest point of its orbit during an eclipse, it will be too small to cover the entire disk of the Sun. As a result, observers on Earth experience an annular eclipse instead of a total eclipse of the Sun.

How frequent are solar eclipses?

For an eclipse to be possible, the Moon has to be in the same direction as the Sun when viewed from Earth – there has to be a new Moon. We don't see a solar eclipse every time there is a new Moon because relative

radiation and some ultraviolet rays to pass through.

It is even more important to be protected when looking at the Sun outside of totality through binoculars, telescopes or telephoto lenses, unless the

lens has been fitted with an appropriate filter. A suitable filter can be made using a special, metal-coated sun-foil, which is available from astronomical equipment suppliers.

What do we know about the life of the solar system?

Our solar system is a busy place. In addition to the large planets, there is an immense throng of smaller celestial bodies. They include dwarf planets, moons, asteroids and comets, as well as innumerable smaller lumps of rock – the meteoroids – that are mainly debris left by passing asteroids and comets. These range in size from fragments as small as grains of sand to boulders several metres across.

Our solar system has many inhabitants. Planets and dwarf planets, as well as comets and asteroids, are in various orbits around the Sun. Most planets have attendant moons, and even some asteroids (in the foreground) are orbited by smaller companions.

What is the difference between stars, planets and moons?

Stars are hot celestial bodies that generate energy in their gaseous core and so shine of their own accord. A few years ago, it was still impossible to say with certainty whether or not there were other stars with planetary systems. Recently, planets have been found orbiting more than 100 nearby stars, so it seems safe to assume that many stars in the universe are probably at the centre of planetary systems.

Unlike stars, planets do not emit their own light, they merely reflect the light of the star around which they orbit. We are only able to see the planets in our solar system because they reflect the Sun's light. So the brightness of a planet in the sky depends not only on its size, but also on its position relative to the Sun and how well it reflects the Sun's light.

Finally, moons are those celestial bodies that do not orbit a star, but travel around a planet instead. Some of the 140 or so moons in our solar system are even larger than planets. Ganymede, for example – Jupiter's largest moon (and the largest moon in the solar system) – has a diameter of 5262km, which is considerably larger than Mercury, which has a diameter of only 4880km.

latter course. Pluto, and other small, but still spherical, celestial bodies have since been referred to as dwarf planets.

What is a dwarf planet?

Astronomers have had difficulty determining how to distinguish a dwarf planet from an 'ordinary' planet. In August 2006, they agreed a definition, which applies only to our solar system. A planet, it states, is a body that orbits the Sun, is large enough for its gravity to make it round, and has 'cleared the neighbourhood' of smaller objects. This is not the case for Pluto, which is why it is now classified as a dwarf planet. The same applies to Ceres, the largest object in the asteroid belt between Mars and Jupiter. But the new definition is still disputed. Some astronomers consider it to be too imprecise because it fails to give an accurate definition of exactly what a planet's 'neighbourhood' is and, also, when the neighbourhood is to be considered 'cleared'. They point out that there are small, celestial bodies whizzing about all over the solar system.

This illustration shows the major celestial bodies in our solar system to scale. On the left is the Sun, next to it are (from top to bottom) the planets Mercury, Venus, Earth, Mars, Jupiter, Saturn, Uranus and Neptune, with their major moons. Right at the bottom is the dwarf planet Pluto with its moon.

How many planets orbit our Sun?

There are now only eight planets in our solar system, namely Mercury, Venus, Earth, Mars, Jupiter, Saturn, Uranus and Neptune. Little Pluto was thought of as a planet until 2006 but, by then, astronomers had come across increasing numbers of similar celestial bodies beyond Neptune's orbit, at least one of which was larger than Pluto. Scientists were faced with having to make a decision. Should the largest of those small bodies also be listed as planets – or should Pluto lose its status as a planet? Following a period of intense discussion, they decided to take the

What decides the shape of celestial bodies?

Gaspra is a typical example of an irregularly shaped asteroid.

The shape of a celestial body depends on both its composition and its mass – and on its gravity. A gaseous or liquid celestial body will always form a sphere because its gravity acts equally in all directions. On the other hand, the inner cohesion of a body made from solid rock can be stronger than its gravity, which is why small asteroids are not generally spherical but often have strange, irregular shapes. When an asteroid's diameter exceeds about 800km, its gravity is sufficiently strong to cause large elevations to collapse under their own weight, and so causes the object to take on a more nearly spherical shape. A good example of this is the dwarf planet Ceres, which has a diameter of 970km. Celestial bodies with particularly high rotational speeds are not exactly spherical, but slightly elliptical. This is because the rotational velocity is greatest at an object's equator, causing a slight bulge.

Why do celestial bodies rotate?

The reason lies with the laws of physics and a phenomenon known as 'the conservation of angular momentum', which means, in simple terms, that rotation cannot simply disappear. It is thanks to this that an ice-skater can increase the speed of pirouettes by pulling in his or her arms. Something similar happens when a gas cloud condenses to become a star with a planetary system. The gas cloud's initially gentle rotation increases in speed as the cloud becomes smaller. Only a small fraction of the rotational motion can pass to the star itself, otherwise it would be torn apart by rapid rotation. This may be prevented by the formation of planets, which absorb most of the gas cloud's original angular momentum. Only 0.5 per cent of the solar system's total angular momentum has passed to our Sun.

Do other planets have seasons?

There are seasons on Earth because of the tilt of the Earth's axis in relation to its orbit. This causes the Sun to shine more on the Northern Hemisphere for one-half of the year and then more on the Southern Hemisphere for the other half. Other planets can have seasons for the same reason, provided their axes of rotation are also tilted in relation to their orbits. Mars' axis, for example, has a 25° tilt, which is almost the same as Earth's at 23.5°. Seasonal differences on Mars are greater because of Mars' eccentric orbit and it is possible to see seasonal fluctuations in Mars' polar ice deposits.

In August 2003 it was summer at Mars' South Pole (below), and the ice cap was considerably smaller than during the Martian winter.

Uranus is an extreme example. Its axis has a 98° tilt, so the Sun shines on the northern pole for half a Uranian year (84 Earth years) and then on the southern pole for the second half. The unlit pole is left in complete darkness.

What laws govern the movements of the planets?

In the early 17th century, the German astronomer Johannes Kepler formulated three laws to describe the movement of the planets within the solar system. The first law covers the shape of a planet's orbit, which is always elliptical, with the Sun at one of the foci of the ellipse (see diagram 1 below). The second law describes a planet's speed. An imaginary line joining the Sun and a planet sweeps out equal areas during equal intervals of time (see shading in diagram 2 below). It can be deduced from this that when a planet is closest to the Sun (at perihelion), it travels faster than when it is furthest away (at aphelion). Finally, the complicated third law compares the orbits of various planets. Using the orbital period (T) of a planet and the semi-major axis of its orbital ellipse (a) – see diagram 1 below – if T^2 is divided by a^3 we get a constant that is the same for every planet. This is useful because, if the dimension of the semi-major axis of at least one planet – Earth, for example – is known, it is possible to calculate the size of other planets' orbits when only their observed orbital periods are known. Kepler's three laws turn out to be a natural consequence of the law of universal gravitation, formulated by Isaac Newton and published in 1687.

Are the planets' orbits stable?

Astronomers have been unable to answer this question with any degree of certainty. Although the Sun is the dominant body in the solar system, and the major factor in determining planetary orbits, the planets also attract each other with their much smaller gravitational forces, causing minor orbital fluctuations. Long-term climatic trends on Earth, such as ice ages, are among the consequences of this kind of orbital fluctuation. It is reassuring to know that planetary orbits have remained more or less stable since the formation of the solar system. But not even computer simulations can help us predict whether this will be the case forever.

1. Kepler's first law

2. Kepler's second law

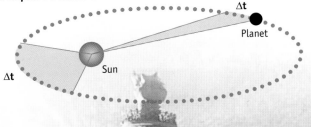

P = perihelion
(closest approach to the Sun)
A = aphelion
(farthest distance from the Sun)
Δt = interval of time
a = semi-major axis

An apocalyptic scene from the 2004 film *The Day After Tomorrow* shows New York locked in the grip of a new ice age. It could become reality if Earth's orbit changed dramatically at point in the future.

Where are the planets to be found in the sky?

Unlike stars, which barely change their positions relative to one another over human lifetimes, planets are in constant motion, which is why our ancestors called them wandering stars. This constant motion is due to their steady progress along their orbits around the Sun. The orbits of all the planets, including Earth, are located in almost the same plane – known to astronomers as the ecliptic. This is why we find all the planets in the sky close to the line of the ecliptic.

Because the Earth also travels around the Sun, and each planet travels at a different speed, strange effects can occur when observing distant planets over weeks or months. A planet can appear to slow down and even move backwards across the sky before reverting to its original direction of travel. But the planet doesn't actually change direction. The effect is caused by the fast-moving Earth catching up and overtaking one of the slower-moving outer planets.

Why is Mercury so rarely visible from Earth?

Mercury is the closest planet to the Sun in our solar system. Since its average distance from the Sun is only 58 million km (compare with 150 million km between the Earth and the Sun), its position in the sky is also very close to the Sun. Seen from Earth, Mercury cannot move any further from the Sun than 28° and most of the time it is even closer than that, or even directly in front of or behind the Sun. From Earth, the planet is nearly always obscured by the Sun's brilliant light. Mercury is only visible for most people when the Sun is not in the sky at the same time – which is briefly at dawn and dusk. Then Mercury can be seen low on the horizon either rising or setting just before or just after the Sun.

What is the link between Einstein and Mercury's orbit?

In contrast to Earth's orbit, Mercury's orbit is a very pronounced ellipse, its distance from the Sun varying between 46 and 70 million km. In addition to the Sun, the other planets exert a gravitational pull on Mercury, causing its orbit to change slightly so that perihelion (the point when the planet is closest to the Sun) steadily changes position – known as the 'advance of perihelion'. If astronomers use Newton's law of gravitation to calculate this advance, the result does not correspond with observation. This is because the description of gravity contained in Newton's law is not entirely correct. Einstein's general theory of relativity provides a better description, and one that allows us to explain fully Mercury's advance of perihelion. In Einstein's theory, huge masses such as the Sun distort space and time. The closer one approaches a mass, the greater the curvature of space and the slower time passes as observed from outside. At great distances from the Sun, relativity theory provides the same elliptical orbits as calculated by Newton's law, but close to the Sun there is a small additional effect – which is why Newton's law cannot give us the correct result.

Mars
Venus
Mercury
Earth

Tiny Mercury is seldom seen from the Earth because it lies too close to the brilliance of the Sun.

What are the evening star and the morning star?

Like Mercury, Venus is an inner planet, which means that its orbit is inside Earth's orbit. This is why, when you look up at the sky, you will never see Venus and the Sun at opposite horizons. So Venus cannot be seen in the sky in the middle of the night, only in the evening or in the morning, when its prominent, brightly shining disc is popularly known as the evening or morning star – both names identifying the same celestial object. But Venus cannot set in the evening as the evening star and then rise again in the morning as the morning star (or vice versa). Venus' position is either to the west or east of the Sun, so that it can only ever be either the morning star, or the evening star.

Why is it so hot on Venus?

The temperature on Venus is about 460°C. This is not only because Venus is much closer to the Sun than the Earth. The conditions observed on Venus are a result of a powerful 'greenhouse effect'. Some 96 per cent of the planet's dense atmosphere is made up of the greenhouse gas carbon dioxide. Some scientists believe that enormous volcanic eruptions occurred on Venus about 800 million years ago, during which almost the entire surface of the planet was covered with fresh lava. It is possible that large quantities of carbon dioxide were released during this event, which could have triggered, or at least intensified, the greenhouse effect observed today.

Venus

Moon

Mercury

Why is Mars red?

Even when viewed with the naked eye, Mars is conspicuous in the sky, not just because it is so bright but because of its strong red colour. The colour comes from the dust that covers almost the entire surface of Mars and which contains large amounts of iron oxide. Put simply, Mars is rusty. Planetary scientists suspect that a large proportion of the iron arrived on Mars in meteorites. In the past, it was thought that for the iron on Mars to have been oxidised, there must have been free water on the planet at some time in the past. Recent studies have shown that the very small quantities of water in Mars' thin atmosphere are sufficient – with the help of powerful ultraviolet rays – to make the iron go rusty. This does not necessarily mean that there have never been large amounts of free water on Mars any time in the past.

Is there, or has there ever been, life on Mars?

Many geological structures found on Mars may have been caused by water – twisting valleys that have the appearance of river courses, stream structures around craters and mountains, the coastlines of large lakes and even the suggestion of dust-covered ice floes. Current findings indicate that 4 billion years ago Mars may have had a thicker atmosphere as well as large expanses of open water – there may even have been an actual ocean in the Northern Hemisphere. Life could have emerged at that point and primitive life forms could have been preserved to this day deep beneath the Martian surface or in active volcanic regions. In the 1970s, biological experiments on board the American Viking space probes provided contradictory results that continue to be debated to this day. For this reason, plans are afoot for future Mars probes to employ more sophisticated methods in the search for traces of life.

What are Martian canals?

In 1877, the Italian astronomer Giovanni Schiaparelli thought he had seen some thin, straight lines on the surface of Mars. He called these 'canali', which was translated into English as 'canals' and so triggered a debate lasting many years surrounding the possible existence of an intelligent civilisation on Mars. Schiaparelli himself believed that the 'canali' had natural causes. The wealthy Bostonian businessman Percival Lowell was fascinated by the canals, so much so that he built his own observatory in Flagstaff, Arizona. Between 1894 and 1916 he mapped 437 canals in total. Lowell was convinced that the canals were an artificial irrigation system devised by the intelligent inhabitants of a dying world. In reality the Martian canals were an optical illusion. The human brain has a tendency to interpret a row of dots as a line. Photographs taken by the first American Mars probes in the 1960s finally cleared the matter up – there are no canals on Mars.

Parts of the wall and floor of the crater *Galle* on Mars shows parallel grooves running down the crater's sides. These may indicate that there were once watercourses on the planet.

Can moons crash?

This kind of thing could easily happen, even in today's relatively stable solar system. Mars, for example, has two small moons called Phobos and Deimos. Both are tiny, irregularly formed objects. Phobos' dimensions are 26.8 x 22 x 18.4km, and Deimos' are 15.0 x 12.2 x 10.4km. Most astronomers agree that these moons were once asteroids from the nearby asteroid belt, which were captured by the Martian gravitational field.

The surfaces of Phobos and Deimos are strewn with numerous craters, and Phobos also exhibits a striking pattern of parallel grooves. So far scientists have not come up with a satisfactory explanation for how they might have been formed. Phobos' orbit is already relatively close to Mars, and it is getting closer by about 1.8m every century. In about 50 million years, the planet's gravitational forces could tear its diminutive satellite apart and turn it into a short-lived ring around the planet, like that around Saturn. Alternatively, Phobos could plummet to the surface of Mars.

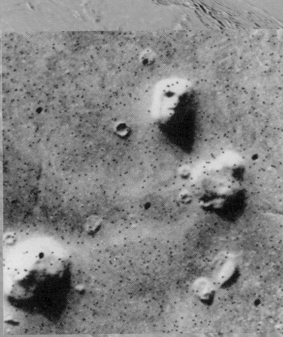

True or False?

There is an artificial face on Mars

In July 1976, the US probe *Viking 1* transmitted images of a mountainous area on Mars that bore a remarkable resemblance to a human face. It was soon recognised as a coincidence caused by light conditions at the time the picture was taken. But when the story was sensationalised by the tabloid press, a dedicated community emerged who believe that the face is a gigantic monument left behind by a vanished Martian culture. Even the Mars *Global Surveyor* probe, which provided detailed images of the area, was unable to alter their conviction. More recent pictures show a mountainous region made craggy by erosion and looking nothing at all like a face.

How are asteroids different from comets?

Asteroids are small bodies chiefly composed of rocks and metals. Most orbit the Sun in the wide asteroid belt that lies between the orbits of Mars and Jupiter. Comets, by contrast, are made up of a loose mixture of rocks and ice, and for this reason they are often called 'dirty snowballs'. Recent research has shown that rocks are the main constituent of most comets. Although the largest asteroids can be several hundred kilometres in diameter, most comets are only a few kilometres across.

Astronomers believe that far beyond the orbit of Neptune the solar system is surrounded by a shell made up of several billion comets. It is known as the Oort Cloud. Comets usually only attract the attention of the general public when gravitational forces at work in the Oort Cloud send them on a path that takes them into the inner solar system. As they approach the Sun, volatile constituents vaporise and form a gas and dust tail that can be more than 100 million km long and clearly visible from Earth.

Are asteroids the remains of a planet that exploded in the distant past?

At one time, some astronomers believed that the asteroids between Mars and Jupiter were the remains of a planet that had been destroyed by a catastrophe. Today, it is generally thought that the strong gravitational pull exerted by Jupiter in the region of the asteroid belt made it impossible for another planet to form there. Asteroids are more likely to be the remains of small bodies – known as planetesimals – from which the Earth-like planets were formed about 4.5 billion years ago.

Nonetheless, most of the asteroids seen today are thought to be fragments of what were once much larger bodies that have been smashed into pieces by collisions between the original planetesimals over millennia. This probably explains the fact that there are several different kinds of asteroids, with 'rocky' asteroids having been formed from the crust of ancient planetesimals and 'metallic' asteroids from deeper internal material.

What is a 'dead comet'?

Every time a comet approaches the Sun it loses some of its mass as frozen gases and ice evaporate and are blown into space. A comet orbiting closer to the Sun than Jupiter has an active life of only about 10 000 years. After that it is 'dead', it no longer has a tail and is easily mistaken for an asteroid. Thanks to recent research, we also know that the dividing line between asteroids and comets is blurred. There are bodies within the asteroid belt that constantly emit dust, triggered by the evaporation of ice, just as is the case with comets.

In young solar systems, collisions in asteroid swarms like this probably provide the material for planet formation.

What else do we know about comets?

In the distant past, comets were thought to be phenomena that occurred entirely within Earth's atmosphere – vapours from inside the planet that soared up into the heavens. It wasn't until the 16th century that Tycho Brahe was able to demonstrate that comets are independent celestial bodies that travel along orbits far beyond the Earth.

Why does the comet's tail always point away from the Sun?

Strictly speaking, a comet has two tails – a thin gaseous tail made up of electrically charged particles (atoms and ions), as well as a broader dust tail which, as the name indicates, is made up of dust particles. The radiation emitted by the Sun applies pressure to the dust particles, which is why they move away from the Sun. The radiation pressure is not sufficient to explain why the gas tail also points away from the Sun. To explain that, astronomers came to the conclusion some 50 years ago that there had to be a particle current emanating from the Sun – the solar wind – which sweeps the tail of gas along with it.

How can comets tell us about the birth of the solar system?

Comets are remnants from the period when the solar system was formed, which is why they are of major interest to astronomers. In the Oort Cloud, far away from the Sun, comets have been preserved unchanged for billions of years in a cosmic deep freeze. Any comet travelling from there to the inner solar system is like a messenger from the past, one that provides us with an insight into the composition of the solar nebula from which our solar system was formed.

Comet Hale-Bopp made an impressive sight in the night sky during its 1997 visit.

The planets – a journey through the solar system

The eight planets in the solar system are very different from one another. There are the inner planets Mercury and Venus, closest to the Sun; the gas giants Jupiter, Saturn, Uranus and Neptune; mysterious Mars; and, the only planet that certainly harbours life, our Earth.

Do all planets have moons?
There are only two planets in our solar system that do not have a moon, and these are Mercury and Venus. The number of moons each of the other planets has varies. While Earth only has only one Moon, Jupiter is orbited by at least 63 satellites. Astronomers keep discovering more, so the actual number may be even larger.

Do other planets have atmospheres?
Mercury is the only planet in our solar system that doesn't have an atmosphere. But even the atmospheres of other planets are quite different from Earth's. On Mars, for example, the atmosphere is very thin and made up chiefly of carbon dioxide, while on Venus the surface atmospheric pressure is about 90 times that on Earth.

Does water exist on other planets?
So far, scientists have found liquid water only on Earth. Some suspect that there may be frozen water in the perpetually shaded regions around Mercury's poles. Frozen water is also present near the poles on Mars, and it is highly probable that some deposits of water are hidden under that red planet's surface.

How tall is the tallest mountain in the solar system?
At more than 24km high and 600km wide, the extinct volcano Olympus Mons on Mars holds the record. Unlike Earth, Mars lacks plate tectonics, so sections of its crust do not move and volcanic hot spots stay in the same position. It is because they have remained in one place that Martian volcanoes have risen to great heights.

Does vulcanism occur on other planets?
Astronomers have discovered old volcanoes on Venus, as well as on Mars. So far, no active vulcanism has been observed on any other planet, although it has been detected on Jupiter's moon Io. Atmospheric gases on Venus indicate that there may be active volcanoes on that planet as well.

Why are the solar system's outer planets so different from the inner planets?
Planets are formed within rotating discs of gas and dust that surround young stars. With Sun-like stars, rocky, Earth-like planets form in the hot, inner area of the disc. Gas giants and ice planets, on the other hand, are formed further out, in cooler parts of the disc.

Why are Jupiter and Saturn so large?
In the inner solar system, the young Sun's radiation blew away any leftover gas at an early stage, but it was preserved for a longer period in the outer areas. This is why those planets that formed in the outer regions of the solar system had more time to attract

Saturn's spectacular ring system is 274 000 km wide, but less than 1 km thick.

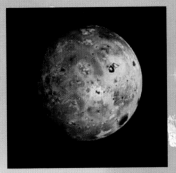   

Io, Europa, Ganymede and Callisto are among the 63 moons so far discovered in orbit around the planet Jupiter.

the gas present in their surroundings, and so were able to grow for much longer.

What is the Great Red Spot on Jupiter?

For more than 300 years, a reddish oval spot has been observed in Jupiter's cloud-streaked atmosphere. This is the Great Red Spot, a huge cyclonic storm that is about twice the size of the Earth.

Is there an ocean on Jupiter's moon Europa?

Magnetic field measurements indicate that an ocean about 100km deep is concealed under a 1km thick layer of ice on Jupiter's moon Europa. It may be that the giant planet's gravity heats the inside of the moon by 'squeezing' it, maintaining any water in a liquid form. The American space agency NASA is currently working on plans to send a space probe to this moon in an attempt to find out exactly what is happening there.

What are the rings around Saturn made of?

Saturn's rings are made up predominantly of ice particles, but they also contain dust and rocks. The size of the particles in the rings ranges from less than 1mm to several metres. Some astronomers think that Saturn's rings were formed as a result of a collision between one of the planet's moons and a wandering asteroid.

Do other planets have rings as well?

The planets Jupiter, Uranus and Neptune have also been found to have rings. These are not as prominent as the rings around Saturn, and they consist chiefly of dust and ice particles from the various planets' moons.

How was Neptune discovered?

The observed orbit of Uranus was found to deviate slightly from the path predicted by calculations. The reason had to be a more distant planet, and it was possible to calculate the position of this eighth planet on the basis of the orbital disturbance. The new planet was first observed in 1846, and given the name Neptune.

Are there any undiscovered planets in our solar system?

In addition to Pluto, there are many more dwarf planets beyond Neptune's orbit – and more are being discovered. It is unlikely that another large planet will be discovered further out in the solar system because astronomers have not observed any more inexplicable disturbances in the orbits of known planets and comets.

Where does the solar system end?

The solar wind meets the gas between the stars at a distance of about 150 times that between Earth and the Sun. This region is often called the boundary of our solar system. But the Sun's gravitational influence extends even deeper into space. It is assumed that our solar system is surrounded by a cloud of comets that stretches up to a light-year into space.

What makes the Earth and Moon special?

There are eight planets in our solar system and most have moons, but there is something special about the Earth and its Moon. The fact that there is life on Earth is one factor that makes it special, but there is another – astronomical – reason as well. No other planet has such a large moon relative to it's size. This is why some scientists do not consider the Earth and its Moon to be a typical planet-moon system, but prefer to think of it as a 'double planet'.

Some 4.5 billion years ago a massive cosmic collision took place that resulted in the formation of the Moon.

How was the Moon formed?

It is probable that the large moons of Jupiter and Saturn were formed from the solar nebula, at the same time as their planets. The smaller moons, on the other hand, are more likely to be captured asteroids. The history of the formation of Earth's Moon is quite different. According to current theories, our Earth collided with another, Mars-sized, proto-planet 4.5 billion years ago, shortly after the Earth had been formed. This caused a large part of the Earth's crust – and most of the other planet – to be hurled into space. This matter initially formed a disc around the Earth, and it

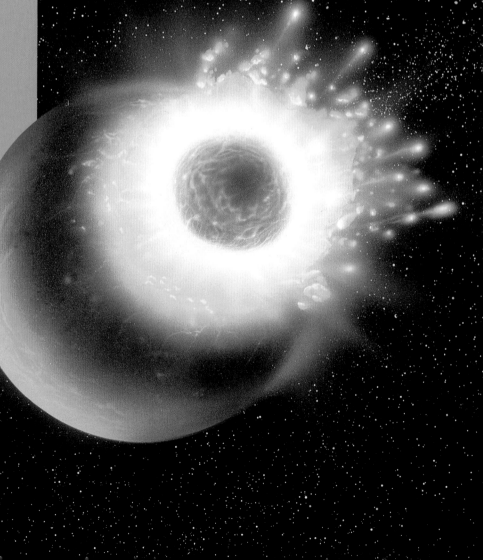

was out of this material that the Moon eventually formed. The newly formed Moon was probably only between 20 000km and 30 000km away from Earth and, because of its proximity, caused massive tidal fluctuations on the surface of the young planet. These tidal forces acted to ensure that, first, the Earth's rotational speed slowed down and, second, that the Moon moved further away from the Earth. The distance between Earth and Moon is now 384 400km on average, and this distance continues to increase by 3.7cm a year.

What is the man in the Moon?

Several prominent dark areas are visible with the naked eye on the surface of the Moon. These, with the aid of a little imagination, can be made to look like a face, popularly known in children's stories as 'the man in the Moon'. Astronomers christened these dark areas mare (plural, maria), which is the Latin word for sea. They are not seas but vast plains that were once flooded not by water but by molten lava. The lava solidified into a rock known as basalt, which is darker than the rocks that make up the rest of the Moon's crust. There are noticeably fewer craters in the lava plains than on the rest of the Moon's surface, and this has led scientists to conclude that the maria are younger than the Moon itself. They were formed about 4 billion years ago, when several large meteorites struck the Moon.

In European literature, the man in the moon is usually depicted as an old man carrying a bundle of sticks upon his back

How do we know...

... there may be water on the Moon?

The Moon is a completely dry, airless world, and strong sunlight in the vacuum of space would cause any free water to evaporate immediately. But there may well be water ice on some parts of the Moon's surface. Because of the orientation of its spin axis, there are no seasons on the surface of the Moon, as there are on Earth. During the course of a year the amount of sunlight reaching areas close to the lunar poles varies very little, with some parts being permanently shielded from the Sun by crater walls. In these areas, with temperatures of –230°C, it is possible that ice – brought to the Moon by comets early in the history of the solar system – could still be stored in the soil. Results obtained from several lunar probes seem to suggest that large amounts of ice are present close to the lunar poles, although this has not been confirmed by recent radar surveys conducted from Earth.

Why do we always see the same side of the Moon?

The Moon turns once on its axis in the time it takes for it to orbit the Earth, which is why the same side always faces us. This characteristic of the Moon's orbit – known as synchronous rotation – came about as a result of the tidal effect of the Earth on the Moon. The Moon does not have any oceans or seas, and no water to ebb and flow, but the Earth's gravitational field does cause the surface of the Moon to rise and fall. This effect has gradually slowed the Moon's rotation to a point where it now matches its orbital period. Although we always see the same craters and seas, thanks to the elliptical shape of the Moon's orbit, our angle of vision shifts slightly, which means we are able to see just over half – about 59 per cent – of the Moon's surface from Earth. Nobody knew what the far side of the Moon looked like until space probes were able to photograph it.

Why are so many craters visible on the Moon?

Most of the craters on the Moon were formed in the early history of the solar system, when innumerable remnants left over from the formation of the planets were still hurtling about in space. These small celestial bodies frequently crashed into the Moon, and into the Earth as well. This dramatic phase ended about 3.9 billion years ago, and since then the solar system has become a less dangerous place. The impact craters that were formed during that period have been preserved on the Moon, where there is no atmosphere, while those on Earth have almost disappeared as a result of erosion by wind and water, as well as movement of the Earth's crust. The impact craters that can be seen on Earth today are for the most part much more recent.

The impacts of innumerable meteorites have left the Moon's surface pockmarked with vast numbers of craters.

True or False?

Only the waxing Moon is visible in the evening

Almost without exception, all the objects in the solar system travel in their orbits in the same direction, including the Moon and the Earth. Viewed from a spot above the North Pole, they move anticlockwise. From Earth, the Moon seems to approach the Sun from the west and move away from it in an easterly direction after the new Moon. So a waxing (increasing) Moon is always to the east of the Sun and can be seen in the evening sky after sundown, and a waning (decreasing) Moon to the west of the Sun is seen in the morning sky before sunrise.

Why are astronauts able to jump so high on the Moon?

With a diameter of 3475km, the Moon is only one-quarter the size of the Earth. The difference is even greater when we compare masses, with Earth having 81 times as much mass as its satellite. For this reason gravity on the Moon is considerably less than it is on Earth. A human being – or any other object – on the Moon weighs only one-sixth what it does on Earth. This helped visiting astronauts move about in their heavy space suits, although the reduced weight meant the astronauts lost some of their surface grip so that their feet slipped away more easily, making walking on the Moon quite arduous. The astronauts also discovered that the moon buggy's wheels lost their grip easily, so they had to drive with extreme caution. They had to take great care not to accidentally damage their space suits. This concern about life-preserving space suits – and the lack of grip between their boots and the ground – is probably what stopped any of the astronauts from attempting to establish a new interplanetary athletics record with some really high jumps. The Moon's low gravity is also the reason why churned up dust takes such an unusually long time to sink back to the ground.

Why does the Moon have phases?

Because the Moon orbits the Earth, the angle between the Moon and the Sun changes constantly as seen from the Earth. If the Moon, as seen from the Earth, is exactly opposite the Sun, we see the half on which the Sun is shining, and we have what is termed a full Moon. If, on the other hand, the Moon is at right angles to the Sun, it is only half lit when viewed from Earth and we have a half Moon. If the Moon is in that part of its orbit where it is in the same line of sight as the Sun, it will show us its unlit side – a new Moon.

The term new Moon, as used by astronomers, is not identical with the appearance of the new Moon, which is of great importance, especially in the Islamic calendar. This is when the thin lunar crescent first appears, and it occurs between one and four days after an actual new Moon, depending on location and time of year.

What is a lunar eclipse?

At a full Moon, the Moon (as seen from Earth) is exactly opposite the Sun, with the Earth in the middle. The Earth's shadow falls in the direction of the Moon, which is why, when the three celestial bodies are aligned, the Moon can sometimes cross Earth's shadow. When this happens, the Moon goes dark as it is eclipsed. We speak of a total eclipse of the Moon when the Moon travels completely into the Earth's umbra – the darkest part of its shadow. An observer on the Moon would then see the Sun as being completely obscured by the Earth. An observer on the Moon would see the Sun only partly covered by Earth if the Moon enters the Earth's penumbra – its partial shadow. Because some sunlight still reaches the surface of the Moon when any part of it is in the Earth's penumbra, this is known as a partial eclipse of the Moon.

Why don't lunar eclipses happen every month?

The Moon's orbit is at an angle of 5.1° to the Earth's orbit. This is not much, but is sufficient to ensure that the Moon can travel up to 37 000km above or below the centre of the Earth's shadow. A lunar eclipse is only possible when the full Moon crosses the Earth's orbit, and this only happens twice a year. Unlike solar eclipses, which can only be seen by observers watching from small portions of the Earth's surface, a lunar eclipse can be seen from anywhere on the side of Earth facing the Moon.

Why is the Moon red during a total eclipse?

Even during a total lunar eclipse the Moon does not become fully black, but instead it glows a dark and ghostly copper colour. This is because the Earth's atmosphere refracts (bends) some sunlight so that it falls into the shadow area on the Moon. For an observer on the surface of the Moon this would look as if the Earth were surrounded by a red glow. Because blue light is scattered more strongly in the Earth's atmosphere than red light – the reason we have red sunsets and sunrises – it is mainly red light that reaches the Moon.

During this lunar eclipse, only part of the Moon was in the Earth's umbra.

Visitors to the Panthéon in Paris – and many museums around the world – can observe the rotation of the Earth indirectly by means of a Foucault Pendulum.

How do the Earth and the Moon move through space?

It is often said that the Moon orbits the Earth, but that is an oversimplification. First, the Moon does not move around the Earth, rather both are moving together around their joint centre of gravity. Because of the great difference in mass, this centre of gravity falls within Earth. Second, the Earth and the Moon together orbit the Sun. If we look at the movement of the three bodies together, the Moon no longer seems to go around the Earth. Instead, the Moon accompanies the Earth on its orbit around the Sun. Sometimes it is on the inside of the Earth's orbit and sometimes on the outside. Sometimes it is leading the Earth and sometimes following. Because the distance between the Moon and the Earth is much less than the Earth–Sun distance, the Moon's orbit around the Sun is barely distinguishable from a circle. It is not a spiral, as it is sometimes depicted, since the Moon never travels in the opposite direction to the Earth.

How do we know it is the Earth that is moving and not the sky?

In 1851, the French physicist Jean Foucault conducted an experiment in which he allowed a 2m long pendulum to swing backwards and forwards close to the ground. During the course of several hours of steady motion by the pendulum, he noticed that its direction of movement was gradually turning. Foucault had been at pains to ensure that no external forces could influence the pendulum, so the only possible solution to the riddle of what he was seeing was that it was the ground beneath the pendulum – the Earth itself – that was turning. This

was the first time anyone had produced experimental proof that the Earth is rotating. Foucault subsequently repeated the experiment, using much longer pendulums, in the Paris Observatory and in the Panthéon, where the experiment was eventually opened to the public.

Why doesn't the Earth's rotation hurl us off the surface of the planet?

The Earth rotates very rapidly, its speed at the equator being almost 1700km/h. To keep all of the inhabitants of the Earth firmly on the ground and rotating along with the planet, requires the gravitational pull of the Earth's immense mass. It is gravity that keeps our feet on the ground. In fact, there is a reduction in weight due to rotational speed that means that we weigh around 0.5 per cent more at the North and South Poles than we do at the Equator. If there was a planet that actually rotated so fast that it was impossible for its inhabitants to be held down by gravity if they stood at its Equator, the planet itself would fly apart.

The Earth and its Moon in a composite photograph taken by the *Galileo* space probe.

What dangers threaten us from outer space?

Our Earth is an island of life in the hostile ocean of space. Space is a cold vacuum full of dangerous radiation and high-energy particles, as well as objects of all shapes and sizes – from specks of dust to rocks, asteroids and comets – hurtling around in all directions. But reassuringly, the Earth's magnetic field and its atmosphere help to protect us from many of these dangers.

A magnetic field (green) protects Earth from particles that emanate from the Sun or from outer space.

What are the magnetic field and the atmosphere protecting us against?

Earth is surrounded by a strong magnetic field that operates like an invisible shield protecting us from the majority of high-energy particles that come from the Sun and from deep space. It is able to do this because the particles are usually electrically charged and can be diverted by a magnetic field. Meanwhile, the atmosphere protects us from dangerous radiation. The ozone layer absorbs some of the ultraviolet radiation that causes skin cancer. Even the more energetic X-rays and gamma rays are blocked by the atmosphere. The atmosphere also protects us from the smaller pieces of rock that are constantly bombarding Earth from space. They simply burn up and never reach the surface.

Why doesn't the Earth's atmosphere just blow away into space?

Like any other kind of matter, air is subject to the Earth's gravity. In order to escape from the planet, the gas molecules that make up the atmosphere would have to move with sufficient speed to overcome gravitational attraction. The speed of a molecule depends on the temperature of the gas of which it is a part, and the speeds achieved in the temperatures that prevail in the Earth's atmosphere do not come close to those required to permit the molecules to escape into space. So Earth is able to retain its atmosphere. Not all planets and moons are able to do this. The strength of a body's gravitational field depends on its size and mass, which is why small celestial bodies like Mercury and our Moon have no atmosphere. In contrast, large objects, even some moons like Saturn's Titan, can retain substantial atmospheres.

Are exploding stars a danger for Earth?

If a star were to explode in our immediate vicinity, the radiation released would represent a grave danger to life on Earth. Fortunately, the closest star with sufficient mass to explode as a supernova at the end of its life is Betelgeuse, which is more than 400 light-years away in the constellation Orion. Given the distance, the Earth's magnetic field and its atmosphere should be able to provide us with adequate protection when such an explosion occurs.

Far more dangerous would be the explosion of a high-mass star made from primordial gas (with a low heavy element content). This is believed to lead to the creation of gamma ray bursts – bundles of intense, high-energy radiation. If these were to strike Earth, the ozone layer could literally be torn away – even if the explosion itself took place many thousands of light-years distant – and for years afterwards life on Earth would be exposed to dangerous ultraviolet radiation. Current scientific opinion is that potentially dangerous gamma ray bursts of this kind only ever occur in young galaxies. At the grand old age of 10 billion years, our Milky Way is far too old for such an event.

Huge amounts of energy are released when a star explodes. Only planets around stars far from the supernova are safe.

What are shooting stars?

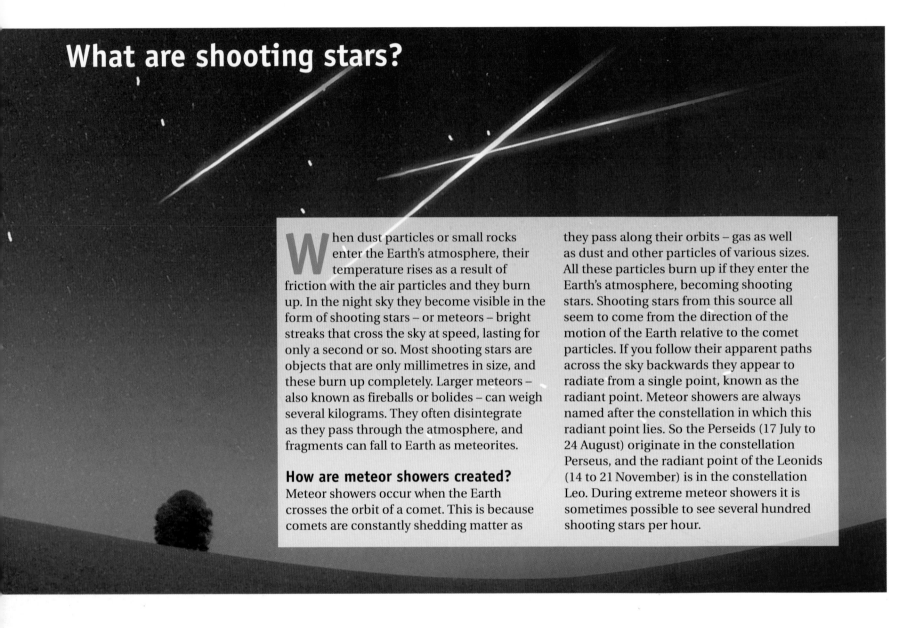

When dust particles or small rocks enter the Earth's atmosphere, their temperature rises as a result of friction with the air particles and they burn up. In the night sky they become visible in the form of shooting stars – or meteors – bright streaks that cross the sky at speed, lasting for only a second or so. Most shooting stars are objects that are only millimetres in size, and these burn up completely. Larger meteors – also known as fireballs or bolides – can weigh several kilograms. They often disintegrate as they pass through the atmosphere, and fragments can fall to Earth as meteorites.

How are meteor showers created?

Meteor showers occur when the Earth crosses the orbit of a comet. This is because comets are constantly shedding matter as they pass along their orbits – gas as well as dust and other particles of various sizes. All these particles burn up if they enter the Earth's atmosphere, becoming shooting stars. Shooting stars from this source all seem to come from the direction of the motion of the Earth relative to the comet particles. If you follow their apparent paths across the sky backwards they appear to radiate from a single point, known as the radiant point. Meteor showers are always named after the constellation in which this radiant point lies. So the Perseids (17 July to 24 August) originate in the constellation Perseus, and the radiant point of the Leonids (14 to 21 November) is in the constellation Leo. During extreme meteor showers it is sometimes possible to see several hundred shooting stars per hour.

Can you buy meteorites?

To own a rock from outer space is an idea that fascinates many people. Meteorites are easy to purchase, with many examples being offered on Internet websites such as the Meteorite Market (www.meteoritemarket.com). Special meteorite fairs are often held in parts of Europe and in the USA, and these are also good places to buy specimens. Prices vary enormously, ranging from as little as a couple of pounds up to hundreds or even thousands of pounds, depending on the type, size, rarity and condition of the meteorite.

How do we know...

... that Earth puts on weight with every day that passes?

The amount of matter that falls onto the Earth from outer space is difficult to measure, but it is estimated to be several tonnes a day. This may sound like a lot but, compared to the mass of the Earth, it is relatively little. Even over a period of 10 billion years, the increase would be less than one-millionth the total mass of the Earth.

Did the water on Earth come from comets?

Many scientists today believe that comet strikes during the early history of the solar system had positive as well as negative effects. Comet strikes may have been destructive, but it is also possible that they brought water to the Earth, and possibly even complex carbon molecules as well. The simultaneous arrival of these ingredients from outer space would help to explain the surprisingly rapid emergence of life on Earth.

Could life have originated in comets?

Some scientists would go even further and claim that life itself was formed inside comets. All the necessary ingredients were there, and radioactive decay could have produced the heat necessary to melt ice, producing a supply of liquid water within a comet's core.

Radical claims such as these can only be checked by examining cometry matter, which is precisely the task given to *Rosetta*, a European space probe that is due to approach comet 67P/Churyumov-Gerasimenko in 2014. *Rosetta* has a small (90cm) lander, named *Philae*. To make sure it doesn't bounce back into space in the comet's almost imperceptible gravity, the lander will attach itself to the ground by means of harpoons. *Philae* will then be able to use a small drill to remove samples from the comet's crust, which will then be analysed in the lander's small, automated laboratory.

In September 2001, the probe *Deep Space 1* flew to within 2200km of the comet Borrelly.

As part of the spectacular *Rosetta* mission, in 2014 the probe's tiny lander will become the first space craft to land on the surface of a comet, gather samples and analyse them.

Missiles from outer space

A gigantic object hurtles towards Earth from outer space – a terrifying scenario that has been played out repeatedly on film and TV.

An aerial view of the epicentre of a blast that occurred when what is thought to have been an asteroid exploded high above Siberia in 1908.

Siberia: 30 June 1908. Suddenly a massive explosion rips apart the silence hanging over plains near the Stony Tunguska River, and a huge pillar of smoke and fire rises to the sky. The thunder of the explosion is heard hundreds of kilometres away, and seismic tremors are registered all around the globe. Trees across an area of more than 2000sq km are snapped like matchsticks and a forest fire rages beneath the centre of the explosion.

This real catastrophic event was probably caused by a 30m-wide asteroid exploding 10km above the Earth's surface. The affected region was almost uninhabited so, although several reindeer herds fell victim to the explosion only two people died. Had the asteroid exploded above a city it would have been a tragedy of major proportions. The force of the blast was comparable to 10-20 megatons of TNT – the equivalent of 50-100 times the explosive power of the Hiroshima atom bomb.

Astronomers have calculated that an event like the one at Tunguska involving an asteroid is likely to happen somewhere on Earth every couple of hundred years. This is because of the innumerable asteroids and comets that are travelling through our solar system, one of which could at some point crash into the Earth.

Are comets dangerous?

In medieval times, the appearance of a large comet in the sky caused fear and alarm because they were considered to be the harbingers of plagues and wars. As recently as 1910, when Earth crossed the tail of Halley's Comet, many people feared that the world was about to end because astronomers had detected substances like sulphur and cyanide in the comet's tail. Today we know that the matter in a comet's tail is much too thinly distributed to pose any danger to us. A comet is only dangerous if it crosses Earth's path. In 1994, the comet Shoemaker-Levy crashed into Jupiter, a vivid demonstration that such catastrophes are not events restricted to the past. It is extremely unlikely that humankind would have survived if Shoemaker-Levy had collided with the Earth rather than Jupiter.

What happens when an asteroid or comet hits Earth?

Boulders from space as small as 1m across can cause explosions that are comparable to the blast that resulted when the atom bomb was dropped on Hiroshima. On average, events of this kind are registered once a year by reconnaissance satellites circling the Earth. The shockwaves do not reach the Earth's surface because the explosions occur in the uppermost layers of the atmosphere. Many small fragments of rock from space strike the Earth as meteorites, but an asteroid needs to be at least 50m across in order to have a chance of reaching the ground or even the lower levels of the atmosphere. Such an event above a populated area would be devastating. An asteroid landing in the sea close to the coast would also have dramatic consequences as it would generate huge waves that would wash ashore. A global catastrophe would require an asteroid with a diameter in excess of 1km. The enormous amount of dust such a strike would inject into the upper layers of the atmosphere would be particularly destructive. Such an event would lead to a global winter that would last for several months, resulting in the collapse of global agriculture and other food production.

Did an asteroid cause the extinction of the dinosaurs?

There have been several mass extinctions in Earth's history. The best known occurred at the transition between the Cretaceous and Tertiary periods, 65 million years ago, when around one-third of the planet's animal and plant species, including the dinosaurs, disappeared. Experts agree that this event was caused, or at least triggered, by the impact of an asteroid or comet. An impact crater up to 300km across, formed at this time, has been found on Mexico's Yucatan Peninsula.

Can humans protect themselves from asteroids and comets?

Experts disagree about whether it would be possible to defend the Earth from rogue comets and asteroids. Unlike weapons experts, who see this as a new business opportunity, astronomers remain unimpressed and maintain that protective measures may not always be necessary. Depending on the size of the body involved, an impact in a large ocean, far from a coast – or in an unpopulated land area – would probably not pose a major threat to life. If calculations predicted an impact by an Earth-bound asteriod in a heavily populated area, scientists believe there would still be plenty of time to prepare a defence. They expect to have warning of an approaching asteroid several decades in advance of an impact, and there are already programmes in place to record all potentially dangerous asteroids in order to make long-term forecasts of their trajectories. Comets are more of a problem, because their emergence from the outer reaches of the solar system cannot be predicted. Their orbits are also hard to calculate because surface eruption of gas can change their course.

What sort of protective measures could be taken if a cosmic boulder were actually to threaten Earth? Blowing it up, as in the 1998 movie *Armageddon*, would not be possible with the means at our disposal today. Nor would it be advisable, since thousands of fragments crashing into the Earth would be even more destructive than an intact asteroid. Given sufficient advance warning, a small course correction would be enough to divert an asteroid and make it miss our planet. This might be achieved by attaching a solar sail or a rocket engine to the surface of the oncoming threat.

Some 30 000 years ago, a meteor about 30m in diameter fell in Arizona's desert at a spot appropriately known as Meteor Crater. The result was the 1200m-wide, 170m-deep hole.

Why do we need satellites?

Satellites are far from an indulgent luxury for the use of the rich nations. The everyday lives of almost everyone on Earth would be affected if satellites were to disappear. They have become the foundation of modern telecommunications. Without satellites, there would be no global positioning system (GPS), on which ships and aircraft depend. Satellites provide data for weather forecasts and monitor everything from forest fires to marine pollution. They are also vital in science. Without telescopes in space, astronomers would have a greatly restricted view of the cosmos.

The Giove-A satellite was the first to be launched into space as part of the European Galileo navigation system.

Who built the first rockets?

Rockets driven by black powder (gunpowder) have been part of the arsenal of European armies for centuries. Invented by the Chinese and brought to Europe by Mongolian warriors, an 1807 on Copehagen saw the British army fire more than 25 000 rockets. Further developments in rocket technology were driven by three men: Konstantin Tsiolkovski in Russia; Robert H Goddard in the USA; and Hermann Oberth in Germany. At the forefront of their endeavours was the search for a better fuel – gunpowder was not powerful enough to propel a rocket into space. In 1926, Goddard built his first liquid fuel powered rocket. It only flew for 46m, but this approach proved to be the way forward. The first ballistic rocket – the *V2* – was developed during the Second World War for German armed forces by Wernher von Braun and his team.

When was the first satellite sent into space?

In October 1957, the Soviet Union used a rocket to launch the first man-made satellite into an orbit around the Earth. This caused some dismay, especially in the USA where there had been a general conviction that the Americans were ahead in the race to conquer outer space. For 57 days, *Sputnik I* sent out its famous 'beep-beep' signal before burning up in the atmosphere. It took the Americans until January 1958 to launch their first satellite. *Explorer 1* was the first real scientific satellite, and counted among its achievements the discovery of the Van Allen radiation belts – bands of energetic charged particles that girdle the Earth.

The American rocket pioneer Dr. Robert Goddard was among the first to experiment with liquid fuel for rockets.

How many satellites are there now?

More than 5800 satellites have been sent into space since 1957. About 3100 satellites currently orbit the Earth, although only about 1000 of these are still operational – the remainder should really be regarded as space junk. Nowadays, space agencies try to give satellites targeted crash landings at the end of their useful lives, but at the beginning of the space age this kind of precautionary measure had not been considered. After several years, low orbiting satellites crash of their own accord because the friction caused by the extremely thin outer atmosphere of the Earth gradually slows them down.

What is a geostationary satellite?

The speed of a satellite in a high orbit is less than it is in a low orbit, and the distance that a satellite has to travel on its path around the Earth also increases with its altitude. All of this means that the orbital period of a satellite increases with increasing altitude. At an altitude of 36 000km, the orbital period is precisely 24 hours. For a satellite in an orbital path that takes it along directly above Earth's equator, this means that it will always be positioned in its orbit above precisely the same spot on the surface of the planet, since the Earth also completes one full revolution in 24 hours. This kind of orbit is called a geostationary orbit. Geostationary orbits have obvious advantages for telecommunications satellites, since the fact that the satellite is always in the same place means that the dishes sending and receiving signals can point at a fixed position.

Why don't satellites fall out of the sky?

Satellites travel around Earth without extra propulsion. They remain in orbit despite the Earth's gravitational pull because they travel so extremely rapidly. If there were a satellite that orbited very close to Earth's surface, it would have to travel at 7.9km/sec, or 22 000km/h. The need for speed reduces with increasing altitude. At a height of 400km – at about the same altitude as the orbit of the International Space Station – the speed has fallen to 7.7km/sec.

Why can't balloons and aircraft travel into space?

A balloon rises upwards because the gas inside it is lighter than the air that surrounds it. An aircraft's dynamic lift – which is what allows it to fly – is produced when air flows around the machine's wings. Furthermore, most aeroplane engines need oxygen from the air in order to burn their fuel. Both balloons and aircraft require air around them in order to stay aloft. Because there is no air in outer space, balloons and aircraft would be unable to fly in that environment. This is why spacecraft need to be powered by something that does not depend on either the lift provided by passage through the air, or on oxygen that the air provides. The fuel that powers rocket motors includes its own oxygen.

Space is beyond reach for any machine powered by an engine that relies on using oxygen present in the Earth's atmosphere.

Is space junk dangerous?

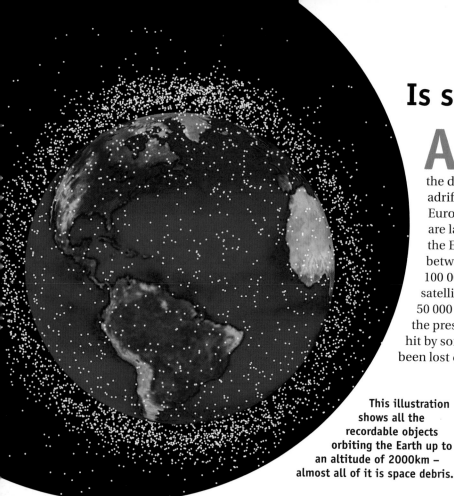

A large proportion of all the junk in space was produced by the almost 200 explosions that have taken place in the Earth's orbit in the past 50 years. These include the controlled explosions of old rocket stages and the destruction of leftover fuel. Also among the debris are dead satellites set adrift, tools lost by astronauts, paint flakes and much more. American and European radar systems have registered more than 10 000 pieces of junk that are larger than 10cm and fly at altitudes of up to 2000km above the surface of the Earth. Nobody knows precisely how many less easily registered particles of between 1cm and 10cm there are, but the experts fear there may be more than 100 000 of them. All these objects flying around pose a permanent danger to satellites, manned spacecraft and space stations, because at velocities of 50 000 km/h, a 1cm-long piece of metal is capable of punching a hole through the pressure hull of a spacecraft. In July 1996, the French spy satellite *Cerise* was hit by some debris and badly damaged. Nobody knows how many satellites have been lost due to collisions with space junk.

This illustration shows all the recordable objects orbiting the Earth up to an altitude of 2000km – almost all of it is space debris.

Why do we need telescopes in space?

T here are many disadvantages to observing the sky from the surface of the Earth. Turbulence in the atmosphere causes images to become blurred, which is why large, Earth-bound telescopes are not able to exploit their theoretical resolution fully. Space telescopes, such as the Hubble Space Telescope, can provide clear images because they are orbiting 569km above the Earth and its atmosphere. The atmosphere – especially the water vapour it contains – absorbs a large part of the radiation that reaches Earth from space. It is only by means of visible light and those radio wavelengths between a few millimetres and several metres, that astronomers can get a reasonably clear look into space from the ground. To study using other types of radiation – infrared, ultraviolet, gamma rays and X-rays – their instruments are best moved into space.

How do we know...

... that satellites are visible in the sky?

After nightfall and before sunrise on Earth there are some orbiting satellites that may be lit by the sun. These are visible as slowly moving dots of light, which may suddenly go out when they enter the Earth's shadow. Relatively few satellites reflect enough sunlight to rival the brightest stars. Those that do include the International Space Station (ISS), which can be as bright as the planet Jupiter. Iridium flashes – which last a few seconds and are extremely bright reflections from Boeing's Iridium communications satellite system – are a particularly spectacular sight. Some of these Iridium flashes are even bright enough to be visible during daytime – if you know where to look. For more information on satellite visibility, see the Heavens Above website (www.heavens-above.com).

How does GPS work?

The global positioning system (GPS) is made up of many satellites that are constantly transmitting very precise time signals to Earth. These are produced with the aid of atomic clocks. Because the satellites are positioned at various distances from the receiver, it is possible to calculate the position of the receiver from the satellites' orbits and the time delays experienced in receiving the signals – provided the signals are received from at least four satellites. The problem is that GPS satellites travel through space at 14 000km/h. Because of their speed, the clocks on board the satellites – as predicted by Einstein's theory of relativity – run slower than clocks on Earth. A second effect is caused by the Earth's gravitational field being a little weaker in orbit than it is at ground level. This causes the clocks on board the satellites to run a little faster than those on Earth. Unfortunately, these two effects do not cancel each other out and the navigation system has to make precise allowances for the temporal distortions caused by gravity and motion. Since the divergence amounts to several kilometres a day, these corrections are essential if GPS is to get us to where we want to go.

A radar image of the Strait of Messina, which separates Sicily from the rest of Italy.

True or False?

Satellites can see through clouds

Satellites don't just observe the Earth using visible light, they also make use of other types of radiation. Each of these radiation bands provides insights into different areas of interest. The infrared band provides information about temperatures, while microwaves are particularly useful when it comes to observing precipitation. Using radar, satellite sensors can even penetrate clouds. Radar observations also provide the best data on subtle differences in height of the Earth's surface. This is how satellites can detect the minute movements of the Earth's crust that can act as indications of impending earthquakes or volcanic eruptions.

Have all the planets been visited by space probes?

The planet closest to the Sun, Mercury, has had two visitors from Earth. The first was the American *Mariner 10* space probe in 1974. A second visitor, also American – the *Messenger* space probe – made its first flyby in January 2008. The European *Venus Express* is at present orbiting Venus, and Mars is being observed by an entire fleet of probes, which include the European *Mars Express* and the American *Mars Reconnaissance Orbiter*. The last probe to visit Jupiter was the American *Galileo* craft, which burnt up in the giant planet's atmosphere in September 2003. Saturn is currently being orbited by the American probe *Cassini*, and the American *Voyager 2* probe flew past Uranus and Neptune in 1986 and 1989 respectively. Every planet in the solar system has had a visit from at least one terrestrial space probe.

The European *Venus Express* space probe has mapped the planet's atmosphere over many orbits, and has covered the lower atmosphere for the first time.

Which was the first planetary probe?

After their success with *Sputnik I*, the Soviet Union attempted to beat the USA in the race to send unmanned space probes to investigate the planets. In October 1960, two Mars probe launches failed. Although an initial Venus probe was successfully launched in February 1961, it didn't get beyond Earth's orbit because the final rocket stage failed to ignite. A second attempt several days later was more successful, but only seven days after the launch, radio contact with the probe was lost. Then, in July 1962, the USA made their first attempt to send a probe to Mars, but the launch failed. The next Soviet attempt to reach Mars, also in 1962, failed. Then an another Venus probe failed to leave Earth's orbit. In the end, it was the Americans who 'won' when, on 27 August 1962, they launched *Mariner 2*. In December that same year it flew within 34 773km of Venus and into the history books as humankind's first successful planetary probe. It was another four years before the Soviet Union celebrated a successful planetary mission, when *Venera 3* crashed into Venus in 1966.

What will power the space probes of the future?

Today, most space probes are accelerated to high speeds by means of chemical rockets, and then – except for a few steering manoeuvres – they travel to their distant destinations without any additional propulsion. Flights to the outer planets using this technique generally take several years. In order to reduce these times, future probes will also be accelerated during their flight. Ion drives, which have already been used in some probes, are considered to have a particularly promising future. This technology uses inert gases, such as Xenon, as fuel. The gas is heated until it ionises – that is, until the electrons have been stripped away from around the atoms of gas. With the help of magnets, the inert, electrically charged gas ions are then shot out of the back of the drive. The propulsive force achieved is small – equivalent to less than the weight of a postcard – but ion drives can run without interruption for months on comparatively small quantities of fuel, unlike chemical rockets.

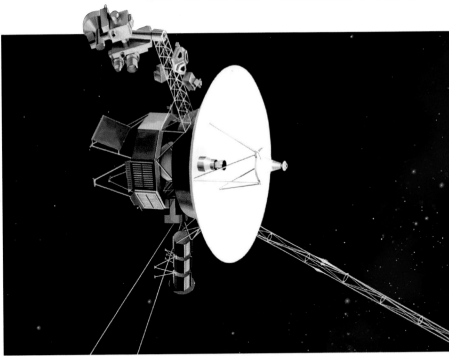

The NASA *Voyager* spacecraft, launched in 1977, has so far penetrated further into space than any other probe.

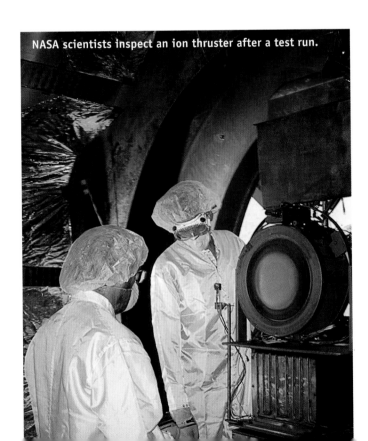

NASA scientists inspect an ion thruster after a test run.

Has a spacecraft ever left the solar system?

So far, no space probe has completely left the solar system. In 2005 the American *Voyager 1* probe became the first to arrive in a region that many scientists consider to be the edge of the solar system – the boundary of the so-called heliosphere. The heliosphere is a gigantic bubble blown by the solar wind – a current of electrically charged particles that is constantly emitted from the Sun. At the edge of the heliosphere, the solar wind collides with interstellar gases. Each year, *Voyager 1* travels a further 525 million km away from the Sun – a distance that is three and a half times that between the Earth and the Sun.

How long will *Voyager 1* stay in radio contact?

Even today – over 30 years after its launch on 5 September 1977 – *Voyager 1* is still collecting information and transmitting it to Earth. The energy of the detected signal is now 20 billion times weaker than the battery power of a digital watch. Despite this, NASA scientists hope to maintain contact with the probe until about 2020, after which its atomic battery will cease to provide sufficient energy.

Why bother sending people into space?

This question is at the heart of an often heated debate – with supporters and opponents of manned space travel both making convincing cases for their respective points of view. Unmanned space probes are much cheaper than manned craft because they do not require life support systems. And exploration of the solar system would not have been possible without robot space probes. But there are many areas where humans are superior to machines, no matter how clever the machines seem to be.

A 17m long arm allows astronauts to work on the ISS.

Who was the first human in space?

Four years after the 'Sputnik shock' of 1957, the USA had to accept another humiliating defeat in the space race. In April 1961, the Soviet Union sent the first human being into space aboard the *Vostock 1* rocket. The cosmonaut Yuri Gagarin and his

Almost seven years after his spectacular trip into space, Yuri Gagarin died in an aircraft training flight.

space capsule circumnavigated Earth only once, and after a flight lasting 1 hour and 48 minutes he landed safely back on Earth in southern Russia. The flight was entirely remote controlled and Gagarin would only have needed to intervene in the event of an emergency. It wasn't until February 1962, after two successful sub-orbital flights, that the USA succeeded in sending their own astronaut, John Glenn, into orbit.

Why are astronauts in space weightless?

Weightlessness does not mean that there is no gravity, only that the gravity is not perceptible. In fact the gravity in an orbit close to Earth is not much less than it is on Earth's surface. If we could construct a tower tall enough to reach that height, a person standing on the top of the tower would not be weightless. We feel our weight because the ground pushes up on us, balancing the force of gravity. As there is no ground below, when an astronaut is floating in space, he or she does not feel the force. This is also true if the astronaut is in a satellite. This does not provide any firm ground either as it orbits the Earth with the astronaut. Gravity is nevertheless present, and keeps the spaceship and its astronaut in Earth's orbit, just as it does the Moon.

True or False?

Weightlessness can make you ill

Weightlessness can cause astronauts to suffer from space sickness as a result of disturbances to the vestibular system in the inner ear. This is the same effect as seasickness. About 60 per cent of astronauts suffer from space sickness on their first mission. Medication is available, although symptoms tend to disappear after a couple of days. More serious are the results of spending long periods in weightlessness. It can cause muscular atrophy, calcium depletion in the bones, reduced cardiac function and blood count changes. Despite the intensive training programme undertaken by astronauts going aboard spacecraft and space stations, these effects can only be partially avoided.

How many people have been to the Moon?

The US Apollo space project sent a total of only 12 people to the Moon in the years between 1969 and 1972. First to reach the Moon, on 20 July 1969, were the *Apollo 11* astronauts, Neil Armstrong and Edwin Aldrin. On 14 December 1972, the commander of the *Apollo 17* mission, Eugene Cernan, became the last human being to have stood on the lunar surface. Although the Americans sent nine manned flights to the Moon, only six landed. *Apollo 8* was the first manned spacecraft to circumnavigate the Moon. During the flight of *Apollo 10*, tests were carried out on the lunar module while in orbit around the Moon, but the astronauts did not land. *Apollo 13* was severely damaged on its outward flight to the Moon and as a result the astronauts had to abandon their plans for a landing as they struggled to survive. After *Apollo 17*, the lunar mission programme was cancelled for financial reasons, although the original plan had been for another three manned landings.

When will people fly to the Moon again?

In 2004, US President George W. Bush announced a new 'vision for space exploration', according to which American astronauts would again set foot on the Moon's surface within 15 years. This time, the plan was not for single, short visits like the Apollo missions, but for the establishment of an enduring presence on the Moon. The goal is to establish a manned lunar station. In the meantime, a new space capsule named *Orion* and a new carrier rocket named *Ares* are being developed for the Moon flights. The first manned flights using this new system are planned for 2014, with the first Moon landings expected in 2018.

After Neil Armstrong, astronaut Edwin 'Buzz' Aldrin (left) was the second human being to set foot on the Moon.

The Earth rising above the Moon in an image taken during the Apollo 11 mission. There is a simple explanation for the lack of stars.

Why are there no stars visible in the *Apollo* Moon photographs?

To photograph stars in the night sky with a normal camera requires exposure periods that are at least several seconds long. The same is true on the Moon. The photographs taken by the *Apollo* astronauts were not taken at night, but during the day – the Sun was always shining in areas where Moon landings took place. So the astronauts were not taking night shots, but ordinary daylight shots with correspondingly short exposure times of about 1/125 second. Even though the lack of an atmosphere on the Moon means that the sky is dark during the day, the short exposure times are not enough to allow stars to show against the blackness.

Were manned flights to the Moon simply clever fakes?

Could it be that thousands of people at NASA, at aerospace companies and in the American government have been party to a giant conspiracy and remained silent until now? This seems highly unlikely. In fact, the supposed proofs put forward by those who believe in 'The Great Moon Landing Hoax' are easily refuted. Is the American flag fluttering in the wind? No, it's moving because the astronauts caused it to move when they raised it. Shouldn't shadows on the Moon be jet black? No, because the surface of the Moon reflects sunlight and scatters it into the shadow. There are sensible, logical explanations for all the supposed anomalies pointed out by Moon landing sceptics. Of course, as final proof – if any were needed – there are the 382kg of moon rock that the *Apollo* astronauts brought to Earth. These have a chemical composition quite unlike anything found on this planet.

What do we need a space station for?

The conditions for performing extended experiments in weightlessness under constant human supervision exist only in a space station. It is possible to carry out some experiments on board unmanned satellites, but having astronauts work on processes allows more flexible and more useful experiments. In the long term, it is hoped this will lead to the manufacture of new materials and medicines in space, although sceptics believe that production in space would be too expensive and so of little interest to industry. Space stations are particularly indispensable when it comes to preparing for flights to other planets. This is because the effects on the human body of long periods in space can only be researched by actually having people spend extended periods in space – and the only practicable place for that is aboard a space station.

The International Space Station (ISS) in mid-2005, photographed by astronauts aboard Space Shuttle *Discovery*.

What can astronauts do to protect themselves from radiation in space?

Protecting astronauts from cosmic radiation during long space missions – a trip to Mars, for example – would require a shield of matter with a density of around 1kg/sq cm, which would involve something like a 10m thick layer of water. The enormous weight of such a shield would vastly increase the costs of a mission to Mars, which is why space travel experts are searching for alternatives. It may be possible to produce an extremely strong magnetic field to act as a shield, but the strengths required would call for the installation of superconducting electromagnets aboard spacecraft.

How long would it take to travel to Mars?

The European Space Agency (ESA) is planning a manned Mars mission as part of its Aurora programme. It is due to take place in 2033.

With the rockets currently in use, a flight to our neighbouring planet would take around six to seven months – provided the launch took place when Mars and the Earth were suitably aligned. This latter factor is a major problem facing a manned trip to Mars. The astronauts would have to wait on Mars for anything up to 18 months before Earth would again be in a suitable position for the return flight. This means that a manned Mars mission would require a total travel time of at least two and a half years. The journey could be completed much faster with something like a powerful nuclear engine. This would not only cut travel time to Mars to just two months, but it would also reduce the need to take into consideration the relative positions of the two planets. But a nuclear engine that would enable this kind of extremely rapid trip has to date only made it as far as the engineer's drawing board.

Why is a long flight to Mars so dangerous?

Individual astronauts have spent about the same periods of time aboard orbiting space stations as it would take to travel to Mars. But a Martian flight is a more serious proposition than spending time – however long – in a near-Earth space station. If there was a medical emergency on the outbound leg, for example, it would be impossible to return to Earth, since much of the route to Mars is a one-way street.

The Mars-bound travellers would also have to take with them all the provisions needed for the journey, whereas a space station can receive regular supplies from Earth. In addition, there are the well-known dangers posed by the slow bone and muscle deterioration caused by living in weightless conditions. The greatest of these is that posed by radiation in space and on the surface of

Mars. Astronauts in near-Earth space are protected from most of this by the Earth's magnetic field, but in outer space and on Mars – which has only a very weak magnetic field – this would not be the case.

How do we know...

...that you'd only be a third of your weight on Mars?

At 6787km, the diameter of Mars is only about half that of Earth (12 756km). The red planet is also only about one-tenth the mass of Earth, which is why the pull of gravity is much less on the surface of Mars than it is on the surface of the Earth. As a result, the weight of a person – or any object – on Mars is only 38% of what it would be on Earth. A person who weighed 60kg on Earth would weigh only about 23kg on the red planet.

What does terraforming mean?

To shape another planet in Earth's image is the declared goal of terraforming visionaries. The planet in our solar system that would be most suitable for this kind of treatment is Mars. This is because the now parched red planet seems to have once possessed a denser atmosphere and free water on its surface. Attempts to 'terraform' Mars could include the production of large volumes of carbon dioxide to bring about an artificial greenhouse effect and so reproduce conditions that once existed on the planet. Another idea would be to deploy huge mirrors in space around the planet to melt the frozen water and carbon dioxide locked up in the Martian soil and so bring about climate change. This kind of transformation of an entire planet would take millennia – and would be a gigantic and costly experiment with highly uncertain outcomes.

This vision of the future prepared by proponents of terraforming shows a factory on Mars designed to emit carbon dioxide. The aim of the plan is to trigger an artificial greenhouse effect.

Visitors can try out astronaut training equipment in Huntsville's space museum.

including a lift-off during which they must bear several times their own weight. Added to that, unavoidable stresses can drive the pulse rate up to more than 200 beats a minute – not something for those with weak hearts. Anyone planning a trip into Earth's orbit must be in the very best of health. Cardiovascular diseases, bone or muscle disorders and auto-immunity diseases are all disqualifying criteria. The health requirements for flights to the edges of space are much less onerous, with the only disqualifying criteria being certain heart diseases.

When will I be able to travel into space?

In principle, outer space is already open to ordinary people – the only requirement being great wealth. Each year the Russian space agency conveys one or two paying space tourists to the International Space Station for a week-long stay. Most of us would need to save for a long time before we could pay the US $22 million fee needed for that particular package holiday. Would-be space tourists must also complete a six-month training programme. For those who would be satisfied with a couple of minutes of weightlessness at an altitude of just over 100km, there will soon be a cheaper alternative that will only require an investment of around US $212 000. The British company Virgin Galactic is offering rocket flights to the edge of space, with the first flights expected to take place in 2009. Bookings are now open, but you will need a US $22 000 deposit.

How physically fit does a space tourist need to be?

Before flying to the ISS, space tourists have to undertake several months of cosmonaut training, which includes extensive medical check-ups. After all, the travellers have to be able to cope with some unusual experiences,

When will we be able to stay at the first orbital hotel?

The American hotel entrepreneur Robert Bigelow hopes to open a type of inflatable hotel in Earth's orbit for affluent space tourists as early as 2010. In July 2006, his company – Bigelow Aerospace – successfully placed a smaller prototype, named *Genesis 1*, into Earth's orbit using a converted Russian cold war ballistic missile. A second, larger prototype followed in June 2007, and in 2010 *Sundancer* – a 180 m³ structure with all the necessary life support systems – is due to be launched. There is a problem, because so far no suitable spacecraft has been found to get customers to and from the space hotel. This needs to be resolved as quickly as possible, and so Bigelow Aerospace and Lockheed Martin are currently investigating whether Lockheed Martin's *Atlas 5* rocket can be adapted for manned flights.

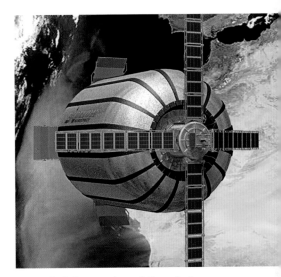

An artist's impression of *Genesis 1*, the precursor of a space hotel, which was launched in 2006.

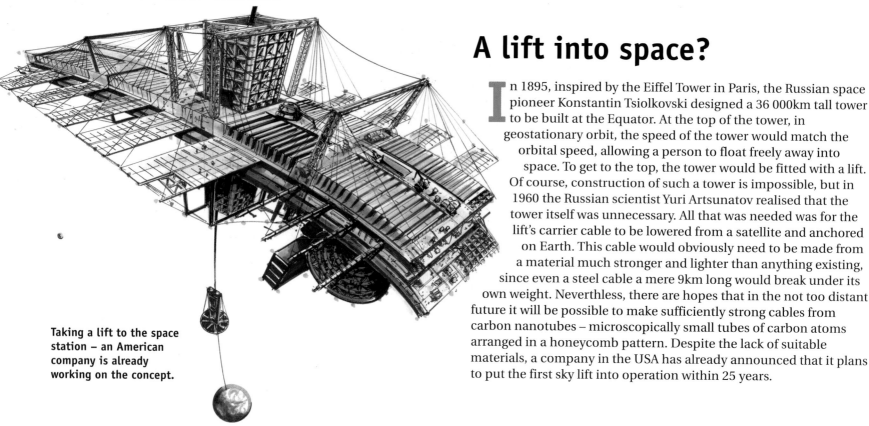

Taking a lift to the space station – an American company is already working on the concept.

A lift into space?

In 1895, inspired by the Eiffel Tower in Paris, the Russian space pioneer Konstantin Tsiolkovski designed a 36 000km tall tower to be built at the Equator. At the top of the tower, in geostationary orbit, the speed of the tower would match the orbital speed, allowing a person to float freely away into space. To get to the top, the tower would be fitted with a lift. Of course, construction of such a tower is impossible, but in 1960 the Russian scientist Yuri Artsunatov realised that the tower itself was unnecessary. All that was needed was for the lift's carrier cable to be lowered from a satellite and anchored on Earth. This cable would obviously need to be made from a material much stronger and lighter than anything existing, since even a steel cable a mere 9km long would break under its own weight. Neverthless, there are hopes that in the not too distant future it will be possible to make sufficiently strong cables from carbon nanotubes – microscopically small tubes of carbon atoms arranged in a honeycomb pattern. Despite the lack of suitable materials, a company in the USA has already announced that it plans to put the first sky lift into operation within 25 years.

What is the future for space travel?

Forecasts are difficult to make. In the 1960s, after the *Apollo* Moon flights, plans for a permanent, manned Moon base were already being made, and flights to the entire solar system seemed to be within the reach of astronauts. But then the Apollo programme was cancelled, and for many decades manned space travel was limited to flights into Earth's orbit. In the 1970s, many believed that the space shuttle ushered in the era of reusable spacecraft. The shuttles have since proved to be too complicated and too dangerous, and NASA is now returning to an approach that makes use of expendable rockets. Nevertheless, if current plans are carried through, astronauts will return to the Moon by the end of the next decade and, if all goes well, the first manned flights to Mars may follow in about 20 years.

Why isn't it possible to fly faster than the speed of light?

The speed of light in a vacuum (299 792km/sec) is an absolute limit imposed by the laws of physics. Imagine a spacecraft travelling through space at half the speed of light. If an astronaut aboard this craft shone a beam of light from the front of the ship, would a stationary observer outside see the beam spreading out from the spacecraft at one and a half times the speed of light? Surprisingly, the answer is no. Whether measured from inside or outside the

spacecraft, the speed of light remains the same. This apparent contradiction can only be resolved with the help of Einstein's theory of relativity, according to which time appears to pass more slowly and distances are shortened at high speeds. Since time and space aboard the rapidly moving spacecraft appear relativistically distorted, a measurement of the speed of light made by the astronaut will give the same result as one made by the stationary observer outside. Another consequence of the theory of relativity is that the mass of an object appears to increase with its velocity, so by the time a craft reached the speed of light its mass would be infinitely large.

How long would it take to journey to the stars?

Today's technology may allow speeds of perhaps one-hundredth that of light, which means that a journey to the nearest stars would take several centuries. But if it ever becomes possible to fly at velocities approaching the speed of light, then journey times could be greatly shortened. For an astronaut aboard a spacecraft travelling at 99.9% of the speed of light, only 4.5 years would pass on a journey to a star 100 light-years away. For everyone else left at home on Earth, 100 years would pass and the astronaut would return to find a world quite different to the one he or she left 200 years earlier.

Will we ever be able to travel through 'wormholes'?

Because we are restricted to velocities below the speed of light, the kind of space travel to distant stars and galaxies we see depicted in science-fiction films is impossible. The realisation that this is the case, and that humans may never be able to visit even a fraction of the vast universe, is a situation that even some physicists find unsatisfactory. This has led to a search for loopholes in the laws of physics that may make intergalactic space travel possible.

One such loophole involves the possibility of so-called 'wormholes' in space. These strange entities are a result of the theory of relativity, and the suggestion is that distant regions of the universe may be connected by a kind of space-time tunnel. A number of theories have predicted that wormholes will be unstable and so unsuitable for use as shortcuts by future astronauts. As soon as they have formed – if they do form – they would collapse again. One suggestion is that wormholes could be stabilised by matter with negative energy and negative gravity. Unfortunately – like wormholes themselves – this kind of matter only exists in theory so far.

A 'wormhole' serves as a shortcut for a journey to the Andromeda Galaxy – the sort of scenario depicted in science-fiction films. It may be an attractive idea, but most scientists doubt that such journeys will ever be possible.

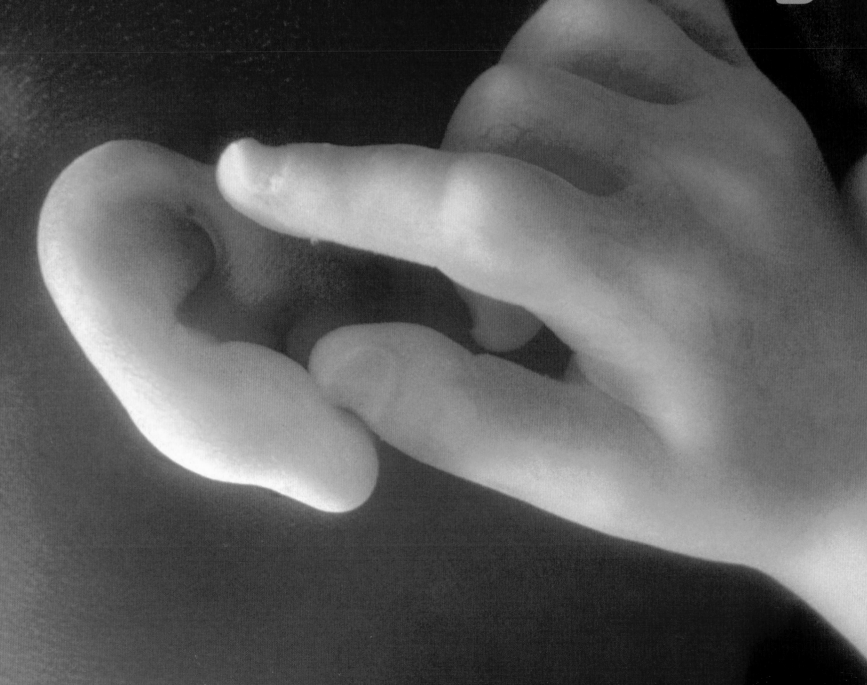

The miracle that is a
human being

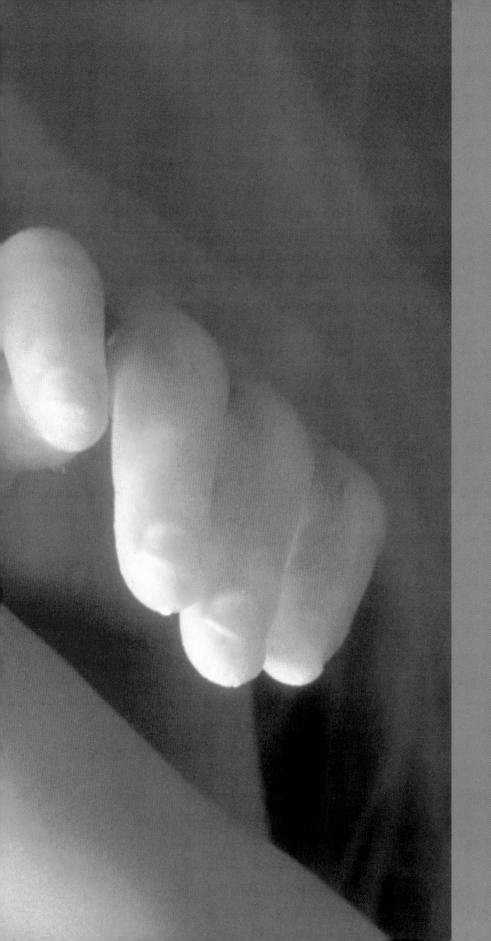

Humans are **amazingly complex** living creatures – the product of millions of years of evolution. Our most conspicuous external characteristics are that we walk in an upright position and lack body hair. However, it is the **large human brain** and its achievements that make us unique among the other creatures that share the planet with us. As far as we know, no other animal can **develop theories and think** about them, or translate thoughts into language and then **record them for posterity** as we do.

Are genes alone responsible for the way we look?

Our bodies develop according to a pattern laid down in our genes, so our appearance is based primarily on the genes inherited from our mother and father. But our bodies are exposed to a wide range of environmental influences, especially during childhood and even in the womb. These also affect the way we look.

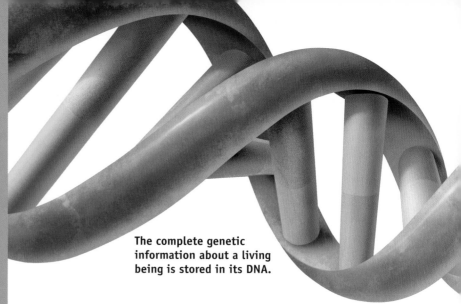

The complete genetic information about a living being is stored in its DNA.

How alike are humans and the great apes?

Chimpanzees and bonobos are our closest relatives, followed by the other anthropoid apes – gorillas, orangutans and gibbons. There is a difference of less than 1.5 per cent between the 3.2 billion DNA building blocks that form the genotypes of humans and apes. This difference is ten times smaller than that between mice and rats and ten times greater than that between any two unrelated humans.

Now that the genomes of humans and chimpanzees have been decoded, the search is on for genes that explain characteristics that are uniquely human. During the course of evolution, there have been several changes in our genetic make-up that have had major effects on the functioning of whole gene groups. How these genes are regulated and how they influence one another accounts for the differences between humans and apes, rather than the type of gene.

Is there a language gene?

Language is one of the key abilities that sets human beings apart from the apes. Scientists have discovered a gene that is crucial to the development of speech, and people with a mutation of this FOXP2 gene cannot speak.

Comparing FOXP2 genes in humans and apes, researchers found two small differences in the DNA sequence. These result in the formation of slightly different FOXP2 proteins, which act as genetic switches and affect other unidentified genes necessary for speech. So the ability to speak does not depend on one single 'language gene'.

Apes cannot speak, but they do have a simple communication system based on sounds and gestures.

Why are people today taller than they used to be?

The maximum height to which a person can grow is dictated by his or her genes. The height a person reaches depends mainly on living conditions during childhood and adolescence. Over the past 120 years, the average height of Europe's population has steadily increased because living conditions have improved. Development in previous centuries tended to occur in waves.

Between the 9th and 11th centuries, when the climate was mild, food plentiful and the population density low, men were about 173cm tall – almost the same as they are today. Because of poor nutrition, a dense population and an increased risk of infectious disease, average height subsequently decreased until, in the 17th and 18th centuries, it was only 167cm.

Why are men taller than women?

On average, the men in developed western countries are about 12cm taller than the women. At the start of human evolution, a tall, strong man was a guarantee of food and shelter for women and children. Although height has little influence on professional success or social status today, tall men continue to be considered attractive, so are more likely to find a partner and produce larger-than-average families. And men prefer shorter women, who reach sexual maturity earlier. These preferences in choice of partner are genetically determined, so despite emancipation and equal opportunities for men and women, height differences between the sexes will be slow to change.

Men are on average taller than women – and it's all down to evolution.

Are there genes that make us fat?

If a person is overweight, about 60 per cent of the blame can be attributed to their genes. Several genes influence the appetite by releasing hunger and satiety hormones. The proportion of energy intake that the body stores as fat, rather than allowing it to be burned, is also partly genetically determined.

A healthy diet and physical activity can reverse any trend towards obesity, unless a rare genetic defect is responsible. The mutation of a hormone or its binding site in the brain can produce feelings of constant hunger that lead to obesity. Only medical treatment can help in these cases.

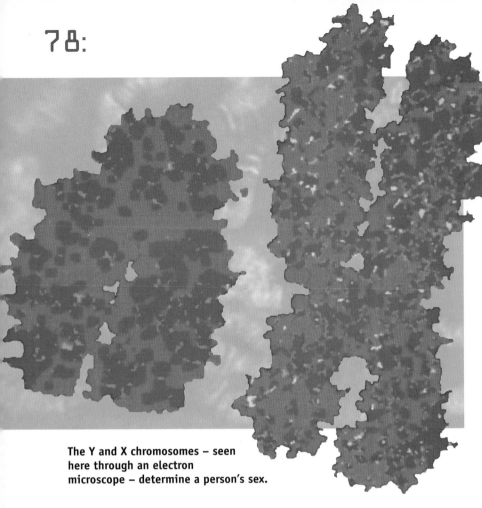

The Y and X chromosomes – seen here through an electron microscope – determine a person's sex.

How many genes determine a baby's sex?

Whether a sperm with an X or Y chromosome fertilises the egg cell determines the sex of a baby. If it is an X chromosome, a girl will develop with each of her body cells containing two X chromosomes; if it is a Y chromosome, the child will be a boy with an X and a Y chromosome in each body cell.

Many genes contribute to the development of internal sexual organs and external sexual characteristics. The SRY gene is particularly important to the Y chromosome. In the seventh week of pregnancy, this gene gives the starting signal for the male foetus to develop. It calls in several

other genes and together they gradually form all the male characteristics. If the SRY gene is missing, female organs develop. If a Y chromosome is present but the SRY gene is faulty, the result will be a sterile XY female.

What are the chances of a boy or a girl?

Equal numbers of sperm with an X or a Y chromosome are formed in the testes. After merging with the egg cell – which always has an X chromosome – there should be an equal probability of producing male (XY) or female (XX) offspring. In fact, 51.2 per cent of babies are male.

Parents may succeed in determining the gender of their child. This can be done because the two types of sperm have different characteristics. Male Y sperm are more mobile, while female X sperm survive longer. When intercourse takes place on the day of ovulation, the faster-moving Y sperm have an advantage and the chances of conceiving a boy are greater. When intercourse takes place a few days before ovulation, a girl is the more likely result.

What is a human hermaphrodite?

In every cell of his body a man has the X and Y sex chromosomes, and so develops testicles and other male characteristics – a penis, a scrotum, facial hair and a deep voice. A woman carries two X chromosomes in her genotype and therefore develops ovaries, vaginal labia, breasts and a high voice. A hermaphrodite – or intersex person – cannot be clearly assigned to either sex. Such individuals have testicles as well as ovaries, and usually two X chromosomes. True hermaphrodites are extremely rare. More common are intersex individuals who have either ovaries and masculine characteristics or testicles and feminine characteristics. The reasons for these aberrations remain unexplained.

Male or female? Sexual characteristics are determined by a number of genes, with the SRY gene triggering the development of a male foetus.

Are identical twins exactly alike in every detail?

Identical twins come from the same fertilised egg cell. As a result, they carry the same hereditary information and share many characteristics. But neither the physical nor the psychological characteristics of a person are determined by genes alone. While still in the womb, each twin will develop individual differences. It is possible to distinguish between twins by their fingerprints and iris patterns, as these features depend on chance environmental influences during development in the womb.

Factors that influence the development of identical twins include the position each foetus occupies in the womb and how well each is supplied with nutrients. If one twin is better nourished, the result can be the birth of two babies with different sizes and weights.

Twins' immune systems do not develop in the same way, since chance plays a major role in the formation of antibodies. One twin can be more prone to particular diseases than the other.

The DNA in some genes changes during our lives, and these changes vary between individuals. This means that the activities of certain genes can be reduced or increased – one twin could develop a greater or lesser susceptibility to cancer. As they grow older, twins will become more distinct from one anotheras they experience these epigenetic changes. For example, one twin can develop a greater or lesser susceptibility to cancer.

Identical twins have identical genomes – but this does not mean they are exactly the same.

Can a man with unfit sperm father a child?

Only sperm that are sufficiently motile – able to swim – can find and penetrate female egg cells. If a man produces too few or insufficiently motile sperm, a special artificial insemination procedure can be carried out. Known as intracytoplasmic sperm injection, the technique uses a micromanipulator to inject a single sperm into an egg.

With the help of this technology – first used in 1992 – fertilisation is possible even when a man has a low sperm count. The procedure requires a minor operation to remove sperm directly from the testes and immediately transfer them to the egg cell.

Why do some humans have pale skin?

Skin colour is determined by genes that control the production of the pigment melanin in skin cells. Skin will be darker or lighter according to the amount, density and distribution of melanin. After early humans lost their body hair, strong skin pigmentation provided protection against the ultraviolet rays of the African sun.

As regions of the Earth with less sunshine were populated, this protection was less important and people whose skin had mutated to lighter shades were no longer at a disadvantage. It is possible that the opposite was true: people with lighter skins were at an advantage because of genetic changes and this enabled them to become dominant in Europe.

It is believed that light skin is an advantage in parts of the world with less sunshine as it allows more ultraviolet radiation to be absorbed so the body can manufacture Vitamin D. Inuit, from the far North, have relatively dark skin because they are able to meet their need for Vitamin D in another way – by living on fish. This theory fails to explain why many of the inhabitants of South-East Asia have pale skins, despite being subjected to strong sunshine.

Skin colour is not a racial characteristic. A single mutation can so affect thee production of melanin that skin colour changes from black to white.

Why do some people have freckles?

The skin protects itself from strong solar radiation by producing increased amounts of brown melanin pigments. The skin of individuals with blond or red hair does not brown evenly. Instead, small brownish spots – known as freckles – are formed. As with hair and eye colour, freckles are a genetically determined feature. Several genes are responsible for how much melanin pigment is produced and how it is distributed in the skin cells.

Freckles can become paler when there is little solar radiation, or be minimised with the help of bleaching cream, but they do not disappear altogether. Even after laser therapy to destroy pigment cells, new freckles will reappear eventually. Hormonal changes during pregnancy can cause freckles to darken, but pigment production usually returns to normal after giving birth.

People with different skin colours hardly differ from one another genetically.

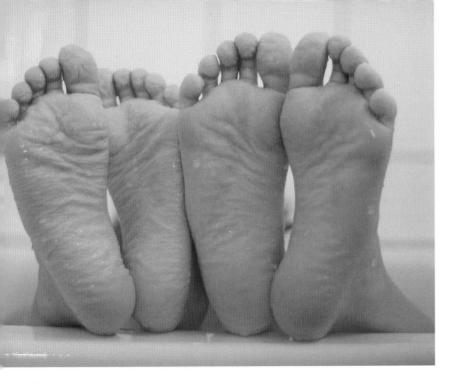

What are the white half-moons on our fingernails?

Fingernails and toenails are made up of several layers of hardened cells. The nail plate (the hard part) is less than 1mm thick and transparent, allowing the pink nail bed beneath, with its plentiful supply of blood, to show through. Only the half-moon-shaped area, known as the lunula, at the bottom of each nail is white and opaque. This is where the nail's root or matrix is located and it is here that new cells are produced continuously. These new cells push forwards, dying as they go, and the stored keratin inside causes them to harden and finally become transparent.

Fingernails grow at a rate of about 1mm a week, toenails more slowly. The little white half-moons vary in size, and if the matrix is hidden beneath the cuticle and skin on the fingers the lunulae can't be seen at all.

A bluish discoloration of the lunulae is a sign of a shortage of oxygen due to poor circulation, possibly caused by heart or vascular disease. The small, white flecks that are seen on nails are the result of minor injuries and nothing to worry about.

Why does skin wrinkle in the bath?

Most skin is covered by a thin, waterproof film of oily sebum. This substance, secreted by the sebaceous glands, keeps the skin supple and prevents moisture loss and water penetration. When skin is in contact with water for relatively long periods of time – in the bath or at the swimming pool – this film of sebum dissolves. The process is accelerated by the use of soap or other skin cleansers.

Once the film of sebum has dissolved, water can penetrate the gaps in the outer layer of the skin, which consists of dead, hard skin cells filled with the protein keratin. The skin expands and swells up. Because the outer layer is attached to the impermeable subcutis beneath, which is made up of living cells, creases form and the skin wrinkles.

A lot of hard skin builds up, especially on the fingers, toes and heels, where everyday activities place additional demands on the skin. Here, the layer of hard skin can be between five and 20 times thicker than in other places. These are the areas most prone to water penetration and where the skin wrinkles most noticeably.

When we get out of the bath or pool, the absorbed water quickly evaporates, the skin becomes smooth again and a new layer of oily sebum is formed. The exception to this is the soles of the feet and palms, which don't have sebaceous glands.

How do we know...

...that some people have no fingerprints?

The familiar pattern of dermal friction ridges can be seen across the palms of both hands and the soles of the feet. People with certain hereditary skin diseases do not develop these ridges. The skin stays smooth and does not leave behind any usable impression. Such disorders are rare, with one of the diseases – Naegeli Syndrome – affecting only one person in three million. Most people without fingerprints also have fewer sweat glands and may have other skin abnormalities.

A fingerprint is a patch of sweat on which the dermal ridge pattern has been reproduced. Scene-of-crime experts use chemicals to make these prints visible. Every one of us has different fingerprints, which don't change over the course of our lives. Since 1901, fingerprinting has helped the police track down criminals.

Why do albinos have red eyes?

People with albinism have a genetic defect that affects the production of the pigment melanin. Depending on which genes are affected, the brown pigment may be missing from only the irises of the eyes, or it may also be missing from skin and hair. In dim light, the eyes of people with albinism appear a greyish-blue. In brighter light, and depending on the angle of the light, the blood vessels in the concave rear surface of the interior of the eye become visible, making the eyes appear red.

The lack of pigment means that albinos are without any natural protection from light, which makes their eyes sensitive and can affect their sight. The genetic defect responsible for albinism can occur in people from all ethnic groups.

Can the signs of disease be seen in a person's eyes?

Doctors can find indications of some physical diseases while examining a patient's eyes. Examination of the concave interior of the eye provides information about vascular diseases and diabetes. Measuring the arteries and veins in the eye can enable doctors to estimate a patient's risk of heart attack or stroke. The shape and size of the pupils can supply evidence about blood pressure as well as the risk of a stroke. A yellowing of the normally white eyeball may indicate jaundice.

Iridology or iridodiagnosis – a technique used by practitioners of complementary medicines – is not recognised by orthodox medicine. Iridology divides the iris into different sectors, each of which corresponds to a part of the body. The shape and pattern of the pigmentation in a sector is believed to indicate the presence of a disease or susceptibility to a variety of complaints.

What determines eye colour?

The amount of melanin stored in the iris will result in an eye colour ranging from pale blue to dark brown. People with blond hair produce small amounts of melanin and their eyes reflect more of the blue light that strikes it, causing their almost colourless iris to appear blue. As the quantity of melanin increases, eyes colour changes from greenish to dark brown.

A person's eye colour is determined by several genes, but the precise effect of each of these remains unknown. Genes for brown and green eyes prevail over those for blue, with most blue-eyed people living in northern Europe. The overwhelming majority of people have brown eyes.

About four people in every million have a right eye that has a different colour to their left. This condition, known as heterochromia, can be genetic or the result of disease.

True or False?

All newborn babies have blue eyes

Generally, only pale-skinned babies have blue eyes. Little or no pigment is stored in their irises because the eyes are not fully developed at birth. The blue eye colour in a baby's early months is caused by the reflection of the blue component of any light that falls on the iris, which at that stage is still colourless.

Dark-skinned people produce more melanin and their babies have enough pigment stored in the iris at birth to make their eyes brown, becoming darker as they develop.

Why did humans lose their hair?

The extensive loss of body hair probably gave early human beings living in open country several advantages. As humans became more efficient hunters, they needed to cover long distances on foot at speed. Bare skin helped to increase the cooling effect of sweat. Less hair also meant that the risk of infestation by disease-carrying parasites – lice, fleas, ticks, mites and bugs – was reduced. Bare skin may have become an important characteristic when it came to choosing a mate. If less hirsute partners were considered more attractive, natural selection over tens of thousands of years will have resulted in bare skin.

The disadvantage of having no hair to conserve body heat was partly offset by the development of a layer of fat under the skin. And as people learnt to use fire, make clothes and build shelters they became less susceptible to adverse changes in weather or climate.

Some scientists disagree. They explain the loss of body hair as an evolutionary phase during which humans lived on seashores and river banks, feeding on mussels, crabs and fish. It is known as the aquatic ape hypothesis. A covering of long, thick hair did not suit this way of life. This hypothesis is supported by the direction of growth of our body hair. Unlike that of anthropoid apes, it still facilitates the easy flow of water when swimming.

The lack of thick hair is one obvious difference between apes and humans, but describing humans as the 'naked ape' is not accurate. We are almost completely covered in hair, but it is very sparse.

Thick hair may distinguish us, but it is not accurate to describe humans as 'naked apes'

What determines our hair colour?

Human hair colours are produced by pigments passed into the hair root by pigment-producing cells known as melanocytes. These melanin pigments are divided into light and dark types, and which predominates is determined by a gene that has several variants. Where hair contains a greater proportion of the light pigment phaeomelanin, it is blond or red; where the hair cells have stored mostly the darker pigment, eumelanin, it is brown or black.

Children of northern and central European parents often have fair hair in their early years that darkens as eumelanin production increases later in life. In old age, the cells of the hair root produce insufficient melanin and the hair turns grey or, if it lacks pigment altogether, white. Tiny air bubbles in the central medulla of the hair fibre can also make the hair appear white.

Are blonds an endangered species?

The gene variants responsible for blond hair are recessive. A child is only blond if he or she has inherited the gene variant responsible from both parents. If the child has one gene for blond and one for dark hair, then the dark hair colour will prevail. Over 90 per cent of the world's population has black hair. Only in northern and central Europe and in countries settled by European migrants do populations include large numbers of fair-haired people.

As global integration increases, the number of blond people will decrease. This does not mean the blond gene is disappearing. It remains in the human gene pool in the same proportions because many people with dark hair still harbour this gene. In the future, some dark-haired parents may be surprised by the birth of a blond child.

Can hair turn white overnight?

Dark hair cannot suddenly turn white because it would have to lose its pigments in a short time to do so, and that is impossible.

In middle-aged adults, pigmented hair grows alongside grey or white hair. In cases of a rare immune deficiency disease that leads to sudden hair loss, pigmented hair can fall out very rapidly, leaving only the white hair behind. This gives the impression of hair suddenly turning white.

There is no proof that stress can trigger this kind of hair loss. But as there is a correlation between the nervous and immune systems, the possibility cannot be ruled out.

As people age their hair gradually turns grey because it stores fewer pigments.

A trick that never fails to amaze – the hair hang is all about physics.

How can performers hang by their hair?

The trick in performing a 'hair hang' lies less in conquering pain than in physics. Performers are suspended by all of their hair at the same time so the load is widely distributed. A fakir can withstand the pain from the points of his bed of nails for the same reason – the nails are numerous and close together.

For the hair hang, the decisive factors are how tear-proof the hair is and how firmly rooted it is in the scalp. Depending on hair colour, there are between 85 000 and 140 000 hairs on a human head. Redheads have the fewest, blonds the most. Even with red hair and a body weight of 100kg, each hair has to support a maximum of only 1.2g. This is well inside its capability, as a single hair will only tear under a load of around 100g and considerably more force is required to pull one out by its roots.

Most hair-hang performers weigh only about 50kg, so they are able to take partners, tables and chairs into the air with them. The important thing is for each hair to have an equal load to bear. The strain on the hair hardly varies, provided the performer is raised into the air slowly and smoothly and only turns on his or her own axis, rather than swinging back and forth.

Is it true that some men cannot grow a beard?

In Hollywood movies about the Wild West, you won't see beards on the faces of the Native Americans – their cheeks are always smooth. But Charles Darwin observed Native American men who tore out their sparse facial hair, and the male native inhabitants of South America are also assumed to have plucked their beards with tweezers, suggesting some slight hair growth at least.

The vigour of facial hair growth does vary between ethnic groups. While European peoples have comparatively dense beards, Asians and the natives of the Americas tend to have little facial hair. This characteristic is genetically determined, although it finds a variety of expressions.

Some scientists believe that the sparse beard growth typical of Inuit men developed as an adaptation to life in a cold climate. In below-zero temperatures, beards become encrusted with ice, and this can damage the skin of the face.

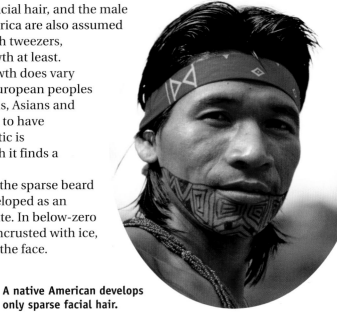

A native American develops only sparse facial hair.

What makes hair either curly or straight?

As with colour, the form of hair is genetically determined. East Asians have straight hair; Europeans have straight, wavy or curly hair; Africans usually have hair that is tightly curled. Different hair types have different structures. Seen in cross-section, straight hair is round, while wavy to curly hair is oval to kidney-shaped.

Different hair forms can also be distinguished by their chemical make-up. Each hair consists mainly of long, fibrous molecules of the protein keratin. In straight hair, most of these molecules are held together by sulphur bonds, making the fibres rigid. If additional, elastic fibres from less closely linked keratin are present, the hair curves to create waves or curls.

When hair is permed, chemicals are used to break down the sulfur bonds. Once that is done, it is possible to create waves and curls, which can be fixed in place by a second chemical process.

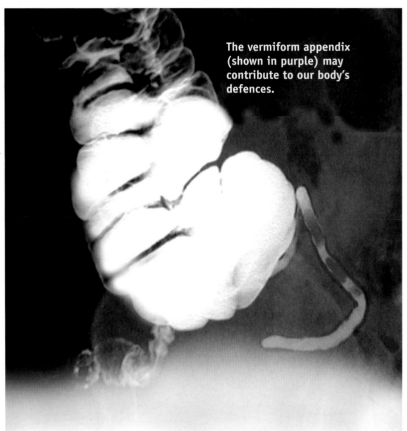

The vermiform appendix (shown in purple) may contribute to our body's defences.

Are some parts of the body completely useless?

Parts of the body that no longer serve any purpose usually tell us something about human development over millions of years. These features were important to our ancestors, but their original function has been lost during evolution. Examples include the coccyx, which once served as a tail; the appendix; the muscles of the outer ear, which some people can still use – but only to waggle their ears; body hair, which no longer protects us from the cold; wisdom teeth that often fail to fit in the jaw; and, the plantar muscle in the soles of our feet, which is used by the apes for gripping.

Processes that take place during the development of the human embryo are responsible for some redundant body parts. The embryo develops male and female characteristics before its gender is determined, which is why men have nipples and women the vestiges of a sperm duct.

What is the function of the appendix?
In rabbits and other herbivorous mammals, the appendix is the largest part of the large intestine. The koala's appendix is 2m in length. In these animals, the appendix serves as a fermentation chamber, where otherwise indigestible cellulose from the plants they eat can be broken down. The human appendix no longer has this function. A leftover from our vegetarian past, this 7cm-long, sack-like protuberance is situated below the junction between the small and large intestines.

While the appendix is routinely removed surgically without causing any apparent ill effects, it may make some small contribution to our body's immune defences. Like the tonsils on either side of the throat, it contains lymphatic tissue, which is why it is sometimes known as the intestinal tonsil.

Why doesn't everybody get wisdom teeth?
Wisdom teeth are the molars furthest back in the mouth, in both the upper and lower jaws. They usually appear between the ages of 16 and 35. When and how wisdom teeth appear varies, and about one in four people never get them at all. There is a genetic reason for this: some people's genetic profile has given them a more 'modern' set of teeth, while the rest are still carrying around an evolutionary relic.

For our forebears, a full set of 32 teeth was crucial if their powerful jaws were to be capable of chewing up tough food. As humans have evolved, our jaws have become smaller but the number of teeth in them has not reduced. This is why there is often no room left for wisdom teeth. The result is often inflammation and impaction, and the affected teeth have to be extracted.

True or False?

The tongue web serves no purpose

The *frenulum linguae*, or tongue web, is a fold of mucus membrane running between the middle of the underside of the tongue and the floor of the mouth. It appears to hold the front part of the tongue in position but is a redundant leftover from the embryo's development. It is formed at the same time as the tongue, which grows in two longitudinal halves that fuse together, leaving a membranous fold along the seam marking the junction. If this process goes wrong, the *frenulum linguae* can be too short, causing the condition ankyloglossia. This can limit the movement of the tongue and cause speech defects. A very short tongue web is generally cut surgically.

Does the navel continue inside the abdomen?

The navel marks what was once the entry point for the 50 to 60cm-long umbilical cord. Three blood vessels running through the umbilical cord – two arteries and the umbilical vein – link the mother's blood supply to that of the unborn child. When a baby is born, the umbilical cord is cut and the remaining stump clamped on to the baby's abdomen. At this time, the blood vessels in the umbilical cord are still intact. But, like the external section, the continuation of the cord atrophies inside the newborn. All that is left in the end is a strand of connective tissue running along the front of the abdominal wall towards the liver.

The fact that not all navels look alike has less to do with the way the umbilical cord was cut than with the musculature of the abdomen. In some people, the navel looks like a hole, while in others it is a protruding lump. The hole does not continue into the abdomen.

In the womb the umbilical cord forms a vital link between the mother and the growing infant.

Do we have senses for everything?

Our five senses enable us to receive vital information from our environment, but they deliver only a tiny part of the information available. We can't see ultraviolet or infrared light and our ears are deaf to ultra- and infrasound. Many animals can smell, taste and feel things far better than we can, and we have no sense organ to enable us to detect magnetic fields.

How does our colour vision work?

The impression of colour occurs in the brain when we look at an object that reflects only one part of the visible spectrum. This light enters the pupil and then falls upon the retina of the eye, where light-sensitive receptor cells respond by transmitting nerve impulses through the optic nerve to the brain.

There are two groups of receptor cells: 125 million rod-type cells, which enable us to distinguish light and dark at low light levels; and, six million cone-type cells, which come in three different types and are responsible for colour vision. Each of the three types of cone cells contains a different kind of pigment: one is sensitive to

Listening to music and seeing colours 'in the mind' simultaneously is possible for a small number of people whose sensory impressions are linked.

blue light, the others to green and red. The human eye can detect a spectrum of colours stretching from violet to dark red. We see mixed colours when more than one type of cone cell is stimulated at the same time. When all three types of cone cells are stimulated at once, the brain receives an impression of white. Everything looks grey at twilight because the cone cells are not active in low light. Under these conditions, the more light-sensitive rod cells allow us to see, even if what we see is only in black and white.

Why do some people claim to 'see' sound?

There are people in whom one sensory impression is connected to another. When people with this condition listen to music, sounds or words, they see colours simultaneously. Other senses can be involved in this merger, which is known as synaesthesia.

Synaesthesia is not an illness. It has been proven that people who are affected by it are not imagining things. Brain scans – which make active areas of the brain visible so that researchers can identify them – show that when people with synaesthesia hear music, both their hearing and sight centres are being stimulated. Researchers assume that the condition is a result of different nerve centres in the brain being stimulated at the same time.

This phenomenon affects roughly one in a thousand people and occurs far more frequently in women than in men. Recent research has shown that many more people than previously thought may experience much milder forms of synaesthesia without really being aware of it. Individuals experience this form of perception in very different ways, and many find it pleasurable. A similar but temporary fusion of the senses can also be produced in people who have taken LSD or other drugs.

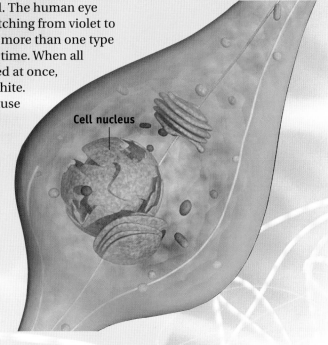

Cell nucleus

The cone-type receptor cells react to blue, green or red light, which they convert into nerve impulses that are transmitted to the brain.

Where do 'floaters' come from?

Floaters are little spots and streaks that hover like midges in front of the eyes, moving when the eyes move. They are caused by clouding in the normally clear gelatinous vitreous humour that makes up most of the eyeball. The vitreous humour shrinks as we get older, and protein fibres begin to detach themselves. These come together and produce shadows on the retina that float across the field of vision. This kind of clouding of the vitreous humour can sometimes affect younger people. Some people find floaters a nuisance, others are less troubled by them. In severe cases, sufferers should consult a doctor who will investigate whether they are caused by an abnormality, such as a detached retina.

Why can't we see a star in the sky when we look directly at it?

When we stare hard at an object, an image of it is created at a central point on the retina – the so-called yellow spot. This part of the retina contains only cone-type receptor cells, which enable us to see objects in sharp outline and in colour when sufficient light is available. The weak light of a star is not enough to stimulate the cones. Conversely, the rod-type receptor cells closer to the edge of the retina continue to function even in dim light. So some stars are only visible when observed indirectly – when we look a little to one side of them. As soon as we look straight at them, they disappear.

Could we still hear without external ears?

The auricle, or external ear, funnels sound waves into the auditory canal. Actual hearing takes place with the help of sensory cells in the inner ear. Without auricles, we would still be able to hear but not as well. The auricles intensify sounds and help us to determine which direction sound is coming from. Our two ear holes enable us to distinguish between left and right because they are at different distances from sounds coming from either side. Sound waves reach one ear, then there is a slight delay before they reach the other. This delay lets us know which direction the sounds are coming from. The distinctive shape of the auricles, with their grooves and hollows, function to identify sound direction much more precisely. The cartilage folds reflect incoming sound waves, producing many small echoes. These indicate whether the sound is coming from above or below, in front or behind.

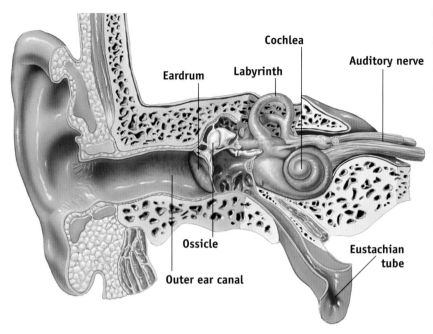

Sounds conducted via the eardrum and the ossicle are transformed into nerve impulses by the cochlea in the inner ear.

Why do older people find it difficult to hear high-pitched sounds?

Our hearing is most sensitive to the sound of human speech, which produces sound waves with frequencies between 1500 and 5000 Hz – although it is possible to detect frequencies between 16 and about 18 000 Hz. As parts of the middle and inner ear deteriorate with age, hearing capacity decreases. Because the sensory cells responsible for hearing higher frequency sounds wear out more rapidly, it is there that hearing loss is first noticed. The hair cells that receive high-frequency sounds are found in an area at the front of the cochlea, while low-frequency sounds are received via a larger area in the rear section. As well as age-related hearing loss, noise, circulatory disorders and chemicals can cause auditory cells to die. Once lost, they cannot be restored.

Is earwax useful or a nuisance?

Earwax is secreted by modified sweat glands in the outer ear canal. Its job is to clean the ear canal and protect it against infection. Dust, bacteria and fungal spores, as well as dead skin cells, are carried out of the ear by the constant production of this fatty secretion. Earwax also contains antibacterial ingredients and bitter substances to deter small insects from entering the ear.

To maintain the protective function of earwax, you should not try to remove it with soap or cotton buds. Nevertheless, an excessive build-up of earwax can cause a blockage of the auditory canal, leading to temporary hearing loss.

A single gene decides whether earwax is wet or dry. In Europe and Africa, the wet variety is most common.

What have our ears got to do with balance?

The inner ear contains not only the cochlea with its hearing cells but also the closely linked labyrinth, the organ that gives us our sense of balance. It consists of three looped, fluid-filled canals arranged at right angles to one another. Balance, like hearing, is controlled by small hair cells in which mechanical pressure triggers nerve signals that are transmitted to the brain. We hear because sound waves cause the eardrums to vibrate and these vibrations are passed, via the ossicles, to fluid in the cochlea. There they bend the hairs of the auditory cells. Our sense of balance is affected when we move our head, changing the position of calcium particles inside the labyrinth, which in turn affects the hair cells in the part of the inner ear concerned with balance.

How do we know...

... there are more than four taste sensations?

Sweet, sour, salty and bitter – these are the four taste sensations that everyone knows. We detect them with the help of sensory cells on the tongue and in the mucous membranes of the mouth and throat. Sweet and bitter substances bind to special proteins – or taste receptor cells – releasing a nerve impulse in the process. Sour and salty flavours are detected without the help of receptors. Scientists now recognise two additional types of receptors. The umami receptor detects protein-rich foods by reacting to the presence of glutamate and other amino acids. A second type of receptor is responsible for detecting the flavour of fat. It is possible that there are more receptors, possibly one for a metallic taste.

The human tongue is only able to detect a few flavours but, with the help of olfactory cells in the nose, people can enjoy an incredible variety of subtle taste sensations.

What is the connection between taste and smell?

The nose has more to do with the flavours we experience when we eat and drink than the tongue. We are only able to experience the full flavour of food or drink with the help of olfactory cells in the mucous membrane of the nose. You can confirm this by simply holding your nose as you eat. There are sensory cells in the mouth, throat and especially on the tongue, but these only react to a few taste sensations, including sweet, sour, salt and bitter.

As we eat or drink, molecules reach the olfactory cells in the mucous membrane of the upper nasal cavity. These olfactory cells are sensitive and can distinguish between hundreds and even thousands of different aromas. The brain combines the flavour and olfactory signals that are received separately to create a single taste sensation.

Why do hot spices burn the mouth?

The hot, spicy taste of foods is not a taste sensation but a feeling of pain. Capsaicin – the chemical compound that makes chilli peppers hot – binds to proteins, or pain receptors, of nerve cells in the mucous membranes of the nose and mouth. The nerve impulses produced in this way pass via the trigeminal nerve into the brain, creating a painful burning feeling. The same receptors also react to heat, so that when heavily spiced food is eaten hot, the effect is even more intense. The pain is offset by the body's reaction, which is to release endorphins – naturally occurring opioids that produce a feeling of wellbeing – which could explain the popularity of hot, spicy food. Another positive effect of hot spices is that they kill pathogens and promote sweating – the latter effect being especially useful for cooling the body in hot climates.

How is a fresh, minty taste created?

The fresh taste and smell of peppermint oil is due to the menthol it contains. However, olfactory and taste cells in the nose and mouth play no part in this sensation. Menthol stimulates the nerve cells in the mucous membranes that normally react to cold. The trigeminal nerve carries the impulses produced by this reaction to the brain, where the impression of cooling freshness is linked to other taste and smell signals.

Do humans react to pheromones?

The existence of human pheromones is unproven, although there are some reasons to believe that pheromones like those found in insects and other mammals are also present in humans. Scientists are searching for the active ingredients and the sensory cells that react to them. Researchers will need to prove that these messenger substances can alter sexual functions or influence behaviour. The most promising discoveries so far have been substances found in the sweat of men and urine of women. These substances cannot be detected by ordinary olfactory cells, but they trigger measurable brain activity in the opposite sex. We don't know whether the vomeronasal organ found in the nasal septum in some mammals – where it plays a role in detecting pheromones – also functions the same way in humans.

Can we imagine pain?

The causes of some kinds of pain are unknown. In the past, these were explained as symptoms of psychological disorders. We now know that chronic pain can lead to changes in the central nervous system. Sufferers develop a sort of pain memory, so that they continue to feel pain even when the original trigger has disappeared. This is why some people develop hypersensitive reactions to harmless minor irritations.

The phantom pain that amputees feel in a missing body part may be the result of a region of the brain recalling earlier pain signals. In the absence of these signals, other signals produce the same feelings of pain. Medication or electrical stimulation can help the brain to forget previously experienced pains.

Hindus walk over red-hot coals at a religious festival in the Andaman Islands. It is now possible to take part in rituals such as this much closer to home.

Why can people walk over red-hot coals without burning their feet?

Walking barefoot over burning embers is a psychological experiment. Fire-walking is popular not only among practitioners of esoteric religions but also as part of mind-training programmes designed to improve self-confidence. The experiment is not without risk, and many who attempt the feat come away with burns to show for the experience. The temperature of the coals is between 650°C and 900°C, and 60°C is sufficient to burn skin.

The trick is to walk quickly on dry feet, and then to cool the soles of the feet in grass or a bucket of water. Thick, horny skin will help but the feet should only briefly touch the hot surface. Embers glow, but they are not a good heat conductor so at least a quarter of a second's contact is necessary for the heat to pass to the skin. The soles of the feet need to be dry to prevent bits of burning material from sticking to them. The important part of fire-walking happens in the head. Concentrated preparation is required to remove fear and prevent slow or hesitant steps.

Are men or women more sensitive to pain?

Not only do women experience pain more frequently than men, they are actually more sensitive to it. This has not been fully explained, but women have greater sensitivity to touch, which could be due to differences in their skin structure. Research has found that the facial skin of a woman contains twice as many pain receptors as that of a man.

Male and female sex hormones play a major role in how each gender experiences pain. Women experience pain differently at various stages during their menstrual cycle. In the late stages of pregnancy, hormones cause the body to release increased numbers of pain-relieving neurotransmitters to make the pain of giving birth more bearable. As oestrogen levels fall after the menopause, sensitivity to pain also decreases.

In males, the sex hormone testosterone has a clear influence on pain, with experiments showing that raised levels reduce pain.

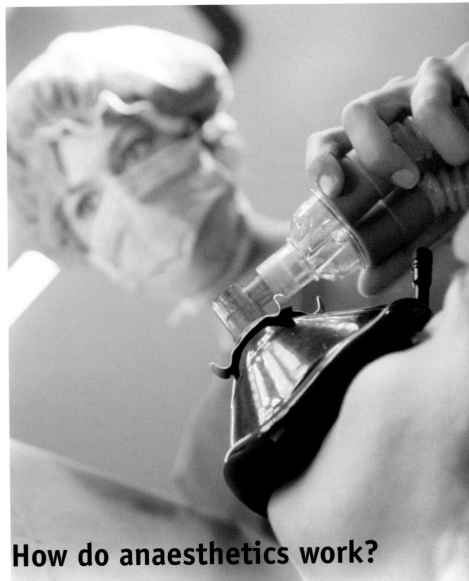

Why does the 'funny bone' tingle when it is struck?

The 'funny bone' is a touch-sensitive place on the inner side of the bone of the upper arm, near the elbow, where a barely protected nerve runs directly under the skin. This is the ulnar nerve, which transmits information from the skin and various muscles of the forearm, right down to the little finger. When something hits this part of the arm, the nerve begins to produce impulses uncontrollably. The sensation is something like that of an electric shock, the blow sending a slightly painful, tingling feeling down the forearm to the little finger. The immediate unpleasant sensation usually disappears after a few seconds, but severe bruising can cause longer-lasting after-effects. The term 'funny bone' may be a pun on the word humerus – the medical name for the long bone in the arm – along which the ulnar nerve runs.

How do anaesthetics work?

A general anaesthetic causes unconsciousness, which relaxes muscles and results in freedom from pain. Depending on a patient's age and state of health, doctors administer anaesthetics by injection or inhalation – in the latter case feeding an anaesthetic gas like sevoflurane through a mask or a tube inserted into the windpipe. This temporarily deactivates nerve cells in the thalamus region of the brain and the patient loses consciousness.

There are alternative pain-relieving and muscle-relaxing medications that act on other parts of the nervous system.

Local anaesthetics interrupt pain messages to the brain only from the site of an injury, or a specific area of the body. Hence, pain-free operations can be performed while the patient remains conscious.

Why are many physical reactions involuntary?

Vital bodily functions like breathing, heartbeat, digestion and temperature control are regulated unconsciously. They are governed by the autonomic nervous system. Since these processes have to be kept going non-stop, while we are awake or asleep, it would not make sense for them to be controlled voluntarily. It is only because they are fully automatic that important reflexes occur quickly and reliably.

True or False?

The body is the same temperature all over

The human body has its own thermostat, which is set at 37°C, but this temperature is only constant in the brain and core of the body. Depending on the outside temperature and the level of muscular activity, arms and legs can be a few degrees warmer or colder than the set point. Slight fluctuations in body temperature are also linked to our daily cycle of waking and sleeping. As we become tired, our body temperature drops, only to rise again when we wake after a sleep.

Why do our stomachs sometimes rumble?

If some time has passed since our most recent meal, the stomach contains only gastric juices and swallowed air. The silent, slow and regular contractions of the stomach walls become stronger and produce a gurgling sound. These hunger contractions subside as soon as we eat something. The intestines can produce similar noises when the food pulp and gas it contains are being moved steadily down its length.

Why do we sweat?

The human body has between two and three million sweat glands distributed over its surface. Their job is to release fluid when the body becomes overheated. Sweat is made up of salt, protein and urea dissolved in water. As it evaporates on the skin, it produces a cooling effect that helps to reduce body temperature. Extreme heat or physical effort produces more than 1 litre of sweat per hour.

Hot and spicy food can also cause us to sweat, as can fear and stress when messages pass to the autonomic nervous system that controls perspiration. Damp, sweaty hands that can grasp things better than dry ones, are a part of our natural fight-or-flight response.

How heavily an individual perspires is genetically determined, but there are also gender differences.

Some disorders can cause excessive sweating, with up to five times as much sweat being produced as normal.

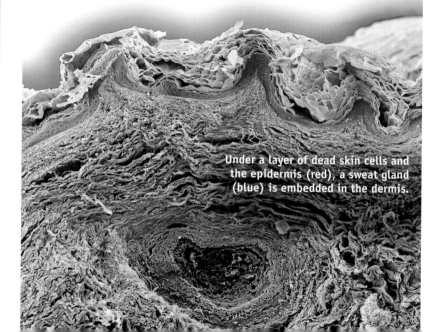

Under a layer of dead skin cells and the epidermis (red), a sweat gland (blue) is embedded in the dermis.

Are swimming and walking innate skills?

Newborn and very young babies possess a number of reflexes, some of which they lose quickly. These include the stepping, swimming and breath control reflexes. When a baby is plunged into water it reacts by closing its mouth and windpipe while making swimming movements. These automatic reactions disappear at three to four months, and children learn to swim when they are three or four years old.

The automatic stepping movements, which babies make when held upright with feet touching a flat surface, are also lost after three months. Only much later are the leg muscles and sense of balance sufficiently developed to enable the child to stand upright and walk. Other babyhood reflexes, such as swallowing, coughing and sneezing, are retained permanently.

Why do we blush when we are embarrassed?

We turn red when the blood vessels in our skin expand. This can occur after physical effort or when it's hot, because it enables the body to lose heat. It also occurs if we are angry or in an unpleasant or embarrassing situation. Stress hormones cause blood pressure to rise, which increases the blood supply to the skin, showing up particularly on the face and neck. Blushing caused by embarrassment only lasts a few seconds and is often barely noticeable. Some people have a stronger reaction and the increased blood flow continues for longer periods. The fear of blushing can make the problem even worse. This phenomenon – which specialists call erythrophobia – can be treated by psychotherapy, relaxation therapy, medication or, even, as a last resort, surgery.

Why do we sleep?

A person must sleep to stay physically and mentally fit. One reason for this is that sleep helps the body to recover, since heartbeat, breathing and many organs work more slowly during sleep than they do when we are awake. By contrast, the brain is highly active during sleep. It uses the time to assimilate experiences, consolidate what has been learnt the previous day, and perhaps to erase useless information.

Our internal clock regulates how tired we get and how long we sleep, causing the brain to release neurotransmitters that induce sleep. These include the sleep hormone melatonin, which the pineal gland produces in larger quantities during the hours of darkness. The need for sleep is determined by our genes, and varies considerably from one person to another. Anything between four and 12 hours a day is considered normal, provided a person feels well.

The need for sleep varies between individuals. While a few can happily go for a run before dawn, far more struggle to stay awake if they don't get as much sleep as possible.

Are we born to be owls or larks?

Each of us has an internal clock that controls our daily cycle of sleeping and waking. Several genes determine when our physical and mental activities increase and decrease over each 24-hour period. Daylight is the most important of the environmental clues our body uses to set its biological clock to local time and synchronise it with the naturally occurring light–dark cycle. This mechanism does not operate the same way for everyone. Some people's clocks seem to be slow. They are not ready for action until late in the morning, but can keep going until well into the night. Whether you are an early-rising lark or a night owl is largely innate, but it is influenced by the amount of light an individual is exposed to during the day. This is dependent to a large extent on whether more time is spent in an enclosed space or out in the open. Age also makes a difference. For many young people, their most active phase is delayed until night-time, which is why an early start on a school day can sometimes cause problems.

the body in the bloodstream. In their turn, endocrine glands and internal organs react to these neuro-transmitters to ensure that harmony is maintained between hormones and the body's biological rhythm.

Is yawning infectious?

Unborn babies yawn in the womb. Children and adults don't just yawn when they are tired or bored, but also before becoming active. We yawn when we go to sleep and when we wake up. It is still a mystery why we yawn. Experiments with oxygen masks have shown that it has nothing to do with a lack of oxygen. It is likely that yawning is a social signal that invites others to join in with whatever we want to do, whether it's some communal activity or going to sleep. Around half of those who see another person yawning will do the same, so yawning really is infectious. People who find it easy to empathise with others are among the most affected.

Where is our internal clock?

There are many clocks ticking away inside our bodies. Every organ, even each individual cell works to its own rhythm. There is also a central pacemaker that synchronises all of the activities of the body and harmonises them with the 24-hour day–night cycle. This pacemaker consists of two clusters of nerve cells, each the size of a grain of rice, found in the region of the brain called the hypothalamus. Known as the suprachiasmatic nuclei, these clusters of cells are linked to sensory cells in the retina, which measure the intensity of daylight. Cells in the internal clock coordinate the dark–light cycle with the independent rhythms of particular gene activities. The body's inbuilt clock is driven by protein-producing clock genes, which switch on and off at a regular rate. Depending on the time of day, varying quantities of neurotransmitters are released and distributed around

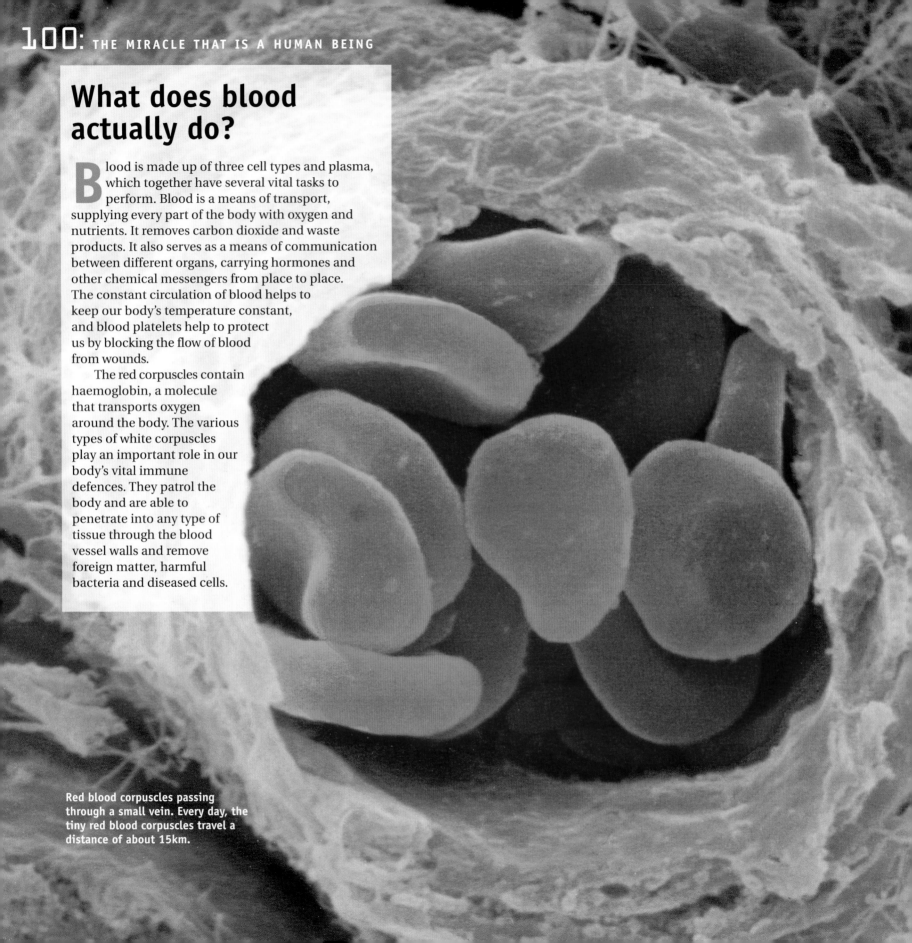

What does blood actually do?

Blood is made up of three cell types and plasma, which together have several vital tasks to perform. Blood is a means of transport, supplying every part of the body with oxygen and nutrients. It removes carbon dioxide and waste products. It also serves as a means of communication between different organs, carrying hormones and other chemical messengers from place to place. The constant circulation of blood helps to keep our body's temperature constant, and blood platelets help to protect us by blocking the flow of blood from wounds.

The red corpuscles contain haemoglobin, a molecule that transports oxygen around the body. The various types of white corpuscles play an important role in our body's vital immune defences. They patrol the body and are able to penetrate into any type of tissue through the blood vessel walls and remove foreign matter, harmful bacteria and diseased cells.

Red blood corpuscles passing through a small vein. Every day, the tiny red blood corpuscles travel a distance of about 15km.

Why is blood red?

Red blood cells contain the protein haemoglobin, which gives it its red colour. The haem group of protein molecules – or the iron attached to them – is responsible for the colour.

Haemoglobin's task is to bind with the oxygen in the lungs, transport it to the rest of the body and then release it. To do this, the haem group enters into a loose chemical bond with oxygen, which is easily released in surroundings where oxygen is in short supply. Haemoglobin is bright red when full of oxygen. Without oxygen, the molecular structure changes and with it the absorption of light that dictates its colour. This is why blood that is deficient in oxygen appears darker.

How many blood groups are there?

Division into blood groups is based on inherited characteristics found in blood components. There are several blood-grouping systems. From a medical point of view, the most important are the ABO and Rhesus systems. The ABO system distinguishes four blood groups, which are known as A, B, AB and O, depending on certain distinguishing marks on the surface of the blood cells. The blood groups A and O predominate in central Europe, and only seven per cent of people belong

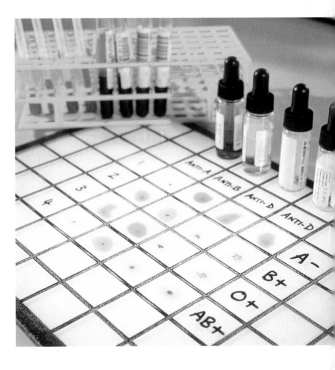

Because mixing blood from different groups causes defensive reactions in the body – such as particles clumping together – it is essential that a patient's blood group is known before a blood transfusion can take place.

to blood group AB. Division by the Rhesus system is based on other surface characteristics of the red corpuscles, known as Rhesus factors, of which some 40 are known.

There are 19 or more blood grouping systems, which were mainly used in paternity testing and forensic medicine before the introduction of DNA analysis.

Where did the Rhesus factor get its name?

The pathologist Karl Landsteiner was awarded the Nobel Prize for Medicine in 1930 for his discovery of the ABO blood group system.

In 1940, the Austrian doctor and his colleague, Alexander Wiener, were engaged in research into the characteristics of blood groups in Rhesus monkeys. They found that, when injected into other animals, the Rhesus monkeys' blood triggered the formation of antibodies that attached themselves to previously undiscovered structures on the surface of the red corpuscles. Further tests showed that the same structures – or antigens – were present on human blood cells. Since these were discovered for the first time in Rhesus monkeys, they became known as Rhesus factors. It was later discovered that the Rhesus factor was a protein occurring in many different forms. Some 85 per cent of people in central Europe are Rhesus positive, which means that their red blood cells contain the Rhesus factor D, the most important and also the first to be discovered.

What are antibodies?

Antibodies are an important part of the immune system. These proteins – also known as immunoglobulins – help defend the body against foreign matter, especially bacteria, viruses and parasites. They are produced by B lymphocytes – white blood cells, the specific purpose of which is to make contact with foreign matter and then release one or more of a number of antibodies into the blood. There are various types of antibodies, with immunoglobulin G occurring most commonly in blood. It has two bonding sites that can recognise an invading molecule, or antigen. When a bacterium carrying this antigen on its surface enters the body, the antibodies recognise and bind to the antigen so that the intruder is either paralysed or destroyed by immune cells.

Immunisation against a specific disease prompts the immune system to produce immune cells and antibodies in the blood that, if the body becomes infected with the disease in the future, will try to destroy the pathogens.

What does healthy really mean?

Sixty years ago, the World Health Organization defined health as 'a state of complete physical, mental and social wellbeing'. Most people could be said to be unhealthy for most of their lives – even in developed countries. Today, health is regarded more as a subjective awareness of wellbeing. The prerequisites for wellbeing are still sufficient amounts of healthy food and physical activity, adequate amounts of sleep, a healthy environment and good social relationships.

What is the maximum age for humans?

The maximum human lifespan is around 125 years. Some experts believe that this limit could be extended by between ten and 15 years. The oldest person whose age could be proved was a Frenchwoman who died in 1997 at 122 years of age. In 1933, *The New York Times* reported the death of a Chinese man who was allegedly 253 years old, but this and similar reports of people living to such great ages are now considered unreliable. There has been no change in the maximum age over the centuries, which is the same for all ethnic groups. What has increased since the end of the 19th century is the average life expectancy in developed countries. Thanks to the improvement in living conditions, this is now between about 79 and 82 years for both sexes (at birth), and rises the older you get.

Why is it that women live longer than men?

On average, women live seven years longer than men. In parts of the developing world the gap is not as wide. Men of all ages in developed countries run a greater risk of dying than women. This is particularly true of young men and those over 60. It was thought that this difference had biological origins and might be due to sex hormones, but a study carried out in German monasteries and convents found that monks lived almost as long as nuns and other women.

Karate master Uehara Seikichi was still practising his chosen sport, and remained a highly respected teacher, when he was over 95 years old.

A sadhu in the ruined city of Vijayanagar in India. These Hindu monks have renounced worldly things and opted for a life of austerity.

Lifestyle and living conditions play a greater role in life expectancy than genetic differences. The average life expectancy of men is probably shorter than that of women because they run more risks and are more likely to be weighed down by stress and work – unless they happen to live in a monastery. It may also be significant that men pay less attention to their health.

Why is it that ascetics tend to live to very great ages?

Along with genes, nutrition has a major influence on life expectancy. Experiments on animals have proved that a 20 to 30 per cent reduction in energy intake – as part of an otherwise well-balanced diet – slows down the ageing process. This seems to apply to human beings, too. Over the long term, a starvation diet of this kind slows the metabolism and probably reduces the number of free radicals – associated with oxidation damage – created by metabolic processes. This reduces the risk of chronic inflammation, and the heart and blood vessels stay healthier longer. But self-imposed starvation diets without medical supervision are not a good idea. There is a danger that dieters may not consume vital nutrients in sufficient quantities to maintain health.

Is there a longevity gene?

Although the maximum life span for the human species is determined by the human genotype, a person's individual life expectancy is not only a matter of genes. Nutrition and other environmental factors have a strong influence on how long we live. Together with genes, these factors can dictate how many free radicals – which damage cells – are created as the body generates energy, and how effectively these can be rendered harmless and the damage repaired.

In flies, worms and mice, various genes have been discovered that have a particularly marked effect on life span. By modifying the way such 'Methuselah genes' act, scientists have succeeded in prolonging the lives of animals. Gene activation in mice resulted in metabolic changes similar to those brought on by a life-prolonging starvation diet. Corresponding genes are also found in humans. Some researchers believe that, with the help of newly developed pharmaceuticals, it will be possible to extend the human life span.

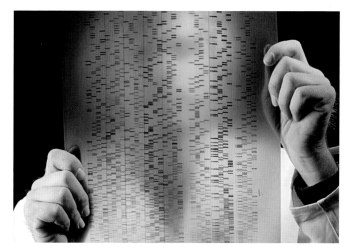

Scientists are studying DNA segments in an attempt to find a longevity gene.

Parents faced with a muddy spectacle like this can feel comforted by the knowledge that at least playing in the dirt helps bolster their offspring's immune system.

Does dirt stimulate the immune system?

In the first years of life, the human immune system has to be exercised through contact with germs in the environment. This is how it learns to tolerate harmless microbes and develop immunity to those that might cause dangerous infections. A relatively germ-free environment means that the immune system does not develop to the full. Children who live in cities suffer from more allergies than those who grow up on farms, because dust and dirt contain bacteria, fungi and cell particles that stimulate the immune system.

Hygiene measures we take for granted also offer protection against dangerous infections. In the home, a sensible level of hygiene is essential but there is no real need for disinfectant.

When do we run a temperature and how does it happen?

A rise in body temperature is the body's way of defending itself when attacked by an infection or other foreign matter. This helps to stop viruses, bacteria and parasites from multiplying, and activates the immune system. After coming into contact with foreign matter, the immune cells release chemical messengers that raise the body temperature set point, which is controlled from a special centre in the interbrain. By increasing heat production and reducing heat emission, the temperature rises to more than 37°C and, in cases of high fever, to over 39°C. This speeds up the metabolism, pulse and respiratory rate. A temperature rarely rises above 42°C, since at this level many human proteins are destroyed. Psychological factors, such as stage fright, can also affect the biological thermostat in the brain causing body temperature to rise.

How do antibiotics work?

Most antibiotics are produced from bacteria found in the soil that are able to fight off infection by other bacteria. These active agents are used in medicine because they can combat disease-causing organisms without damaging the body's cells. Penicillin, for example, kills

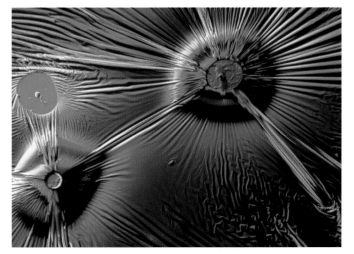

The antibiotic doxycyclin, used to treat respiratory infections and Lyme's disease, and also for the prevention of malaria, shows up as brightly coloured crystals under the microscope.

Do medicines work differently in men and women?

The effectiveness of a medicine depends partly on how quickly it breaks down in the body. Many drugs are broken down by enzymes in the liver, and the level of activity of the genes that control production of these enzymes is often different in men and women. The same dose of a particular medication will work differently in male and female patients – and may also cause side effects to a greater or lesser extent. Aspirin has been shown to offer men more protection against heart attacks than it does women.

Gender differences in hormone balance and distribution of fat and muscle tissue can influence the effectiveness of some medicines. The choice and dosage of drugs will become more gender-specific in future.

bacteria by blocking the formation of their cell walls. Human cells are not affected since they are not built in the same way. Other antibiotics slow down processes such as protein production and DNA replication.

However, antibiotics that are spread throughout the body in the blood also damage beneficial bacteria in the gut. This can lead to intestinal complaints. The widespread and frequent use of antibiotics has also meant that more and more germs are becoming resistant to them. Soon after Alexander Fleming discovered penicillin in 1928 it could kill most staphylococci. Today, more effective drugs are needed.

Why have scientists still not found a cure for the common cold?

Immunisation would be the ideal solution, but because there are more than 200 different cold viruses it is impossible to produce a vaccine that would protect against all types. Some agents, including the immune system messenger interferon, stop viruses multiplying but treatment with these can cause side effects that are worse than the cold itself. New medications are being developed, but in order to succeed they will have to be inexpensive and cure the cold more quickly than it would cure itself by simply running its course.

How do we know...

...there is a kissing disease?

When two people kiss, they exchange bacteria and viruses in their saliva. Fortunately, most of these bacteria are not dangerous but instead work as a kind of oral vaccination, stimulating the immune system when transferred from one partner to another. Bacteria that cause tooth decay and mouth odour will not survive if they move to a new site. Viruses are a different matter. Glandular fever, or mononucleosis, caused by the Epstein-Barr virus, is rightly called the kissing disease. This virus attacks the mucous membranes of the nose, mouth and throat and can be transmitted by kissing or through coughs and sneezes. Most adults have already had the infection – often without realising it – and are therefore immune. The herpes virus, which causes troublesome sores and blisters around the mouth, is also transmitted by kissing.

What causes cancer?

The actual causes vary enormously, but all cancers result from damage to the genes that control cell growth. Large numbers of these genes control the distribution, development and ageing of body cells.

Scientists have known for a long time that ultraviolet rays from the sun can damage the DNA of sunbathers. The carcinogenic (cancer-causing) effects of certain chemicals, such as those present in tobacco smoke, have also been recognised for many years. Infection by viruses, such as those that cause warts and hepatitis, can also encourage

some forms of cancer. When genes become damaged, cells can grow far beyond their usual limits and can become cancerous.

Even without external influences, faults in the replication of DNA, or random mutations, at any stage in an individual's life can increase genetic susceptibility to cancer. It is usually hard to determine what triggered the disease in the first place.

A recent theory suggests that tumours develop after genetic damage to the stem cells found in all body tissue. For therapy to be successful, treatment must destroy these cancerous stem cells.

True or False?

Sunshine protects against cancer

Excessive, unprotected sunbathing can cause skin cancer. The sun's ultraviolet B rays damage DNA in skin cells where cancer can develop, even many years later. Exposure to a sensible amount of sunlight – between 10 and 15 minutes a day – is good for human health and provides some protection against several types of cancer. This is because the ultraviolet B rays help skin cells to produce vitamin D. Vitamin D encourages bone development and reduces the risk of breast, intestinal and prostate cancer, among others.

Has chemotherapy had its day?

After surgery and radiation, chemotherapy is the third pillar in the treatment of cancer. The chemical substances used – known as cytostatics – are toxins that only attack cells that are dividing. Since cancer cells divide with particular frequency, they are the first to be destroyed. Cancer drugs also kill healthy cells in the roots of the hair, bone marrow and the mucous membranes in the mouth and intestine, which is why common side effects of chemotherapy include hair loss, nausea, diarrhoea and weakened immune defences. In older patients, damage can be long lasting.

There are some indications that chemotherapy may interfere with memory and brain functions because it destroys brain cells. Despite the drawbacks, it is thanks to chemotherapy that many forms of cancer, such as childhood leukaemia, can now be cured. Scientists are working on the development of new types of chemotherapy with fewer side effects.

Is cancer treatment more successful in children?

There are a number of differences between cancer in children and the disease in adults. Children tend to become ill mostly because of an innate susceptibility to cancer, or due to some external influence while in the womb. One in 470 children will develop some form of cancer between birth and the age of 15. Most of these fall ill in their first five years of life with leukaemia, brain tumours or cancer of the lymph glands. In adults, lung, breast and intestinal cancers are the most common forms of the disease. Children's bodies find chemotherapy easier to tolerate, so they can be given higher doses of cytostatics. Treatment is successful in one in three children, and the success rate for leukaemia is even higher. This is due largely to the fact that children's bodies find it easier to regenerate cells.

Can a vaccine protect against cancer?

A general vaccine against cancer on the same lines as those used against influenza or other infectious diseases is not possible. There are so many different forms of cancer that there can be no one single overall immunisation. What is feasible is inoculation against infections that increase the risk of certain types of cancer. People can now be vaccinated against hepatitis B and carcinogenic wart viruses. These reduce the risk of liver and cervical cancer. Another way of immunising against cancer is with a form of immune therapy in which, after the disease has broken out, the patient's immune system can be activated to fight against the cancer cells. The aim is to increase the number of immune cells that are able to recognise cancer cells from their external characteristics and then destroy them.

Does gene therapy help combat cancer?

Gene therapy has not been used successfully to fight cancerous tumours. A new gene is inserted into as many tumour cells as possible so that the marked cells can be targeted and killed. As a first step, it is vital to ensure that the therapeutic gene is not also being produced in healthy cells. Deactivated viruses are commonly used to carry the genes and attack cancer cells. In what is known as 'suicide gene therapy', the transferred gene triggers the production of an enzyme that transforms subsequently administered medication into an inhibitor that causes the cancer cells to self-destruct. In another variation, the transferred gene changes the tumour cells so that they are recognised by the body's immune defences and destroyed.

Scientists are trying to find chemotherapy treatments with fewer side effects.

How does our environment pose a danger to our health?

Harmful substances in air, water and food, along with radiation and noise, are environmental factors that can damage our health. The World Health Organization estimates that about one-quarter of diseases worldwide are due to environmental factors. Children whose bodies have not yet developed adequate defences are particularly at risk.

Industrial plants, traffic and heating systems release harmful substances into the air we breathe. These are absorbed by the body and can contribute to the development of cancers and allergies, as well as respiratory, heart and circulatory diseases. This reduces average life expectancy, especially among people living in cities. Vehicles are also a problem in cities, especially when sunlight reacts with exhaust ingredients to create a layer of choking smog. Increasing noise pollution is another problem associated with urban life. It causes premature hearing difficulties and results in a build-up of stress hormones that raise blood pressure, adding to the risk of heart attacks.

When can air make us sick?

Contaminated air is one of the most common causes of death across the world. Common air pollutants include dust, sulfur and nitric oxides, ozone and organic compounds. Forest fires and volcanic eruptions are natural events that can cause toxic air pollution. But man-made pollutants that are released into the atmosphere by motor vehicles, factories, bonfires and heating systems are far more significant. The smaller dust particles are, the more dangerous they become and the more deeply they can penetrate the lungs. Ultra-fine dust particles – some smaller than bacteria – can reach the alveoli in the lungs and from there enter the bloodstream. Fine dust in the air can aggravate cardiovascular disease, as well as making asthma and other respiratory problems far worse. In children, it can cause inflammation of the inner ear and contribute to the development of lung cancer. There is no such thing as a harmless concentration of dust.

In enclosed spaces, carcinogenic tobacco smoke is the main source of dust particles. Because furniture and building materials give off fumes, overall pollution levels can be worse inside the home than out. Sick building syndrome is a consequence of people becoming oversensitive to various pollutants in the home.

There is now proof that environmental toxins can reduce the number of sperm in men.

Does environmental pollution reduce male fertility?

The numbers of sperm in the semen of young men in industrialised countries has been declining for the past 50 years or so, and this has led to reduced fertility. Research has revealed the presence of a wide range of environmental pollutants in human blood, and the results of animal tests suggest that certain chemicals with hormone-like effects could be causing problems. These are used in industry, and include cleaning agents, synthetic softeners and flame retardants, as well as pesticides and pharmaceutical residues. Once released into the environment these substances can enter the body in water or food, or through contact with the skin. More than 100 such chemicals have been found to have a hormonal effect that can influence the development of the male foetus. Increased levels of pollutants in the blood of mothers have also been linked to a risk of malformed sex organs and a reduction in the quality. of the sperm produced.

Is sport good for all of us?

Regular physical activity has been shown to be good for our health. Endurance sports, in particular, improve overall fitness, stimulate the metabolism, reduce stress and prevent cardiovascular diseases. Some types of sport have dubious health benefits. These include hang-gliding, boxing and football, which have a high risk of injury. Other sporting activities can result in damaged muscles and joints if practised without proper training and instruction, the most frequent injuries being to muscles. The level of benefit we gain from keep-fit activities depends on our genes. Scientists believe that this is because they have a different configuration of 'fitness genes' that control the metabolic reactions that produce energy.

Will climbing the stairs every day keep you fit?

You don't have to belong to a sports club to be fit. Even gentle physical exercise is healthy if practised daily. Thirty minutes a day of climbing stairs, brisk walking, swimming or cycling have measurably beneficial effects on blood pressure, pulse rate and blood vessels. Since cardiovascular diseases are the most common causes of death in developed countries, the effect of regular exercise is to prolong life. Because physical activity stimulates the metabolism, it reduces the risk of contracting other diseases and improves general fitness and quality of life, especially as people age.

How important is a daily routine to good health?

Our internal clock – which dictates whether we are early-rising larks or night owls and how much sleep we need – is regulated by our genes. As a result, there are individual variations in how mentally and physically alert each person is over the course of a day. The working of our internal organs also follows a definite rhythm. People who can adapt their daily activities to suit their own personal rhythm tend to lead healthier lives, and the establishment of a regular routine can be a contributing factor to that. A life that runs against the internal

Taking the stairs is better for your health than using the lift or an escalator. It may even help you to live longer.

clock – such as a shift worker's, whose pattern of activity changes frequently – can lead to physical and mental illness. They suffer more than average from disturbed sleep, loss of appetite and digestive problems, and are generally more susceptible to illness. Experiments with animals have shown that a normal daily rhythm disturbed on a regular basis can reduce life expectancy.

Spare parts for people

There are limits to the human body's ability to regenerate itself. Minor wounds to the skin and slight damage to internal organs can heal by themselves, but when injury, disease or the natural wear and tear that comes with age seriously impairs the ability of part of the body to function properly, tissue or organ transplant can often be the only solution. Before this can be done, a process known as tissue typing must be carried out to ensure that donor and recipient cells are sufficiently alike. Only then can the body be prevented from rejecting the transplant. But even where there is a good match, organ recipients must take medication for the rest of their lives in order to suppress their immune defences. This means that transplant patients will always be susceptible to infections.

Are pigs suitable organ donors?

The demand for human donor organs far outstrips supply. Many patients die as they wait for many years for replacement organs. Animals could provide the answer to the organ shortage and pigs would make particularly suitable donors because the size and function of their organs closely matches those of humans. Moreover, animals could be bred and kept in large numbers. But the human immune system would recognise a pig's kidney as alien and immediately reject it. For this reason, scientists have begun to modify pigs genetically, to make their cells more human. To achieve this it is necessary to clone animals that are lacking certain genes or have been given additional human genes. The aim of changing the surface structure of cells in such a way that the human immune system no longer attacks donor tissue has yet to be achieved.

There is also another danger to transplanting non-human tissue. It is difficult to ensure that there are no retroviruses in the animal's DNA which could become active after transplantation and trigger

Skin tissue made in a laboratory is widely used to help patients recover from the damage caused by severe burns or skin diseases.

In a controversial laboratory experiment, scientists transplanted a human ear onto the back of a mouse.

diseases. Pig's heart valves are already being used to replace defective valves in humans, but these are treated prior to use to ensure they contain neither live cells nor viruses.

What replacement tissues can already be grown in the laboratory?

An alternative to transplanting donor tissue or organs is tissue engineering – whereby replacement tissue is grown from the body's own. In Europe, about 25 000 people have been treated with skin, cartilage and bone tissue manufactured in the laboratory. Replacement skin has been used in the treatment of extensive burns for over 20 years. The new skin is created using small pieces of skin are taken from the patient. These cells are then left to grow in a nutrient solution for about two weeks. The process creates pieces of skin about the size of a playing card which can then be placed over the burns. The replacement skin is not entirely satisfactory and looks like scar tissue because it does not produce hair or sweat glands, and it also lacks pigment cells.

An alternative method involves taking skin cells that have been grown in the laboratory and spraying them onto the burn where they gradually grow together to form new skin. Another procedure that has been used for many years is the replacement of joint cartilage with a patient's own cartilage cells. Cells are taken from a knee joint, grown in a laboratory and transplanted into the damaged joint.

The replacement of ears, heart valves, bone or blood vessels poses particular problems. In these cases, the cells are required to form tissue with a particular shape. Bioengineers use porous, synthetic or biological carrier material – such as collagen or fibrin – shaped into the desired form. This is fed with the patient's cells and then incubated in a nutrient solution. The cells attach themselves to this scaffolding substance and grow into three-dimensional tissue. Some types of carrier material later dissolve in the body, while others stay put. Artificial blood vessels are produced by cultivating blood vessel cells over tubes of synthetic materials.

How could stem cells be useful in tissue replacement?

Most tissues are not as easily reproduced artificially as skin and cartilage cells. For the production of heart muscle or nerve cells, for example, stem cells are required. All body tissue contains stem cells and their task is to produce replacement cells. Work with such adult stem cells opens up many new possibilities. In theory, it should be possible to use stem cells to produce any kind of tissue but the necessary cell cultivation technology has still not been perfected.

Intensive research efforts are underway to try and regenerate heart tissue destroyed by a heart attack with the help of stem cells. Studies have shown that stem cells

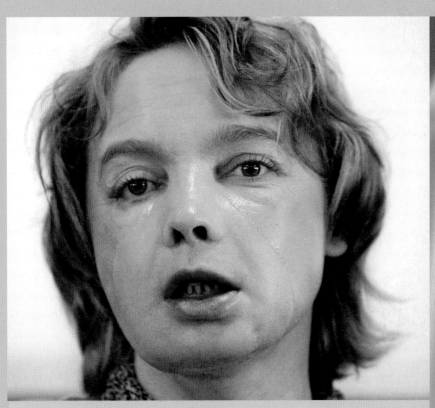

In November 2005 Isabelle Dinoire from France became the first person to undergo a face transplant after the lower part of her face had been mauled by a dog and almost completely destroyed. She still needs to take powerful drugs to stop her body from rejecting this new tissue.

taken from bone marrow and delivered to the heart improve a patient's heart function. Researchers have still to decide whether this is because the transplanted cells are actually turned into fully functioning heart muscle cells or not.

Attempts are also being made to use nerve stem cells to restore dead brain tissue. Patients who are afflicted with Parkinson's and Alzheimer's diseases would probably benefit from this kind of cell therapy.

The only stem cell therapy that has so far become well established is bone marrow transplantation following cancer treatment. This involves replacing the bone marrow cells that have been destroyed with suitable donor stem cells.

While adult stem cells are taken from the patients themselves, embryonic stem cells come from embryos that are only a few days old. The advantage of these is that they multiply more easily and can be transformed into all types of cells. But because the immune system considers them to be alien, there is a problem with rejection. The use of these cells gives rise ethical reservations as well as the possibility that they may grow into tumours.

How does our brain shape our personality?

The brain processes and stores everything that is important for our lives – from our most basic instincts and reflexes to our personal memories and the skills and abilities we have inherited or learned. It is this that makes us who we are.

Playing chess from an early age may not increase IQ, but it will promote certain skills.

When can a person be called intelligent?

What makes human intelligence so special is our ability to use insight and thought to tackle new tasks and cope in new situations. Depending on the extent to which this ability is greater or lesser than average, we talk about different levels of intelligence. An intelligence quotient, or IQ, of between 90 and 110 is regarded as normal. An IQ of between 70 and 90 indicates learning difficulties, while a score below that level points to mental disability. However, within the overall range of intellectual impairment, even an IQ of between 50 and 70 is still classified as mild. Only when a person's IQ is less than 50 does independent living become virtually impossible. At the other end of the scale, people with an IQ of more than 110 are classified as intelligent, while those who score 130 or above are regarded as gifted.

Does intelligence change as we grow older?

The term intelligence does not always mean the same thing. As early as the 1950s, scientists distinguished between fluid and crystallised intelligence. Fluid intelligence is the ability to find ways of dealing with a new situation and to grasp its implications quickly, even in the face of distraction or disturbance. Crystallised intelligence is the end product of accumulated learning. Both forms of intelligence change in different ways over the years. As people get older, their ability to react quickly to new situations may decline, but their fund of knowledge has grown steadily larger over the years.

Is it possible to become more intelligent?

It is not possible to increase intelligence, either with the use of some kind of brain machine or by taking special dietary supplements. It is possible to compensate for low intelligence by exercising the brain and acquiring knowledge and experience. That is why it is important to stimulate less intelligent children – such as those with Down's syndrome – as much as possible.

People of average intelligence can perform better than those with above-average brainpower in particular areas where their accumulated knowledge – their crystallised intelligence – is superior. Chess is regarded as the greatest strategic game. Studies have shown that highly intelligent beginners don't fare so well when matched against less intelligent experts who have been playing for a long time.

Dolphins have a trusting manner that people with autism find easy to relate to.

Are there ways of recognising intelligence in individuals?

If a child starts to speak early, many people see it as a sign of high intelligence. Many gifted children do begin to talk earlier than those with average intelligence, but the reverse can also be true. Some highly intelligent children start late but then surprise their parents by producing whole sentences straight away – as if they had rehearsed what they wanted to say in their minds before saying it.

Average children who start speaking late are not necessarily less intelligent than early starters. Some children – usually boys – simply take longer to develop language skills. Only when a child has failed to speak more than a few words by the age of two should the advice of a specialist be sought.

Are people with autism mathematical geniuses?

Some autistic people have extraordinary gifts. They can work out the square root of 1 657 543 effortlessly, or correctly identify which day of the week 23 September will fall on in the year 4198 without hesitation. Such people are known as savants – French for 'those who know'. But savants are mostly of below average intelligence and are unable to live independent everyday lives. Such people are said to have Kanner's syndrome, but not every autistic person with Kanner's syndrome is a savant. Autism spectrum disorders include Kanner's syndrome and Asperger's syndrome. People with Asperger's – named after the paediatrician Hans Asperger – are often of above average intelligence and are capable of abstract logical thought. Many develop an intense interest in certain things, but often become isolated because of their inappropriate behaviour. Those with Kanner's syndrome – named after the psychologist Leo Kanner – shut themselves off from everything and everyone around them and have a compulsive need to be in the same surroundings at all times.

Is lying a sign of intelligence?

Before children can tell lies, they must be able to put themselves in another person's place. Until the age of about four, a child cannot understand that others might not know about something that they themselves have experienced. Only at that age does a child begin to learn that every person has separate perceptions of his or her own. It is at that point that a child can say, 'I've seen something that you don't know about.' This is also the point at which children can deliberately say things that are untrue.

Why can't we remember being born?

Authentic and conscious memories of events that occurred before the age of one are thought to be unlikely. It is even less likely that anyone can remember being born. The scientific explanation is that too few brain structures are fully developed at the time of birth. Memory seems to be firmly linked to language acquisition. Experiments have shown that people are only able to remember experiences that they were able to describe in words at the time the incident occurred.

What is déjà-vu?

Most of us have had the experience of being in a new situation but having a strong feeling that we have experienced the same thing at some time in the past. The feeling only lasts for a moment or two. Scientists do not dispute that the experience of déjà-vu – French for 'already seen' – exists but, unlike New Agers, who believe that déjà-vu experiences are fragments of memory from

Several parts of the brain are activated when we remember things – memory is not located in a single area of the brain.

a previous existence, scientists consider it a kind of malfunction of the brain. One line of research suggests that when a person has an experience, his or her mind will search for similar experiences that have been stored in the memory archive in an area of the brain known as the hippocampus. During this process, the brain may incorrectly report that there is a matching memory. It is at this point that we feel that we have experienced something before, even though we haven't.

Why do we often relate a tune to a particular memory?

The human ability to hear and react to music is closely connected to other forms of perception. Musical sounds are processed in the rostromedial prefrontal cortex, a region at the front of the brain that mediates between emotional and non-emotional information. This is also where information is processed that is important to the formation of our perception of our own self. This may be why music heard in a particular situation becomes attached to the memory of that situation. Scientists have found that memories of music heard in childhood and youth are more enduring than those heard in adulthood. Work with people suffering from dementia has shown that they can still be reached through music from their childhood and early adulthood.

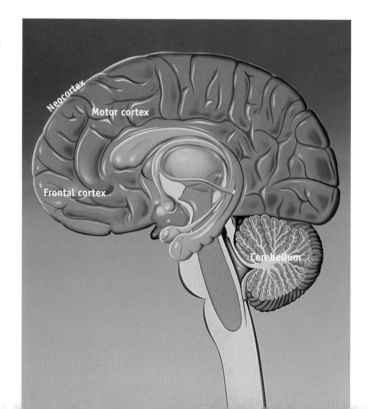

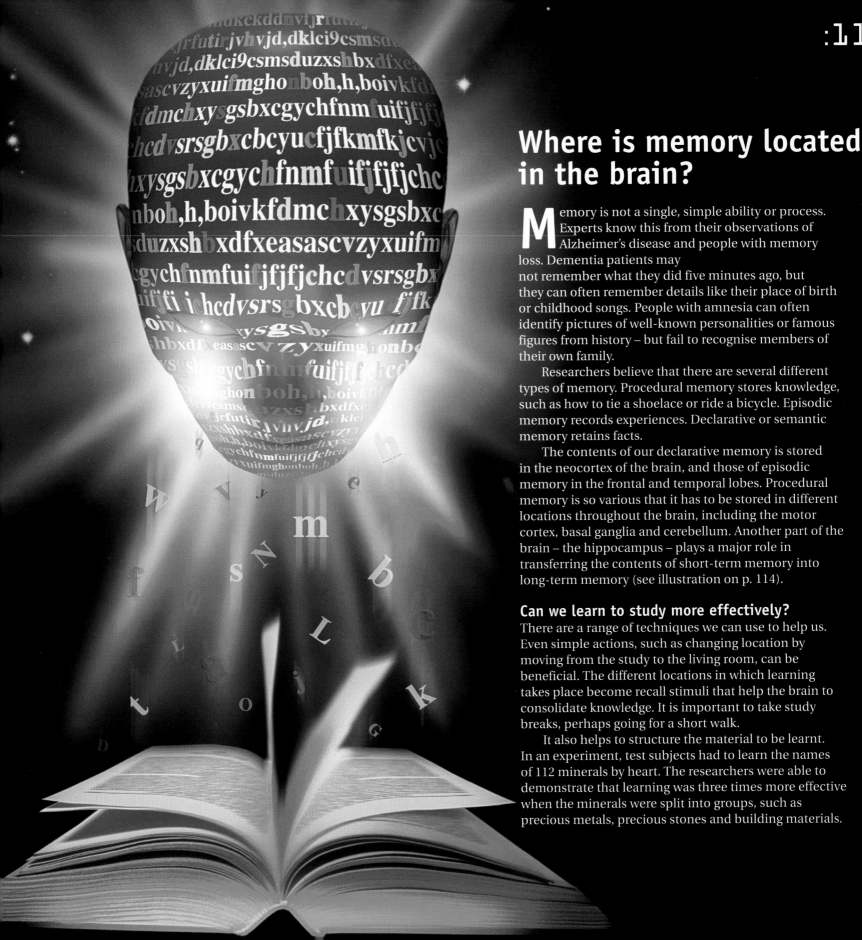

Where is memory located in the brain?

Memory is not a single, simple ability or process. Experts know this from their observations of Alzheimer's disease and people with memory loss. Dementia patients may not remember what they did five minutes ago, but they can often remember details like their place of birth or childhood songs. People with amnesia can often identify pictures of well-known personalities or famous figures from history – but fail to recognise members of their own family.

Researchers believe that there are several different types of memory. Procedural memory stores knowledge, such as how to tie a shoelace or ride a bicycle. Episodic memory records experiences. Declarative or semantic memory retains facts.

The contents of our declarative memory is stored in the neocortex of the brain, and those of episodic memory in the frontal and temporal lobes. Procedural memory is so various that it has to be stored in different locations throughout the brain, including the motor cortex, basal ganglia and cerebellum. Another part of the brain – the hippocampus – plays a major role in transferring the contents of short-term memory into long-term memory (see illustration on p. 114).

Can we learn to study more effectively?

There are a range of techniques we can use to help us. Even simple actions, such as changing location by moving from the study to the living room, can be beneficial. The different locations in which learning takes place become recall stimuli that help the brain to consolidate knowledge. It is important to take study breaks, perhaps going for a short walk.

It also helps to structure the material to be learnt. In an experiment, test subjects had to learn the names of 112 minerals by heart. The researchers were able to demonstrate that learning was three times more effective when the minerals were split into groups, such as precious metals, precious stones and building materials.

Does the brain need sleep?

more rapidly from memory. Research has demonstrated that sleep can even reinforce newly learnt physical movements. People who are learning to dance or ski may find that they are only able to perform new steps or movements effectively after a good night's sleep.

Is it possible to come up with brilliant ideas while asleep?

The German organic chemist August Kekulé von Stradonitz liked to tell the story of how, in 1865, his insight concerning the ring structure of benzene came to him in his sleep. Many other people have also reported occasions when important ideas were gained this way.

Research has confirmed this, with scientifically conducted experiments showing that a good night's sleep can result in people coming up with solutions to problems from the day before. Fewer members of control groups – who were not permitted to sleep at night during the experiment – were able to find the required solution, compared with those who were allowed to sleep. The old saying urging people to 'sleep on it' seems to work. Sleep enables the brain to consolidate knowledge, but also to restructure it and so help find solutions to problems.

Sleep and brain function are closely connected. Sleep consolidates what has been learnt and ensures that new brain structures are formed in babies and small children, which promote further learning. It is not possible to conduct sleep deprivation experiments on babies, so conclusions must be drawn from experiments with young cats. A number of kittens were presented with new stimuli. Some were allowed to sleep for six hours afterwards, while the others were kept awake and provided with further stimuli. The brains of the kittens that had been allowed to sleep exhibited more changes than those of the ones that had been kept awake.

Does it help to continue revising right up to the night before the exam?

Revising on the night prior to an exam does not appear to be harmful – as long as you don't leave everything to the last moment. Allow plenty of time for sleep on the night before an exam. Experiments have demonstrated that memories are consolidated in the brain during sleep, making them much easier to recall. Lack of sleep, or even sleep deprivation, seems to cause learned material to slip

True or False?

Does brainpower diminish if we don't use it?

It really is a case of 'use it or lose it' when it comes to the brain. Some scientists say that there is no such thing as age-related memory loss – the problem is that adults become mentally lazy. Their jobs become routine and their evenings are spent absorbing a constant stream of TV, all of which dulls the brain. There is no agreement on the best way to exercise the mind. Some scientists believe that solving crosswords provides a good workout for the brain, while others consider this to be of little use as it can quickly become another routine. The essential feature of any beneficial activity seems to be that the brain must be constantly confronted with new experiences. Something like learning a new language may be the best sort of workout.

Do older people find it harder to learn new things?

Our ability to learn decreases with age, and people find it difficult to retain new information. This is due to a reduction in inhibition – the ability to ignore or forget unimportant information. In younger people, the memory can better distinguish between important and unimportant facts. When this capacity diminishes, too many bits of information jostle for a place in the working memory, often getting in each other's way. This is why fresh information is more easily lost.

Does physical exercise improve memory?

We need to do more than mental training to ensure that our memory stays fit as we grow older. Studies have shown that people who do not exercise regularly and rarely walk for any distance are more likely to complain of a deteriorating memory as they get older. Exercise activates the brain and supplies it with more oxygen, so people who stay fit and active for as long as possible have a better chance of retaining a good memory.

Can objects be manipulated by the power of the mind alone?

Say the word psychokinesis and most people will think of Uri Geller. In 1974, he appeared on a television show and appeared to bend several spoons without applying any force. For weeks afterwards viewers discussed the possibility that what they had seen was a real phenomenon. Scientific views on this range from scepticism to rejection. The consensus seems to be that it was a sophisticated trick.

A branch of research has emerged that uses the power of the mind in a less magical way – neuroprosthetics. With the help of a silicon chip implanted in the brain, it is hoped that people with paralysis will be able to use their mind to operate a computer or even an artificial limb. Further research is needed to establish which areas of the brain are responsible for particular functions, and which action generates which types of electric currents in the brain.

How did our brain evolve?

With a history dating back about 500 million years, the oldest part of the brain is the brain stem. It is also known as the reptilian brain because even our distant ancestors had one. It is responsible for the body's basic functions – including movement, is slow to learn and is devoid of emotion.

Above it is the midbrain, which acts as a kind of control centre for breathing and a variety of other reflexes. It is also connected to the cerebellum, which is responsible for physical coordination and balance. These parts of the brain have a 250-million-year history.

The youngest part of the brain is the cerebrum, with the cerebral cortex. It is thought to have first appeared only 200 million years ago. The cerebrum is the seat of consciousness, where memories are stored and where thoughts, wishes, intentions and ideas are formed.

Are men's brains different from women's?

On average, men have larger brains than women, although this does not mean that they are in any way intellectually superior. Men often have more nerve cells in the cerebral cortex than women, although the cells in the female cerebral cortex are better connected. The corpus callosum – the structure that connects the two halves of the brain – may be wider in women. Although there are broad general differences, we should not forget that every brain is individual and unique.

Pure mind power is not enough to make objects move – whatever Uri Geller may claim.

The real meaning of dreams

Thanks to modern science, it is now possible to see into a living human brain and observe the processes that occur there during sleep. Sometimes, modern medicine has even been able to bring people back from the threshold of death, and this has made the study of near-death experiences possible. Despite medical progress and huge developments in the instruments used to observe human bodily functions, there are still many gaps in our knowledge about the nature of dreams and memories.

How are dreams and memories formed?

As far as we know at present, for memories to be developed, genes in the brain's nerve cells need to be activated, new proteins formed and nerve connections changed. We know even less about the creation of dreams. For a while it was assumed that they were closely linked to REM sleep – REM being the acronym of Rapid Eye Movement, a stage of sleep which is recognisable by the sleeper's rapid eye movement. REM sleep was discovered in the mid-20th century and has played an important role in explanations of dreams ever since. In addition to the rapid eye movement seen during REM sleep, there are other motor activities as well. REM sleep occurs about every 90 minutes and lasts for about 20 minutes. Experiments conducted in sleep laboratories have shown that people who are woken while they are in a REM sleep stage are more likely to report having dreamt than those who were woken during non-REM sleep.

It has now been found that people dream during other phases of sleep as well, although these dreams have a different quality. They tend to resemble conscious thought sequences, whereas REM sleep dreams tend to be similar to cinematic narrative sequences. The connection between dreaming and REM sleep seems to be looser than was initially assumed. REM sleep has even been observed in the very rare cases of people who no longer dream at all due to brain damage.

What are dreams trying to tell us?

It has been suggested that dreams are a kind of mixture of information made up from originally unconnected experiences and knowledge from different areas of the brain. But sleep scientists have noticed that people don't usually dream at random and that their dreams do tend to have something to do with themselves and their lives. Some people repeatedly dream that they have failed an important exam, for example, or they dream of people and objects that they had dealings with during the day. Many

Dreams are amongst our most mysterious experiences. They can be strange, funny, frightening or stimulating – but asa general rule, they immediately fade from memory when we wake.

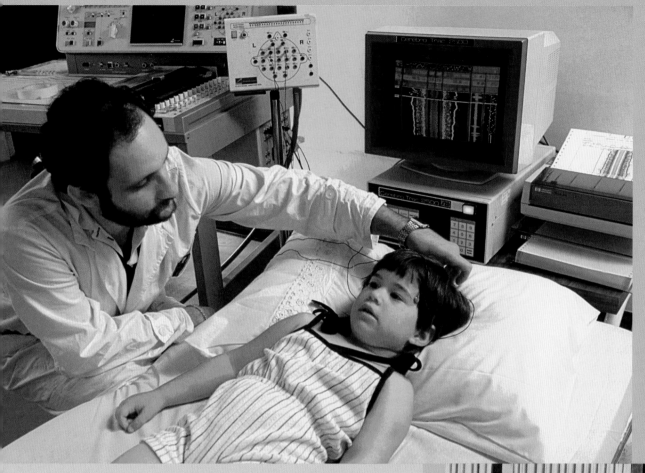

theory in which he claimed that dreams have meaning, and that they are, in fact, the fulfilment of suppressed desires. But, according to his theory, when a suppressed desire appears in a dream it is disguised, with the substance of the dream being presented through a series of subtle shifts and transformations. This is how Freud explained the well-known phenomenon where a dreamer is sure he or she knows a person in a dream but is quite unable to give that person a name. In cases like this, different people from the dreamer's past and present have be transformed into a single composite individual. Shifts occur when images from the dreamer's distant past find their way into dreams about places and events from the preceeding day. According to Freud, it is because of these shifts and transformations that we often only remember fragments of our dreams when we wake the following day.

An important method for finding out more about dreams is sleep research – also known as somnology – in which the various stages of sleep are studied and analysed with the help of instruments. The causes of snoring are also studied in this way and the research can help restless sleepers to be calmer.

researchers have therefore gone back to the idea that dreams are likely to be more than merely neurons stimulated at random.

Which dreams do we remember most clearly?

We usually only remember the fact that we have had particularly intense dreams when we wake up. This makes sense, since we usually have several dreams each night. If we do remember a dream, it is likely to have occurred in the last stage of dreaming prior to waking, but for some people, the dream's contents tend to fade as soon as we try to recollect it. Only nightmares, or dreams of a particularly emotional nature, stick in our memories.

What were Sigmund Freud's theories about dreams?

For Sigmund Freud (1856–1939), dreams were a valuable tool that could be used to help people to understand the working of the unconscious mind. By interpreting his own dreams, as well as those of a number of other people, Freud developed a

Sigmund Freud's influential book *The Interpretation of Dreams* (1900) was a major step on the way to a psychoanalytical understanding of the human mind.

Are humans 'wired' for language?

Humans seem to be programmed for language learning, and they will only fail to speak if prevented from doing so. Evidence of the extent to which language is a human characteristic is supported by the observation that even people with serious mental disabilities – who are unable to learn to read and write – can and do learn to speak a language.

Are there difficult and easy languages?

There are no difficult or easy languages, but foreign languages can be easier or more difficult for speakers of a particular mother tongue. English speakers find the German language easy, because English is closely related to German. English speakers have no trouble deducing the meanings of words like *kaffe*, *schwimmen*, *lernen* and *wasser*. Chinese speakers may not find German so undemanding.

Wherever they are in the world, children take about the same time to learn their mother tongue if not hindered from doing so. The only thing that varies from language to language is the sequence in which particular grammatical features are learnt.

Whether it's English, French, German or Chinese – children learn their respective native tongues with equal ease.

The time to teach children a second language is while they are still in kindergarten, when their brains are at their most adaptable.

most complex tasks, including speaking their native tongue. During this time, a child is in great need of external cognitive stimulation – they have to be taught consciously to perceive and assess their environment. A study of Romanian orphans demonstrated that the minds of the children, who were properly clothed and fed but whose cognitive development was not particularly encouraged, took longer to develop than would normally be the case.

When a child reaches puberty, his or her brain is restructured. This is why it is a difficult period for children and their parents. From the age of about 17, the brain begins to lose its previous adaptability, and the speed at which new things, such as a new language, are learnt is considerably reduced.

Why do children seem to find it much easier to learn languages than adults?

The brain is an adaptable organ. Studies have shown that even an adult brain can change noticeably, although brains are at their most adaptable in childhood. This is also the period when human beings learn to master the

Can we forget our first language?

A person would need to be relatively young at the time they lost contact with their original native language in order to forget it. If someone were to emigrate in their mid-forties, they would be unlikely to forget their native tongue even if they never spoke it again, although they might have difficulty with rarely used words. Children or teenagers who move into an environment where almost nobody speaks their first language can lose their ability to speak their old language. The brain is an organ designed for constant learning, and it is for this reason that a lack of practice can cause it to forget what it had previously learnt.

How do deaf children learn sign language?

It doesn't matter to the human brain whether a language is made up of sounds or gestures. Linguists now know that deaf children learn sign language in much the same way as children who can hear learn a sound-based language. A deaf child with one deaf parent will learn sign language from that parent. If neither parent is hearing impaired, one or both will need to learn sign language in order to communicate with their child. This can result in the development of individual signs that are only understood by family members, but sign languages are considered to be proper languages, for which dictionaries are available. Among the world's many sign languages are BSL (British Sign Language), ASL (American Sign Language) and Auslan (Australian Sign Language).

Deaf people use a finger alphabet for spelling. Illustrated here are the letters D, E, F and G. Many words also have their own signs.

The davui – a Fijian conch shell trumpet – is blown when people need to be summoned for a celebration or to impart important news.

Can we communicate without using a language?

It is not really possible to communicate without a language. When we find ourselves in situations where conventional spoken languages like English or Spanish cannot be used, we have to switch to makeshift languages, where devices such as hand gestures or flags may be used as linguistic markers. Using one's hands to speak – often the last resort for people holidaying abroad – must be regarded as a kind of language, although not one that endures or has any fixed rules.

How do people 'speak' with signal flags?

Signal flags have been in use internationally since 1901, to enable ships' crews of different nationalities to communicate with one another at sea.

A set of signal flags consists of 26 alphabet flags, ten numeric pennants, four substitute pennants and one answering pennant. Each of the 26 alphabet flags represents a letter in the Roman alphabet, as well as having a special meaning when hoisted on its own. The flag for the letter A, means: 'We have a diver underwater, please keep your distance and travel slowly.'

Up to four different letters are raised simultaneously. As is the case with single letters, groups of two, three or four letters also have particular meanings. The flags PN together mean: 'Your navigation lights are not visible.' It is rare for the flags to be used to spell out messages so, to avoid confusion, every spelled message has to start with the signal for 'I am spelling out this message' and end with the signal 'I have spelled out this message'. Signal flags are read from top to bottom of a hoist.

What links language and music?

Many scientists now believe that language and music are closely linked. If human beings didn't have a natural talent for music, their utterances would be entirely devoid of melody and speech would sound clipped, like that of a robot.

Every human utterance has intonation, pauses, pitch and volume. Put together, these properties of speech are called 'prosody' by linguists. Ultimately, the prosody of human speech seems to be based on an innate human musicality, which leads us to produce musical rhythms. No human culture studied so far lacks music.

How are languages born and how do they die?

Researchers are fairly clear about why languages die. This is generally the result of the speakers of one language gradually converting to another, more widespread language (see p. 124). How new languages emerge continues to be one of the great unanswered problems in linguistics. It involves questions about a human protolanguage – a possible ancestor of all human languages – and the way in which children acquire language.

Was there a protolanguage?
We will never know exactly how and where language began. Scientists suspect that the first words uttered were onomatopoeic, like hiss and sizzle. It is also unknown whether there is one single source for language or whether it emerged in different regions independently.

There have been many attempts to reconstruct a protolanguage, and it has been given names, such as Nostratic or Boreic. Supporters of a protolanguage derive their optimism from a discovery made by linguist Franz Bopp (1791-1867). At the beginning of the 19th century, he realised that Indian Sanskrit shared some features with German and Russian. This group of related languages – known as the Indo-European family – includes other languages and it was suggested that they all developed from a single, earlier language, namely Indo-European. Many researchers remain sceptical that all languages are related. Attempts to reconstruct Indo-European involve a meticulous study of ancient texts. To reconstruct a protolanguage, the amount of common ground between all the languages involved is so small that it is barely possible to make any definitive statement. The protolanguage would have been spoken in a period from which we have no language data at all.

True or False?

Native Americans used smoke signals to send messages

Native Americans threw wet grass onto a fire to produce thick clouds of smoke, and then covered the fire with a blanket. They then released the smoke that collected under the blanket at intervals to produce sequences of puffs of smoke that rose into the sky and were visible from a distance. This was not based on a form of Morse code. The meaning of the smoke signals was agreed in advance. This kind of long-distance communication has been known since antiquity, and variations use torches instead of smoke.

Were the Neanderthals able to speak?
Certain anatomical and mental prerequisites are needed for speech, such as a spine able to support the breathing muscles required. According to a recent study, this was the case with *Homo erectus*, who lived approximately 1.8 million years ago. Neanderthals lived only 120 000 to 45 000 years before our time.

Among the mental prerequisites are an ability to imagine something not present and visible. This may be a hand-axe that needs to be crafted or the preparations for a hunt. If there are indications that an early human species used burial rituals, we can be fairly sure they had language. When people bury their dead it is because they believe in a hereafter, and this can only be conveyed through language. However, a great deal of debate surrounds the question of whether Neanderthals buried their dead or not.

Recent scientific findings suggest Neanderthals possessed more intelligence than had previously been thought.

Why do some languages die?

When a language is spoken by only a small number of people, all it takes is a famine, an epidemic, a natural catastrophe or even genocide for it to become extinct. However, the most common reason why languages die out is probably a gradual shift that takes place over several generations, during which a language community favours another language that can be used to communicate more effectively in the circumstances in which they find themselves. This process can be seen occurring today in Africa, where the speakers of minority languages are increasingly shifting over to more widely used languages, like Swahili and Hausa. These are the languages used in schools, hospitals, offices and the media, and as they gain in popularity, the languages of the home villages fall into disuse.

Languages never die as a result of being fatally degraded. That many people are no longer able to use an apostrophe correctly, or know when to use 'whom', is not necessarily a sign of impending doom. It is simply that language is gradually changing, as it has always done.

Why is Latin regarded as a dead language?

In 476 AD, the Western Roman Empire finally disintegrated. The Germanic mercenary leader Odoacer deposed the child-emperor Romulus Augustulus, subjugated Rome to the Byzantine emperor and declared himself Patrician of Italy. Although Germanic peoples formed a powerful upper class in Rome after the collapse of the Empire, the population did not adopt their language. Latin was almost certainly spoken for decades after the collapse, but the infrastructure in the former Roman Empire went into decline. Schools were closed, illiteracy spread and the written word gradually ceased to be of any use to speakers of the language.

The Germanic peoples didn't have writing, so their language could not function as the language of administration. The writing of Latin only continued in monasteries, but outside the monastery walls the spoken language changed more and more as the Romance dialects emerged. When Charlemagne established his empire in 800 AD, nobody had Latin as their native language anymore. Under Charlemagne, Latin was already taught as a foreign language, as it is to this day. Latin may be a dead language, but it is not extinct.

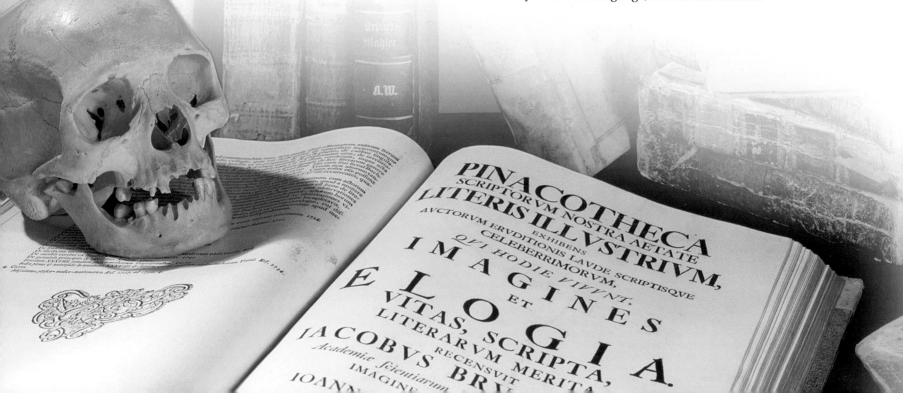

What is pidgin English?

A pidgin is a simple language specially developed so that people from different cultures can communicate, at least in a rudimentary way. New Guinea is an example of a country where pidgins facilitate communication between people of different languages. Pidgins develop when people need to talk to each other to trade or when people have been thrown together and forced to communicate, such as when slaves were brought from Africa to the New World. Groups of slaves speaking the same language were deliberately split up in order to reduce the chance of escape or revolt. Some groups created their own pidgins as a result. Pidgins are not native languages – they have to be learned, and speakers are, therefore, bilingual. Although pidgins are fairly simple languages that can be learned quickly, they have their own grammar, vocabulary and rules of usage. Some linguists think a pidgin becomes a creole, and develops a more complex grammar when it is taught to children as a first language.

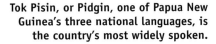

Tok Pisin, or Pidgin, one of Papua New Guinea's three national languages, is the country's most widely spoken.

Why is the English spoken today different from the language that was spoken 100 years ago?

Language is subject to constant change. New words emerge to denote new things. A hundred years ago, the word telephone was just gaining acceptance and nobody could have guessed that one day we would have a mobile. The meaning of words also alters over time. Today, 'nice' can mean agreeable or thoughtful, while its old meanings included foolish, fussy or precise – the latter use preserved in the expression 'legal niceties'.

We are forever looking for new ways of abbreviating our utterances. We talk about getting info or going to uni. These abbreviations eventually gain wide acceptance.

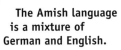

Can a language be frozen?

When a language community forms a minority in a country where there is a different majority language, it is possible for a language to be preserved more or less unchanged. The Amish, who migrated to the United States of America in the 18th century from Switzerland, still speak a form of the dialect they spoke before they emigrated.

The Amish language is a mixture of German and English.

Why are artificial languages like Esperanto not more widely accepted?

Life might be a lot easier if all the people in the world could agree on a single language. But not everyone thinks this sort of globalisation is a good thing. Many small regions around the world are fighting for independence – or at least some autonomy – and their aspirations generally involve maintenance of their region's native language. Adopting an international artificial language – also known as a constructed language – such as Esperanto, is not one of their aims.

It is also difficult to convince people who no longer have to struggle for the sake of their national language of the benefits of learning a new language that – at least initially – would not be spoken by their neighbours. It would be necessary to ensure that a great number of people in neighbouring countries learnt the language at the same time. Although Esperanto adopted the best features from many languages, some linguists point out that sensible features from major non-Indo-European languages – such as Chinese – were not considered.

Are there other artificial languages?

Apart from Esperanto, there is Volapük, an artificial language invented by the priest Johann Martin Schleyer in about 1880. It failed, not least because Schleyer wanted any new words to be invented by himself alone. Another, much younger artificial language is Interlingua, developed by the linguist Alexander Gode. These days, Interlingua is considered to be Esperanto's only serious competitor.

Not every artificial language was developed to improve understanding between nations. The Klingon language was developed by linguist Marc Okrand for the cinema and TV series *Star Trek* in the 1980s. From humble beginnings emerged a language with its own grammar, an extensive vocabulary, dictionaries and a devoted fan-base. The Klingon Language Institute ensures that this language continues to be disseminated and looks after projects that include translating Shakespeare's works into Klingon.

Many of the words in Esperanto's vocabulary sound familiar to Europeans. The word for cat, for example, is kato.

How are languages classified?

Linguists have still not been able to classify every language family and its members definitively, and many minor languages have yet to be assigned to their correct language family. Experts often disagree about whether something is a language or a dialect, and that affects how many members are accepted for a particular family.

The language families that researchers agree have more than 30 member languages include: Niger-Congo, Austronesian; Trans-New Guinea;; Afro-Asiatic; Sino-Tibetan; Indo-European; Nilo-Saharan; Austro-Asiatic; Sepik-Ramu; Arawakan; Tupi; Tai-Kadai; Torricelli; Carib; deaf sign languages; Oto-Monguean; Dravidian; Hmong-Mien; Macro-Ge; Quechuan; Panoan; East-Papuan; Mayan; Geelvink-Bay; Uto-Aztecan; Uralic; and, Tucanoan.

More than 100 language families have been identified. In addition, there are language isolates the membership of which has yet to be determined. Korean is among the better known language isolates.

Writing – preserving our collective memory

The Code of Hammurabi (left) was written in cuneiform script and created in about 1700 BC. It is the world's oldest recorded complete code of laws. The ancient Egyptian texts (above) were written in hieroglyphs, a highly developed and complicated pictographic language that was in use for 3000 years.

W ith the invention of writing, human beings made the transition from a prehistoric culture to a historic one. This is because – in principle, at least – writing enables us to record the past for the benefit of generations yet to come. Considering the revolutionary potential of this new invention, it is perhaps disappointing to learn that it was first used for documenting trade transactions, a task that seems to have taken precedence over the drafting of chronicles to record the thoughts and deeds of our distant ancestors.

Where did the first writing system emerge?
The earliest known examples of writing were discovered in the Sumerian city of Uruk in Mesopotamia. They are thought to have been produced in about 3300 BC – very early in the history of that advanced civilisation. King Hammurabi – under whose rule the renowned Code of Hammurabi was drawn up – did not live until some 1500 years later.

The first writing systems seem to have been pictographic.

The symbol for 'eat', for example, clearly depicts a head next to a bowl, but this literal linking of things and symbols vanished over the course of the centuries. As the use of a blunt reed called a stylus to scratch symbols into clay became widespread, writing became wedge-shaped and gave rise to the name cuneiform script (meaning wedge-shaped).

Cuneiform script makes up what is known as a logographic writing system, whereby a single word is represented by one symbol. Over time, the number of symbols in logographic writing systems tends to become rather large. Chinese, for example, which also has a logographic writing system, uses about 50 000 symbols of which 4000 to 7000 are in common use. Faced with this sort of complexity, it is clear why the development of alphabetic writing systems is regarded as being of equal importance to the invention of the wheel.

What are the most widespread writing systems and why?

Most of the writing systems in current use around the world are alphabetic. They include not only the Latin alphabet but also the Cyrillic, Arabic and various alphabetic writing systems that are used for only one language, such as Amharic or Georgian.

Alphabetic writing systems are phonographic, which means that they record sounds, although sounds and writing cannot always coincide. The Latin alphabet, for example, cannot distinguish between a rolled 'r' and an approximant 'r' (as in raw). The phonographic character of alphabetic writing systems has a major advantage in that it is possible to reproduce an infinite number of utterances with a restricted set of letters. This includes spontaneously created onomatopoeic words like 'plong' or nonsense words like 'glumph'.

Why is it that there are still ancient documents that have not been deciphered?

If centuries or millennia have passed since a language faded from use and only written documents remain, the writing may be impossible to decipher. This is what happened with the North West Indian Indus script from the 3rd millennium BC, or the Minoan writings of the 2nd millennium BC. Unlike the successfully deciphered Egyptian hieroglyphs, no bilingual or trilingual inscriptions have been found for the undeciphered writing, examples such as the crucially important Rosetta Stone, which made it possible to decipher hieroglyphs. It is also often the case that researchers are unclear about which other languages the language of an undeciphered script may be related to. In the case of the Minoan script, we simply do not know what language was spoken by the first Cretan people who used it.

How do we know how the Romans pronounced the word Caesar?

It can seem strange that we have a fairly good idea about how to pronounce an ancient language – such as Latin – for which there are no longer any native speakers. We do know beyond doubt that Latin is an Indo-European language, and many texts written by native Latin speakers from different periods in history have been preserved. So it is possible to make comparisons within Latin. The word 'cuius', for example, is the genitive of the relative pronoun qui, quae, quod. So we can assume that the 'c' in cuius is pronounced 'k' because qu was pronounced like kw. So all that was missing in the word cuius was the 'w' sound.

It is more difficult to make inferences of this sort when it comes to proper names like Caesar, and this is where a knowledge of Greek comes in handy. Caesar was also written about in areas where Greek was spoken, and so we know that Caesar was probably pronounced 'kaizer'. Thanks to the fact that Latin poems and rhymes have been preserved, we know about stresses and vowel lengths in Latin, and because poetry is usually written in a particular metre, we are even able to deduce accent rules. It is also the case that the Romans wrote textbooks in which they provided indications of the correct way to pronounce some words.

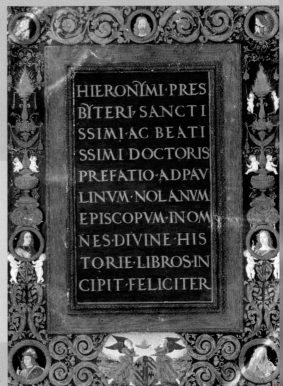

HIERONYMI·PRES
BYTERI·SANCTI
SSIMI·AC·BEATI
SSIMI·DOCTORIS
PREFATIO·AD·PAV
LINVM·NOLANVM
EPISCOPVM·IN·OM
NES·DIVINE·HIS
TORIE·LIBROS·IN
CIPIT·FELICITER

What factors have an effect on our emotions?

Much of what affects our emotions is pre-programmed by the way our brain is designed, and quite a lot is genetically determined. But there is also a great deal that is learnt, and our emotions can be shaped by historical developments.

Can we believe in love at first sight?

This is not simply a cliché from the world of romantic novels and films – it really does happen. Behavioural scientists have proved that the so-called 'first sight' has to last for around 30 seconds, but not much longer than that. The critical 30 seconds is long enough for us to decide whether we are interested in a person or not. It is enough time for a fairly long period of eye contact, or for the exchange of a few sentences. If a woman likes a man, she may offer more 'filler' sounds like 'ah' and 'oh'. Men also tend to do this when they find a woman attractive, unless they are shy.

Why do we talk of men 'paying court'?

In the Middle Ages, many knights went to war in order to secure their livelihood. This changed during the High Middle Ages – in about the 12th century – when sovereigns began to rule over particular territories. Bartering was gradually replaced by a money economy and the population grew. Knights had to adapt to life at court, and to being part of the courtly retinue. Wherever many people live together in limited space, rules of social interaction have to be established to enable communal life to work smoothly. Sociologist Norbert Elias (1897-1990) described the process that began at this point as the 'transformation of the nobility from warriors into courtiers'. The expression 'paying court' goes back to that period of upheaval and change.

Is the search for a partner really all about just one thing?

In addition to primary sexual characteristics such as genitals, human beings have secondary sexual characteristics. These include wider hips on women and a broad, hairy chest on men. When our forebears lived in caves, they may have used these characteristics in their search for a suitable sexual partner. Few people would now claim to have chosen their partner because of the characteristics that made them suitable for procreation, but scientists can identify human behaviours that have been with us since the Stone Age.

How do we know...

...that spring can also have a depressive effect?

As the days become shorter in autumn, many people begin to feel mildly depressed. Doctors and psychologists state that a seasonal lack of light is the cause. The opposite – known as vernal depression – occurs rarely, and why people should become depressed in spring is unknown. One theory is that spring, with its bright sunshine and the cheerfulness of the people around, can cause distress in people who are at an unhappy stage in their lives.

In which areas of the brain are emotions created?

Techniques for observing brain activity, including magnetic resonance imaging, are becoming ever more precise, allowing researchers to determine which emotions are controlled in which regions of the brain. Primary emotions, such as fear or anger, are produced in the amygdala. Unpleasant emotions and ideas are processed in a region of the prefrontal cortex.

Researchers claim to have tracked down the origin of the so-called sixth sense, which appears to originate in the anterior cingulate cortex that runs along the walls that separate the left and right brain hemispheres. A complex emotion like love poses more of a problem, as several areas of the brain are involved, including the amygdala, the hypothalamus (which controls all the body's hormonal cycles), the hippocampus (where short-term memory is located), the prefrontal cortex (long-term memory) and the thalamus (which filters sense impressions). But scientists are still not able to say why someone falls in love with one person and not another.

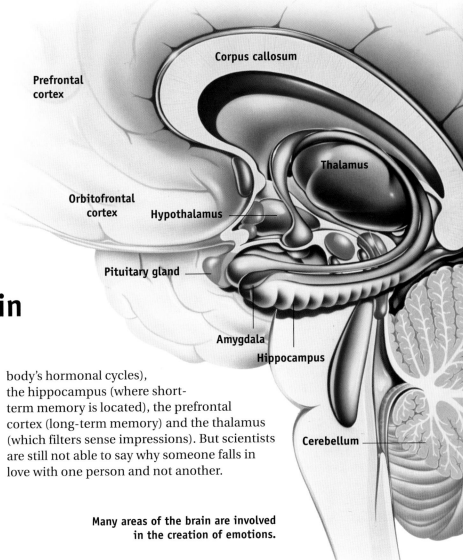

Corpus callosum

Prefrontal cortex

Thalamus

Orbitofrontal cortex

Hypothalamus

Pituitary gland

Amygdala

Hippocampus

Cerebellum

Many areas of the brain are involved in the creation of emotions.

Among the Himba people in Namibia it is still the custom for several related women to look after a child.

Is parental love innate?

Innate (inborn) love generally refers to a mother's love for her child; a father's love is not considered innate. This assumption that a mother's love is innate is now coming under closer scutiny. Psychologists and psychiatrists report that mothers have conflicting emotions about their children. Some mothers certainly exhibit negative emotions towards their children – in extreme cases even hating them. Ethnologists, historians and folklorists point out that the close mother-child relationship that is regarded as the norm in Western developed nations today, is a relatively new phenomenon. Because of the expectations that are a part of this bond, mothers can feel overburdened. Among people living in less urban societies, the norm is often for the young to be nurtured by several female relatives.

When does music cheer us up, and when does it make us sad?

Since antiquity, people have realised that music can make us feel sad, happy, passionate or energetic, and since the Age of Enlightenment attempts have been made to determine which musical characteristics provoke particular emotional states.

Scientists have managed to find only a few clues. A fast tempo is usually associated with joy, cheerfulness and action, whereas a slow tempo is connected to grief or solemnity. Loud music communicates power, intensity and joy, and quiet music tends to provoke feelings of tenderness or sadness. High notes seem to express pleasure or surprise, and low sounds, seriousness.

It is hard for anyone to resist the rhythms at the Carnival in Rio.

When do children begin to interpret facial expressions?

Even at the age of three months, babies find the human face much more attractive than anything else in their environment. From about the age of seven months, a child developing normally learns to interpret facial expressions that reflect emotions, such as joy, anger or fear. If two- to three-year-old children are presented with the words for emotions they are able to allocate facial expressions to the appropriate terms. And by the time they are four or five years old, most children are able to name emotional facial expressions without help.

When do children begin to understand jokes?

Adults tend to think of jokes as puns, double entendres or political gags. For children, someone placing their foot in a shoebox and – apparently – believing it to be a shoe is also a joke. Three-year-olds understand jokes based on inversion, such as Pippi Longstocking's horse on the veranda. At about the age when they are leaving preschool, children are able to understand simple jokes based on the idea of words having several meanings, which is the foundation for understanding verbal jokes. One of the next steps is an understanding of irony, although most children do not achieve this until they are eight or nine years old.

Smoking is bad for your health. Every smoker knows this, and yet few consider themselves to be at risk.

True or False?

Optimists don't fall ill as often as pessimists

It is difficult to be sure whether or not this is true. People who do not fall ill do not turn up in doctors' surgeries, and so are not available for use in tests. Studies looking at healing after serious operations or illness have shown a link between an optimistic attitude and a more rapid recovery. A long-term research project on mortality after cardiovascular disease demonstrated that optimists run a 23 per cent lower risk of dying from a cardiovascular disease than do pessimists.

Why do human beings often ignore risk warnings?

People should be aware by now that smoking can cause lung cancer, too much sunbathing can to lead to skin cancer, and unprotected, casual sex can result in AIDS. Despite this, medical psychologists repeatedly observe even people in a high-risk category assess their own risk as low when compared with people of the same age with a similar lifestyle. Scientists call this 'unrealistic optimism'.

Research into this has discovered that people at risk store a kind of risk prototype in their memory. This prototype may be similar to them in age and social class, but will be someone they classify as a chain-smoker, sun worshipper or sex maniac, whereas they consider themselves to be much more moderate. People feel reassured because they only smoke a few cigarettes in the evenings.

What would be more appropriate would be for people who are at risk to compare themselves with an average person, rather than an extreme prototype. This would result in a much more realistic assessment.

Why can minutes seem like eternities – and hours seem like minutes?

By concentrating on the second hand of a clock and following it with our eyes, we can make time almost stand still. The cause of this phenomenon is that time is not being filled. Psychological experiments have shown that every three seconds a new time window opens in the brain, and the brain asks, 'What happens now?' If what happens now is whatever was filling the previous time windows, a person's consciousness of time stretches and boredom takes hold.

But if the subject takes his or her eyes off the clock second hand for an instant, time immediately starts to move faster in the subjective consciousness. Time is filled again and as a result seems to speed up.

What turns us into addicts?

In principle, anything that gives someone a particular thrill can be addictive. If a person's desire to experience the thrill repeatedly becomes overpowering, he or she can be described as addicted. Cravings are typically for substances such as alcohol or nicotine, although it is also possible to become addicted to harmless or even socially acceptable pleasures, such as chocolate or sport.

What kinds of addictions are there?

The classic addictions – which everyone knows about – are to alcohol, nicotine and drugs. In recent years, psychologists and doctors have ceased to speak of addiction and prefer to use the terms 'dependence' and 'abuse'. These more accurately reflect the fact that some addictions are linked to particular substances and others are not, although both types of addiction share similar behaviour patterns. Researchers recognise the existence of dependence on prescription drugs, gambling, sex and work addictions – in addition to the classic addictions – as well as a new arrival: addiction to the Internet.

Should recovering alcoholics avoid drinking alcohol ever again?

Alcohol dependency therapies always aim to free patients entirely from their addiction. For recovering alcoholics, the aim is that – after detoxification and perhaps a course of psychotherapy – they never touch a drop of alcohol again. The ban even extends to alcohol in food. Total abstinence is always the goal, but a new and controversial treatment – known as 'controlled drinking' – has emerged in recent years. That an alcoholic may have learnt to control his or her drinking is also seen as a success.

Are some people genetically predisposed to become addicts?

This continues to be the subject of research. Studies that looked at the addictive behaviour of adopted and biological children of alcoholic parents and parents who were not addicted to alcohol came to the conclusion that – in the case of alcohol – an inherited predisposition played a role in addiction. Researchers are still to decide to what extent such a genetic predisposition will reduce the chances of successful therapeutic intervention in cases such as these.

As social factors play a major role in the development of an addiction, experts agree that adapting a therapy to suit each affected person individually is probably the most important way of helping someone find a way out of addiction.

Casinos offer a wide range of ways for people to gamble, from fruit machines to roulette.

Why are people sometimes public spirited and at other times selfish?

People tend to behave in a socially responsible or selfless way when they are concerned with the welfare of a group that is close to them, or part of what they regard as their tribe. Kinship appears to play a major role in this. The more closely related people are, the greater their willingness to help or make sacrifices. People are more likely to agree to be organ donors for a relative than they are for a stranger. We are also more likely to help strangers if they look like us, and scientists suspect that this is because we are inferring a genetic similarity. Whether this behaviour is innate or learned is a question for debate.

How do hierarchies develop?

A variety of roles and positions can arise within a social group. If the group is concerned mainly with power and authority, highly defined hierarchies can develop in which one or more people take on leadership roles, while the others allow themselves be led. This can occur as a result of those who are led selecting their leaders, or because an individual – possibly after a power struggle – sets himself or herself up at the top.

In the animal kingdom, this kind of hierarchy often has a particular purpose. It may be the dominant animal's function to protect the pack or swarm from external dangers. When animals live within a hierarchy they do so all the time – as can be seen among bees or wolves. Not all human groups need to be structured along strictly hierarchical lines, and some groups merely have a person in charge of coordinating tasks.

A recent study has shown that men take little time to form a hierarchy within a group. They may decide within minutes who among them is to lead the way, whereas it can take much longer for women to establish a hierarchical social structure.

How does workplace bullying arise?

Workplace bullying is usually the result of a combination of factors rather than a single cause. There may be a sense of unease among a company's staff when restructuring is on the cards. Key positions may disappear or be filled by new personnel who may then be bullied by the people who held their positions previously. A lack of leadership qualities or insufficient information from management about the restructuring process could prevent a senior member of staff intervening.

As well as bullying by colleagues (horizontal bullying), there can also be vertical bullying by staff in senior positions. This is just as common and is often used to dispose of staff who – for reasons of employment law – can't simply be sacked. Anyone can become a victim of bullying, although research shows that women are at slightly greater risk than men.

People who are systematically bullied by their colleagues may break down psychologically under the pressure.

Why do people need other people?

Humans can enjoy being alone for a while but are not suited to permanent isolation. Experiments show that people who have spent long periods in isolation can experience delusions and hallucinations. The need for company is probably innate. This genetic tendency may have developed because human children need to be taken care of by their parents for a longer time than the young of other mammals.

Why is gossiping so much fun?

Gossip helps us to let off steam. By circulating disparaging information about another person, a gossip can reduce the esteem in which that person is held or make them seem ridiculous. The gossip must not question himself or herself, nor must he or she confront the victim with the gossip that is being spread. Gossip can also create a sense of them and us – those that join in the gossip are in, and the subjects of gossip are out. When this function of gossip predominates, it begins to take on some of the features of bullying.

Why do people wear make-up and dress themselves up?

Jewellery, make-up and fashion can be traced back to prehistoric times. Vanity is one reason for personal decoration, but it can also serve to exhibit power and status. For centuries, many places, including Europe, had strict dress codes and people were not permitted to dress above their station. When shoes with long toes became fashionable for men in the Middle Ages, only the highest ranking men were permitted to wear shoes with really long toes. Everyone else had to make do with toes that were only half as long.

Extravagant clothes serve, first and foremost, to attract attention.

Jewellery can be a marker of wealth but it can also indicate that the wearer leads a better life than other people. In ancient China, Confucian scholars allowed the nails on their little fingers to grow as long as possible, and then adorned them by fitting a silver sheath. The signal being sent was: 'Look at me, I don't have to perform any physical work.'

Mozart's genius provided material for several films.

Does a genius have to live at a particular time to be recognised?

Geniuses and the intellectually gifted can find fulfilment in the widest possible variety of fields. This is what marks the difference between someone who is intellectually gifted and an autistic savant, who is only able to do one thing well. If Mozart had lived in the 21st century he might have given rock music a new direction. A genius does not need to have been born at the right time to be recognised but, in addition to possessing genius, it is important that a gifted person is able to communicate his or her ideas in such a way that their value is appreciated by contemporaries. It is then that external circumstances become important. A genius stands little chance of blossoming in a developing country, especially during a civil war, and this is particularly true if the genius is a woman.

Humans by numbers

Scientists from many fields study human beings and their cultural achievements. Their collected findings can be expressed in some impressive figures.

How many genes does a human being have?
The human genome, with its 3.2 billion DNA building blocks, has now been decoded. Only three per cent of DNA contains genes that serve as recipes for building proteins. It was thought that the number of human genes was 100 000, but we now know that it is about 25 000.

How large are human cells?
The human body has more than 200 cell types in a range of sizes. Most have a diameter of between 20 and 50 micrometres (one micrometre = one-thousandth of a mm). At 120 micrometres, the ova are the thickest and spermatozoa are the thinnest, with a diameter of only 3 micrometres. The longest are the spinal nerves, which can form branches of up to 1m in length.

To what length can human hair grow?
The length of a hair depends on its life span and growth rate. Body hair remains short because it drops out after only a couple of months. In contrast, hair on the head generally grows 10mm a month for about seven years. People with hair several metres long are extreme exceptions. Their hair either grows for an unusually long time or at an above average speed.

How rapidly does food pass through the intestines?
The length of time it takes for a meal to make its journey through the 8m or so of the intestine depends on what was in it. Fat and roughage have the most influence on the amount of time food spends in the stomach and the rate at which it is transported on from there. Speeds of between 20 and 40mm a minute are achieved in the small intestine.

On average, it takes two to three days for the undigested elements of our diet to complete the journey from mouth to anus.

How much water does the human body contain?
The human body is made up of 60 to 70 per cent water. The highest water content – 99 per cent – is found in the vitreous body of the eye. Lymph and blood contain 96 per cent and 80 per cent water respectively. The least amount of water – even less than in bones (13 per cent) – is contained in tooth enamel, a mere 0.2 per cent.

How long do the body's cells live?
The shortest lifespan is that of the cells of the small intestine's mucous membrane, which are expelled after only one and a half days – a total of 250g of cells each day. It takes between 40 and 56 days for a cell to migrate from the bottom layer to the horny layer of the skin, and red blood corpuscles die after 120 days. Depending on type, immune cells can live from a few days to several years, while nerve cells can remain sound for a person's entire lifetime.

How many bones are there in the human skeleton?
The human body has about 206 bones, but it is possible for it to have more. Some people, for example, have an additional 13th set of ribs or – in very rare cases – more than 10 fingers and toes. About half of all our bones are to be found in our hands and feet. Newborn babies have about 350 bones, several of which fuse together later in life.

How many sperm does a man produce?
It takes about 70 days for a mature sperm to develop from its precursor cell. About 1000 sperm per second are produced in the testicles, which amounts to 100 million

sperm each day, of which unused sperm are broken down. A millilitre of seminal fluid contains about 60 million sperm. After sexual intercourse they have just three days to reach the ovum before they die.

How many muscles does a human being have?

A human being has more than 600 muscles, which make up 40 per cent of a man's weight and 30 per cent of a woman's. The largest muscle is the broad latissimus dorsi back muscle, and the smallest is attached to the stapes, or stirrup ossicle, in the middle ear.

How many cells does blood contain?

In an adult's 5 l of blood, there are 50 billion white blood corpuscles, 1.5 trillion blood platelets and 25 trillion red blood corpuscles – together representing one-quarter of the body's total number of cells. More than two million red blood corpuscles are produced in the bone marrow every second. If they were laid end to end, they would stretch a distance of 192 500km – five times about the Earth.

What is the output of the heart?

During low levels of physical activity the heart pumps 4-5 litres of blood per minute, or 6000-7000 litres a day. Some 140-300ml go into a heart ventricle, 70-130ml of which are pumped out with every heartbeat. By the time we reach the age of 70, our heart will have beaten about 2.5 billion times and pumped 160 million litres of blood.

How quickly can we react to a sound?

Human beings react more rapidly to acoustic stimuli – such as the crack of a starting pistol – than to any other forms of stimulus. The threshold below which human reaction times cannot go – even following intensive training – is about one-tenth of a second. The minimum reaction time following other forms of stimuli is considerably longer.

How long is 'now'?

The brain seems to operate within periods of time of about 2.5-3.0 seconds in duration. These periods are experienced as being 'now'. If external stimuli remain constant for longer than about three seconds, a human being will feel that something is 'going on a bit'. If these stimuli stay the same for a long time, boredom begins to set in.

What is the highest IQ ever measured?

American Marilyn vos Savant's IQ of 228 is the highest ever measured. She achieved this score at just ten years of age.

Which language is spoken by the greatest number of native speakers?

With over 880 million speakers, Mandarin Chinese has the greatest number of native speakers. In addition to Mandarin there are other Chinese dialects, such as Cantonese. This is why the number of Mandarin speakers is considerably smaller than the total population of China.

What is the largest language family by number of native speakers?

Indo-European is the largest language family, if we are counting the number of people who speak one of the Indo-European languages, which include English, German, French, Italian, Spanish and Portuguese, as well as Russian, Punjabi, Urdu and Hindi. A total of about 2675 million people have one of the languages from this group as their mother tongue.

What is the largest language family by number of languages?

The number of member languages in each of the world's language families varies considerably. The West African Niger-Congo language family, which encompasses some 1514 individual languages, is the largest purely in terms of numbers. By way of contrast, the Indo-European family contains a mere 449 languages.

China, the most highly populated nation in the world, also has the language with the greatest number of speakers.

Mysterious
nature

Human beings have achieved a great deal and discovered many things, but there is still much about the natural world that puzzles the experts. As explorers, biologists and naturalists gradually solve more of these mysteries, they reveal a world of astonishing complexity and beauty. Some of the exciting discoveries made in recent years are featured here.

What strategies do plants use to survive?

Adaptive and regenerative abilities are a plant's most important survival strategies. These enable them to grow in the most extreme locations – from deserts to mountains. Some plants are able to withstand high salt concentrations and periods of drought, while others survive temperatures as low as -80°C. If a plant is injured or suffers stress, its recovery is often surprisingly rapid.

Why are flowers colourful

The spectacular colours of flowers are all tools for the purpose of reproduction. Plants use the bright colours of their blooms to attract pollinators such as insects and birds. Flowers of plants that are pollinated solely by the wind, such as grasses, tend to be rather nondescript as they don't need to look attractive.

Some flowers, like that of the horse chestnut, inform their pollinators when it's no longer worth paying them a visit. Brown stains develop inside the bloom as soon as it has been fully harvested, so pollinators can target other more inviting blooms.

Why are so few flowers blue?

The complex chemistry of flower colours is one reason for the rarity of blue flowers. This is why attempts to breed a blue rose have failed. To produce the colour blue, acidity or alkilinity have to be just right and certain metal-complex pigments – organic molecules that can combine with metals – need to be present. Hydrangeas are only blue if the soil in which they grow contains sufficient quantities of aluminium, a metal that is able to combine with their pigments. Cornflowers are blue because the pigments they contain combine with iron or aluminium ions.

Evolution is another reason for the scarcity of blue flowers. Over time, a close relationship has developed between flowers and their pollinators. Birds are attracted to red flowers, whereas beetles, moths and flies tend to prefer creamy colours. Bees can't perceive the colour red, which is why they tend to be attracted by white and yellow flowers, although they like blue flowers as well. Because the contrast between green foliage and white or yellow flowers is much greater than it is between blue flowers and foliage, this gives the brighter colours an advantage, which is why they occur more frequently.

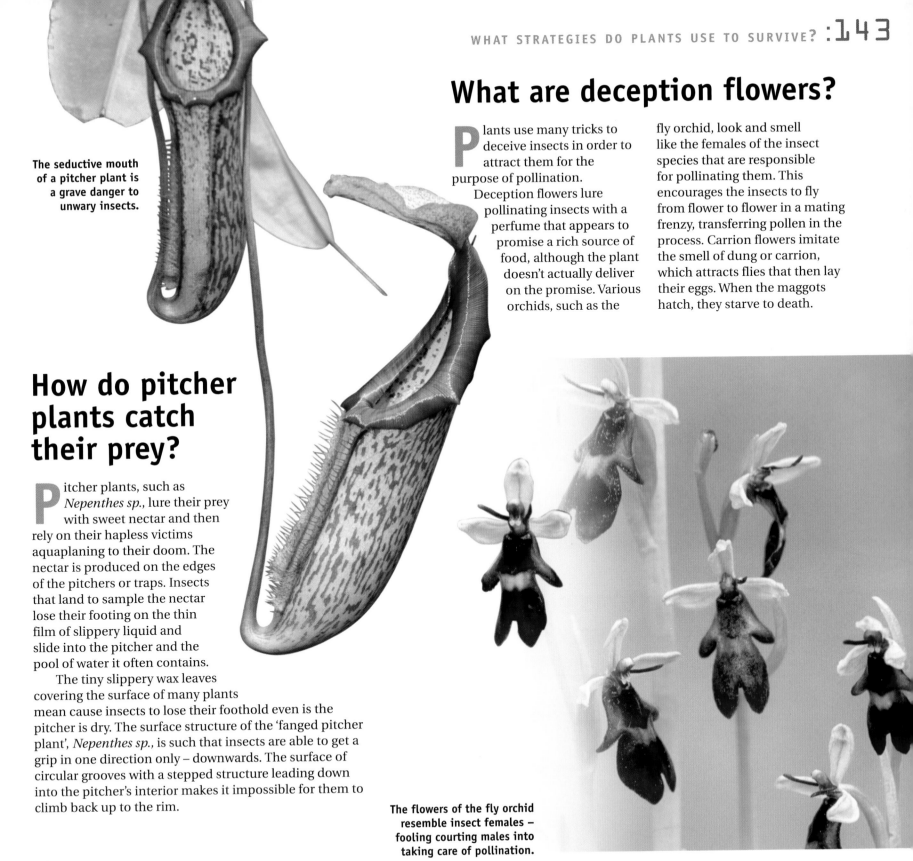

The seductive mouth of a pitcher plant is a grave danger to unwary insects.

What are deception flowers?

Plants use many tricks to deceive insects in order to attract them for the purpose of pollination. Deception flowers lure pollinating insects with a perfume that appears to promise a rich source of food, although the plant doesn't actually deliver on the promise. Various orchids, such as the fly orchid, look and smell like the females of the insect species that are responsible for pollinating them. This encourages the insects to fly from flower to flower in a mating frenzy, transferring pollen in the process. Carrion flowers imitate the smell of dung or carrion, which attracts flies that then lay their eggs. When the maggots hatch, they starve to death.

How do pitcher plants catch their prey?

Pitcher plants, such as *Nepenthes sp.*, lure their prey with sweet nectar and then rely on their hapless victims aquaplaning to their doom. The nectar is produced on the edges of the pitchers or traps. Insects that land to sample the nectar lose their footing on the thin film of slippery liquid and slide into the pitcher and the pool of water it often contains.

The tiny slippery wax leaves covering the surface of many plants mean cause insects to lose their foothold even is the pitcher is dry. The surface structure of the 'fanged pitcher plant', *Nepenthes sp.*, is such that insects are able to get a grip in one direction only – downwards. The surface of circular grooves with a stepped structure leading down into the pitcher's interior makes it impossible for them to climb back up to the rim.

The flowers of the fly orchid resemble insect females – fooling courting males into taking care of pollination.

Mutual benefit: the squirrel gets a meal and the plant maximises its chances of reproducing.

Why are fruits usually sweet?

Fruits are sweet to make them palatable to animals, and eye-catching skin colours make them highly visible. Fruit contains seeds, stones or kernels that are swallowed by animals and excreted in a different location. In this way, the plant distributes its seeds as widely as possible and maximises its chances of reproducing. Depending on the area covered by a moving animal, the seed may be distributed anywhere from s few metres to several hundred kilometres away from where it originated.

Some seeds have to be eaten because they can only germinate after they have passed through the gut of an animal. This strategy ensures that the germinating seed is provided with a good supply of fertiliser, as it is expelled at the same time as a quantity of excrement.

Why do some poisonous fruits look so enticing?

Poisonous plants also depend on animals for the distribution of their seeds, which is why they are never poisonous to all species. The seeds inside the bright red berries on yew trees are poisonous to humans and livestock, but produce no harmful effects when eaten by birds, rodents or even by game animals – which can cause a great deal of damage to the plant. Deadly nightshade is poisonous to humans and livestock, but its shiny black fruit can be eaten by thrushes and other birds without any ill effects.

Why are plums covered in a layer of wax?

The wax acts as a water repellent that almost no moisture can penetrate. It protects the fruit from drying out and acts as a barrier against damaging micro-organisms.

All the parts of plants above ground are wrapped in a waxy layer, although this most obvious with plums and other fruits. This protection against drying out has made plants successful in evolutionary terms.

What is the juice in a coconut for?

Coconut juice, or coconut water to be precise, is liquid food for the coconut tree seedling. The sweet, almost transparent liquid is full of nutrients, containing oil, sugar, water, vitamins and minerals, including potassium, phosphorus and selenium. As the fruit ripens, the hard coconut flesh absorbs the water. The riper the fruit is the less water it will contain. In a medical emergency, as coconut water is sterile it can be used as a substitute for serum and injected directly into a vein if there has been major blood loss .

What does a coconut really look like?
Coconuts are green. They are wrapped in a thick, fibrous layer with a fine green outer skin, all of which is removed as soon as the fruit is harvested in order to save bulk and weight. What remains is the actual coconut seed, which isn't a nut at all but a stone fruit like the cherry. When people speak of green coconuts they mean the six- to

How do we know...
...that just one coconut can be the beginning of an entire island?

Coconut palms grow on isolated islands as well as the mainland, because coconuts can drift across the sea for thousands of kilometres without losing their potential to germinate. Whether they come ashore on a sandbank, a flat atoll or a coral island at the end of their long journey, any little bit of land they do find will provide enough space for them to germinate. As the seedling grows, more and more sand and solid material is deposited around it and so, over time, a small island can form. Because coconuts can spend so much time drifting at sea without being damaged, it is difficult to determine where the first coconuts came from.

seven-month-old, unripe coconuts that are harvested before the hard kernel forms. At this stage the fruit is at its most nutritious. It holds around half a litre of clear coconut water and its flesh is soft enough to be eaten with a spoon. The coconut milk found on grocery store shelves and used in Asian cooking is water that has been pressed out of soaked and pureed dried coconut.

The world's largest seed – which can reach 30cm in length and 18kg in weight – is shaped like a double coconut but is not directly related to the true coconut. It comes from the coco-de-mer palm that grows naturally in the Seychelles in the Indian Ocean.

Is a strawberry really a berry?

The answer is no. In botanical terms, a fruit is the mature ovary of a flower within which seeds develop. The soft outer flesh of a true berry is formed by the ovary's outer wall and contains one to many seeds within. Examples are tomatoes, grapes and even oranges. Technically speaking, strawberries are not true berries but aggregate fruits. Each of the little dots studding the outside is itself a fruit, known as an alchene. A nut is similar to an alchene but has a hard, dry outer ovary wall.

What kind of a fruit is a banana?

Although bananas are sickle-shaped and not round, from a botanical point of view they are classified as berries. Each banana has developed from a single ovary, and its pericarp is fleshy even when the fruit is ripe. Although berries usually contain one or more seeds, the modern, cultivated banana has only the remnants of ovules from which no new plants can develop. This is not a problem, as banana plants grow underground shoots. Bananas are bent because they start out growing downwards and then bend in the direction of the Sun.

Why is vanilla so expensive?

Vanilla comes from a plant that belongs to the orchid family, to a genus with more than 100 different species. Evergreen perennials, they climb up trees, bloom for no more than a few hours and will only develop a fruit if they are pollinated by insects during this period.

Because of its shape and leathery appearance, the fruit of *Vanilla planifolia* is called a pod but the correct botanical term for it is a capsule fruit. It forms from the ovary and several fused petals, bursting open after it has dried to spread its seed.

In the industrial production of vanilla pods, the flowers are artificially pollinated and the green fruits are put through a drying and fermenting process that lasts for several weeks. Only in this way can the famous vanilla aroma develop from the primary chemical component, known as vanillin. The complexity of the production process explains why vanilla pods rank among the world's most eagerly sought after spices.

The Aztecs were using vanilla as a food flavouring more than 500 years ago.

Why does wheat turn golden when it is ripe?

Nitrogen is vital to the survival of both plants and animals. Because it is rarely overabundant in soil, plants recycle it. The chlorophyll molecules that are responsible for the green colour in plants contain nitrogen, so plants break them down when they are no longer needed. This is why wheat turns golden brown when its seeds are ripe and the chlorophyll molecules have been broken down.

Why can wheat grow all over the world?

Unlike rice, wheat can adapt to a wide range of climatic conditions. It needs little water and is able to withstand relatively low temperatures. Human beings have been using wheat for over 10 000 years. The plant has developed great genetic diversity, which is why it is so adaptable. Two particular wheat genes are largely responsible for this diversity: the VRN1 and VRN2 genes. These control growth and reproduction, and determine when the plants are to flower. The VRN2 gene suppresses the production of flowers in winter wheat and so protects the plant from frost damage.

Fields of ripening wheat – like this one in France – display fascinating variations in colour, especially in windy weather.

Why are the young leaves of tropical plants red?

The red colour protects young plants from ultraviolet rays, in much the same way as brown pigments do in human skin. When young plants develop rapidly, the production of chlorophyll – which is the substance normally responsible for protecting plants from ultraviolet radiation – tends to lag behind. Fast-growing plants manufacture additional pigments known as anthocyanins, and it is these that produce the red colour. Anthocyanins absorb the destructive ultraviolet rays and by doing so protect DNA inside the plant's cells. As soon as chlorophyll synthesis catches up, the production of anthocyanins is reduced and the red anthocyanin pigments are covered by the green chlorophyll pigments, which take over the protective role.

Why does grass grow back immediately after it has been mowed?

Grasses become longer and longer but never thicker. The blades of grass are divided by knots that contain the active growth zones, which is why they are good at regenerating. After mowing, the grass sprouts again from these knots, which also direct growth in such a way that the blades stand upright again. These characteristics make grass lawns ideal surfaces for parks and other areas that receive a lot of wear.

Why is grass greener after a cold winter?

Long-term studies have shown that winter weather influences the way that grass grows in the following summer. After a harsh winter, summer grass is greener and lusher than after a mild winter. In a temperate zone, cold winter months ensure that more moisture is kept in the ground. But after a warm winter the earth is drier and grass grows less vigorously when summer comes.

Economically useful plants and genetic engineering

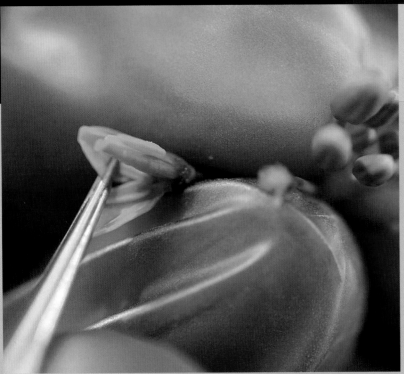

Rape is undergoing genetic modification to produce pharmaceuticals. As it could prove dangerous, it is vital that cross-pollination between pharmaceutical plants and ordinary fodder plants is prevented. This is achieved by ensuring that the pollen of genetically modified plants is sterile.

Ever since human beings settled down and gave up their nomadic lifestyle, they have tried to adapt plants to meet their needs. Targeted breeding has increased yields, reduced susceptibility to disease and produced new varieties. Genetic engineering is now opening up a range of new possibilities, such as the production of transgenic plants – plants that have the genetic material from other organisms inside them.

'Golden rice' illustrates the pitfalls and potential of transgenic plants. This new rice variety was developed to contain two foreign genes: one from the wild daffodil and another from a bacterium. It promised a great deal. The modifications were designed to make the plant produce beta-carotene, a precursor of vitamin A, in its seeds. It was thought that people living in Asia's developing countries would benefit because their predominantly rice-based diets meant there was widespread vitamin A deficiency.

But some of the rice suffered growth disorders, and the vitamin content was so low that a daily intake of up to 2kg of rice would have been required for there to be any benefit to consumers. Despite the setbacks, scientists haven't give up on the idea and believe that rice that contains iron, zinc and vitamin E, as well as vitamin A, will be widely available one day.

Another transgenic plant experiment that attracted considerable attention was a cross between a tomato and a potato. Researchers fused together individual cells known as protoplasts. These new cells contained the genomes for both species, and a hybrid developed that produced tomatoes above ground and potatoes below ground.

What happened to the tomato-potato hybrid?

The hybrid was another failure because the potato yield was meagre and the plant unstable. It failed to produce fruits or seeds so was incapable of reproducing. When cells from different species are merged, the result may be a new strain but a few genes always get lost in the process, something that is difficult to control.

In addition to the technical problems, the reaction of consumers to the new plant was not positive. It has not been alone in attracting a negative response. Many people have concerns about genetic modification, even if only genes belonging to an individual plant are modified or single foreign genes are added.

Why doesn't anyone want tomatoes that don't go squishy?

A tomato was developed in the mid-1990s that didn't go squishy. One of the tomato's own genes – the one that produces the enzyme polygalacturonase – was altered to achieve this. This enzyme starts to dissolve the cell walls within a tomato as soon as it ripens, making the skin softer and less robust, which in turn causes the fruit to become squishy. The plant benefits from the softer skin because this helps it to release its seeds. But, from a grower's point of view, tomatoes keep longer without the enzyme, with ripe, non-squishy tomatoes remaining edible for up to eight weeks. But European consumers thought it lacked flavour, and it is possible that the idea of eating a genetically modified tomato spoiled the fruit's appeal.

Can plants produce medicines and vaccines?

Plants have always played an important role in medicine, and transgenic plants expand the possibilities. By adding a gene, it is hoped that it may be possible to produce vaccines, antibodies, hormones and proteins. Manufacturing these substances with the help of a growing plant would be much cheaper than producing medicines by using animals or cell cultures. Thanks to the so-called 'gene gun' – a machine that shoots tiny particles of gold or tungsten carrying the required genes into plant cells – it is now possible to add foreign genes to almost any plant. Another technique uses a bacterium to introduce foreign genes into plant cells. As early as 1986, the first pharmaceutical tobacco plant was producing human-growth hormones. Today, more than 80 different medicines made using genetically modified plants are marketed worldwide.

Are synthetic plants a possibility?

As well as trying to improve existing plants, scientists have been working on producing synthetic plants that would catch light and use it to turn the carbon dioxide in the air into glucose or other compounds, just as happens in the membranes of plant cells. Researchers have had some success using a mixture of organic materials, photo-sensitive molecules and semiconductors to generate small amounts of energy and produce methane, a gas that can also be used as an alternative source of energy.

New foods, medicines and energy sources already exist, and the possibilities seem limitless. However, there is widespread scepticism among consumers, who will ultimately decide on the technology's fate.

'Golden rice' has not yet succeeded in providing a solution to vitamin A deficiency in the developing countries of Asia. One problem is that the human body will only absorb vitamin A when it is combined with fat – an ingredient that is often missing from poor diets.

Tobacco plants that produce antibodies have been in existence since the 1980s. In 1992, the first genetically engineered vaccine against hepatitis B was produced using this technology.

Do plants speak to each other?

Plants communicate using a chemical language, using scent as the means for sending and receiving information. Some plant vocabularies contain more than 100 chemical 'words', which are used to exchange information about pests or to call for assistance. If caterpillars snack on maize, the plant will produce 'alarm chemicals' that can be detected and recognised by the ichneumon wasp. The insect lays its eggs in the caterpillars, which kills them. Tomatoes that are nibbled by caterpillars produce toxins to combat the pests and emit a scent that warns other tomatoes. The essential oils contained in popular herbs such as sage and oregano also serve as a means of communication. They tell pests to keep away, while attracting pollinating insects.

How do roots and leaves communicate?
In the world of plants, the roots are in charge. They regulate the food supply and decide on the growth and appearance of a plant. They emit the signals that determine whether a leaf should grow or wait for a while – perhaps because water is in short supply. To communicate this information, roots produce particular hormones during dry periods that tell the shoots to curb

This cob of corn may show signs of caterpillar damage, but plants are not entirely defenceless against pests.

Roots do more than simply deliver water – they control the entire plant.

their growth. It is thought that the hormone responsible is part of the large class of carotenoid substances, which also play a major role in determining the colours of flowers.

Can plants learn and remember?
Plants can pass on signals via their cells, some of which can be stored for a period of time in a kind of memory. Young trees are able to adapt their growth to particular watering rhythms, as if they had been able to learn from experience. Plants are also able to make assessments and plan in advance – attributes that are closely related to memory. The parasitic hell-weed tests a number of plants for nutrients before deciding where to settle permanently. During this process it makes precise profit-and-loss calculations – and it does this four days in advance, which is the amount of time it takes for it to get to the actual source of nutrients after an initial sampling.

Can plants move?

Plants do move, although they are not generally able to move away from a particular location. As they grow, stalks may make swinging and circling movements towards the light and grow upwards against gravity. When plants close their flowers or leaves at night, scientists sometimes speak of the movements as 'sleeping'. Clovers fold their leaves at night so that the two external pinnae are wrapped inside the central one. To enable the pinnae to fold in this way, each one is equipped with a tiny joint located at the base of the leaf stem. This is pumped up by hydrostatic pressure, causing the required movement. Biologists suspect that the leaves are seeking protection from nocturnal cold.

Is there such a thing as a walking tree?

The walking palm has 2m-tall stilt roots on which it 'walks' in pursuit of light. It does this by growing new stilts while others die off. With each new root, the palm tree moves a little further in the desired direction. Each 'step' takes more than three months to make, as the aeriel roots of mature palms grow about 70cm per month, enabling them to reach the ground in three months.

The walking palm *Socratea exorrhiza* is supported by 2m-long stilt roots and can grow to over 12m in height.

How do plants protect themselves from inbreeding?

The more varied the gene pool of a plant species, the more diverse are its characteristics and the greater its ability to adapt to environmental conditions. In order to maintain a varied gene pool, many plants protect themselves from self-fertilisation. The yew cannot self-fertilise because each plant is either male or female. The dovesfoot cranesbill uses time to solve the problem, starting life as a male and later entering a female phase. On the iris, the pollen and the stigma are placed as far apart as possible. Self-pollination in the case of plants that have unisexual flowers is often inhibited by special sterility genes, which means that the flower's own pollen cannot germinate on its stigma or form a pollen tube.

The leaves and flowers of a female yew.
The male plant's flowers are more rounded.

How does a seed know when to germinate?

Many seeds go through an initial stage of dormancy to ensure that they don't germinate too early when conditions are unfavourable. The seed will spend this period of dormancy waiting for various signals before germinating.

In the desert, it is important that plants shouldn't start growing until there is sufficient water to support them. The seeds of desert plants store germination-inhibiting substances in their outer coatings. These substances must be rinsed out by strong, long-lasting rain before the seeds can germinate – a short shower isn't sufficient to do the job.

To ensure that autumn-flowering plants don't start to grow until after the winter is over, their seeds are not ready to germinate until they have been exposed to low temperatures and frosts. Some fruits, including tomatoes, contain a substance called abscisic acid, which prevents their seeds from germinating immediately. After winter dormancy, when conditions are favourable, other substances are produced that signal that the time is right for germination.

How far can airborne pollen and seeds travel?

The oldest way for plants to colonise new areas is for their seeds to be carried by the wind. The smaller and lighter a seed is, the farther it will fly if wind conditions are favourable. This is why some seeds or pollen have air-filled cavities to help reduce their relative density.

To travel greater distances, many seeds are fitted with wings, hairs or tiny parachutes. These devices increase air resistance so that, once airborne, seeds take longer to land. Depending on size, shape and weight, every seed has its own landing speed and this can be used to calculate its flight distance during average wind conditions. Dandelion seeds don't descend at more than 26cm/sec., so, given favourable air currents, they can be carried for distances of up to 10km. The pollen of pine trees can fly more than 30km and the tiniest pollens and spores should be able to travel all around the world. In practice, they are more likely to be stopped by difficult weather conditions or other obstacles.

A germinating radish seed, seen through an electron microscope.

Suspended beneath their tiny parachutes, dandelion seeds can travel for long distances through the air.

Lotus flower seeds are able to germinate after more than 100 years.

How long can seeds maintain their ability to germinate?

Seed survival time is genetically determined and can vary between one year and several decades, depending on the species. Grasses can remain viable for up to 100 years and the seeds of the lotus flower can still germinate after more than 100 years.

The most commonly grown vegetables and herbs – such as potatoes, lettuce, parsley, cucumber and carrots – have seeds that keep for between two and seven years.

Grasses, grains and legumes produce seeds that keep for longer. Under normal conditions, legume seeds can still germinate after 30 years, and rapeseed can stay dormant in the soil for more than ten years.

The less moisture a seed contains the longer it can be stored at cool temperatures. Fully ripened and dried soya beans can be kept for almost unlimited periods of time. With the help of cryopreservation techniques – storage in liquid hydrogen at temperatures of -196°C – seeds today are being placed in permanent long-term storage all over the world. Norway is even building a seed storage 'doomsday vault' near the North Pole to ensure food supplies in the event of a catastrophe.

How do we know...

...that some trees need to have a bit of stress in their lives?

Some trees need the stress of a forest fire, because without one they wouldn't be able to reproduce and would become extinct. The seed-producing mechanisms of these trees cannot function until they have been subjected to great heat. In Australia, some of the gases produced in a bushfire provide the signal needed by the cones of banksia plants before they can open. The seeds of the giant sequoia tree also only open once they have been through a fire, while the tree itself is protected by 50cm-thick bark that insulates it against heat. Fires also warm soil that is frozen and thins out the area around a tree. Together with the fertile ash that is left after a fire, these changes produce optimal conditions for germination and growth. Fire also destroys competitors as well as pests and diseased wood.

What do plants really need in order to flourish?

Plants need water, heat, light and carbon dioxide, as well as up to 15 different nutrients – including minerals like hydrogen and phosphorus, and trace elements such as iron and zinc. If one of these ingredients is missing, a plant will struggle to grow.

How much light do plants need?

Too little light is fatal for plants. All plants must receive a minimum amount of light in order to survive. If the light level falls below this critical point, the plant starves to death because it is unable to produce sufficient starch.

Too much light can destroy the sensitive membranes responsible for photosynthesis, and consequently the plant itself. This is why plants are equipped with their own molecular solar protection. Specialised carotenoid molecules – which help to add colour to fruits and flowers – transform excess radiation into heat and disperse it by means of evaporation.

What do plants do during the hours of darkness?

When plants close their flowers in the night their metabolism remains highly active. Since they lose less water at night, some plants open their stomata – tiny pores on the undersides of leaves – wide so that they can absorb large amounts of carbon dioxide. While it is dark, plants also prepare for photosynthesis, which will start as soon as the first rays of light appear. The energy for these nocturnal activities is supplied by the breakdown of starch and sugar molecules, which were produced by the plant's own photosynthesis. There is even a special transport protein that carries the sugar molecules back and forth.

Do plants grow better if we speak to them?

Mozart, rock'n'roll or kind words make no difference at all to plants – they certainly won't grow any faster or produce denser vegetation because of it. In order to make a plant grow more rapidly, it is necessary to increase its rate of photosynthesis so that it produces more sugar, as more energy is required to produce more growth.

The rate at which a plant is able to photosynthesise cannot be affected by sound waves of any type. Plants do not have a central nervous system or any sensory cells capable of being stimulated by sound waves.

The largest number of plant species grows in the jungle

Jungles are home to the planet's richest ecosystems, and more than half of all known plant species. Over 80 per cent of the world's economically useful plants originated in the jungle, including bananas, cocoa, figs, ginger, sugar and vanilla. More species of plants grow in a few kilometres square of jungle than grow in the whole of Europe. We have yet to discover most of the animal and plant species that live in the world's jungles, with estimates of between two and ten million species. A constant climate, high humidity and a year-round average temperature of 25°C are thought to be the main reasons for this diversity.

How do plants react to pollution?

Air pollution places plants under stress, but they do have inbuilt defence systems to protect themselves. If the threat is from ozone, there will be a reaction within a few hours. Pines increase their production of pinosylvin, a substance that would normally protect the trees from fungi and bacteria, while spruces increase their production of catechin, a strong antioxidant. Plants that undergo ozone stress will usually suffer visible damage to tissues, leading to leaf discolouration and leaf drop.

High levels of harmful chemicals such as nitrogen oxides in the air can delay the formation of flowers. Plants concentrate on growth rather than flower formation, which can result in too little time being left at the end of the season for the development of healthy fruit and seed.

Poppies only flourish in a healthy ecosystem. Like other meadow flowers, they are sensitive to harmful chemicals.

Why is it a good sign to see poppies and other flowers growing by the side of the road?

Wildflowers growing in and around fields are a sign of environmentally friendly agriculture and an intact ecosystem. Poppies and certain other wildflowers are so sensitive to change that some are already close to extinction. As many form the base of a food chain, this could have serious consequences. Every species attracts an average of 12 plant-eating and pollinating creatures, which in turn provide food for birds and small mammals. A lot of insects and worms – many of which play a role in pest control – would also disappear if the wildflowers were to vanish. Of the 300 or so known species of European meadow flowers, one-third is now considered to be under threat.

How much carbon dioxide can plants deal with?

Plants use carbon dioxide and water to make glucose, thereby storing the carbon in another form. Climate experts hope this may help to counteract the threat of global warming. As levels of carbon dioxide in the air rise, this will increase the rate of photosynthesis that, in turn, will lead to the formation of more biomass capable of storing more carbon.

But there is little data about the effects of rising levels of carbon dioxide on plants. Recent studies of algae show rising carbon dioxide levels making the plants smaller and less competitive. When normal conditions return, they continue to develop poorly. It is thought that damaging genes are activated when the carbon dioxide levels become too high. Other studies have shown that as carbon dioxide levels rise, plants don't absorb more of it but simply adapt to the altered conditions.

How tall can trees grow?

No tree can keep growing indefinitely – the limit is around 130m. The reason for this is the way water reaches a tree's crown. The supply is made possible thanks to water evaporating through the stoma on the leaves. This produces suction, which moves water from the roots to the crown through a vascular system. Above a height of 130m, this pull becomes so strong that it ruptures the water column. This is because gas bubbles, formed as a result of the extremely low pressure in the water column, interrupt the flow. Botanists describe this process as embolism.

The tallest living tree in the world at present is thought to be a sequoia in Northern California's Humboldt Redwoods State Park. At 113m, this massive conifer is as tall as a 25- to 30-storey building.

In 1872, a mountain ash felled in Victoria, Australia was measured at 132.58m. With the addition of the tree's stump and the material trimmed from the crown of the tree, it probably exceeded 150m in height when growing.

Visitors are dwarfed by giant sequoias growing in America's Yosemite National Park.

Sequoia, California: 113m

Mountain ash, Australia: 150m (felled in 1872)

Silver fir, Europe: over 60m

Statue of Liberty, New York: 93.5m

Where do the world's oldest trees live?

Trees die of old age, like all living things, but they can take quite a bit longer to do so. Even some common conifers have an average lifespan of between 300 and 600 years. At around 1000 years old, oaks and yews are the Methuselahs of temperate latitudes.

The current record for the world's oldest known tree is held by the Rocky Mountains bristlecone pine, found in California's White Mountains. More than 4000 years must pass before many of the trees begin to show signs of dying from old age, with the current record holder still alive after 4700 years. Bristlecone pines are robust, needing little in the way of moisture and nutrients. Inconspicuous and unremarkable, they don't grow much above 12m in height, which is probably why scientists didn't realise until the 1950s that they lived for such a long time. Before their great age was established, the huge sequoias – some of which were between 3000 and 3600 years old – were thought to be the oldest trees on our planet.

Rocky Mountains bristlecone pines take strange shapes over the course of their long lives.

How long do roots live for?

The lifespan of a tree's fine roots is of interest to scientists because it provides an indication of the amount of carbon dioxide that can be absorbed by the forest floor. The faster roots are produced and die, the more carbon is stored in the ground. This is because roots that have already died off continue to retain carbon. Recent research on pine trees and American sweetgums has shown that the lifespan of roots fluctuates between one and nine years – depending on their diameter and the relationships between neighbouring trees. The short-lived roots of the sweetgum were found to be capable of storing much more carbon dioxide than the longer-lived roots of pines.

True or False?

Bamboo is a grass and not a tree

Many species of bamboo grow in dense clumps and some can grow up to 40m tall with a circumference of 80cm. However, the plant's central supports are referred to as stems rather than trunks. Bamboos are actually grasses and belong to the family of *Poaceae*. Like other grasses, bamboo stems do not become thicker with time but grow straight out of the ground with a uniform thickness. Bamboo is one of the world's fastest-growing plants, with stems reaching their full height in just a few months. As well as the tall, woody bamboos a second major group grows much more like 'normal' grasses. These plants do not become woody and rarely grow taller than about 1m.

Bark provides a tree with protection, as well as giving it an individual 'personality'.

Why does a tree need its layer of bark?

Bark is the external part of a tree trunk. Trees would not be able to survive without it because it provides protection from surface injuries, moisture loss, diseases and attacks by fungi and other organisms that might cause damage.

The outer surface of a tree is important because the sensitive vascular cambium is situated directly beneath the bark. It is here that new tissue is constantly being produced – wood on the inside and bast fibres and new bark on the outside. This is also where the system of ducts for carrying glucose from the leaves to all other parts of the tree is situated. The young sapwood found immediately below the cambium is soft and therefore highly susceptible to attack by a range of enemies.

If the bark itself is injured, trees such as conifers open their resin channels, allowing resin to disinfect and close the wound. This acts as a barrier to harmful bacteria. The bark on some trees is so thick and strong that it is able to protect trees from the extreme heat generated by forest fires.

Why don't trees grow on high mountains?

Above a certain altitude trees cease to thrive because the climate is too extreme. The air is too cold, the winds too strong and there is often insufficient water to sustain life. Trees constantly lose water through their leaves due to evaporation, so any interruptions in the water supply can cause them to dry out. In harsh, mountain environments, ground water often freezes, so it isn't always possible for the roots to deliver sufficient water to the rest of the tree. The lack of water causes pressure within the tree to fall to a point where the water column ruptures and fatal air bubbles form. When temperatures are extremely low it is even possible for the water inside the tree to freeze.

The trees seen most frequently at high altitudes on mountain sides are conifers, because they have better protection against excessive evaporation. The timber line – the altitude at which trees cease to grow – varies according to climate and latitude. It typically lies somewhere between 600m (as it is in Sweden) and 4000m (as it is in Nepal). The timber line in the Central Alps of Europe lies at about 1900m.

Trees cannot survive in the harsh climate found at high altitudes.

Are fungi plants or animals?

Beautiful but deadly – *Russula emetica* is poisonous.

Fungi are neither plants nor animals, but belong within a kingdom of their own. Scientists have described over 100 000 fungi species, although some estimates suggest the true number is closer to 300 000 species, with some experts believing the total may be as many as 1.5 million species.

Fungi tend to grow in fixed positions, which makes them seem more similar to plants than animals. This is why for a long time scientists thought that they were part of the plant kingdom. Unlike plants, however, fungi do not photosynthesise and there is also chitin in their cell walls, a substance that is also found in animals but never in plants. Because of these and other physiological and genetic characteristics, it is now thought that fungi are closer to animals than plants.

What do fungi need to survive?

Unlike plants, fungi are heterotrophs – they depend on the organic material of other organisms in order to survive. After bacteria, they are nature's most important decomposers, playing a crucial role in the breakdown of dead organic – especially plant – matter. Fungi have a key role to play in the natural cycle of growth and regeneration. Some fungi species are parasites or live in symbiotic (mutually beneficial) relationships with other organisms. Fungi have spread across the planet, but prefer to live on land, especially on acid soils. All they require is that other organisms live or have lived in the same location. Although most fungi need moisture, only a few species have adapted to life in water.

How do we know...

...that a fungus is deceptive packaging?

The visible surface of a fungus – the part that is generally considered to be the fungus – is merely the fruiting body and a small part of the total organism. The greater part of a fungus is made up of the thread-like hyphae that spread below the surface of the ground. These hyphae intertwine to form a mycelium that can grow to a huge size. A mycelium discovered in Oregon is thought to be the largest living thing on Earth – it is more than 9sq km in area.

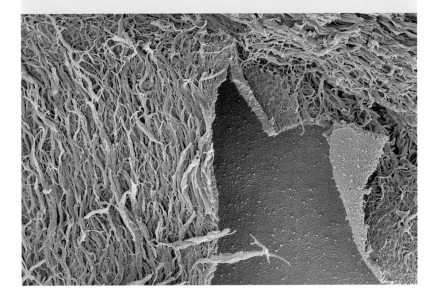

How do plants and animals survive the chill of winter?

Living things have developed a range of strategies to help them survive the cold of winter. Many plants shed their leaves and so begin a period of dormancy or rest. Some animals migrate to warmer climates to escape the cold while others hibernate. There is no let-up in the struggle to survive – life must go on, whatever the weather.

Why do coniferous trees keep their needles throughout the winter?

Conifers do lose their needles, but not all at the same time. Depending on their age and condition, old needles are gradually replaced throughout the year, removing metabolic waste products in the process. Most conifers keep their needles in winter – although there are exceptions, such as larches. Conifer needles have a much smaller surface area than the broad leaves of deciduous trees, which keeps their water evaporation rate as low as possible. The needles are also covered with a special coating of wax, which helps maintain a low evaporation rate.

What makes leaves change colour in autumn?

Before deciduous trees shed their leaves in winter, they go through a nutrient-recycling process. They draw substances from the leaves back into their trunk and roots, because these are vital to the metabolic processes that take place in trees during the winter months. The most important of these substances is chlorophyll, which contains nitrogen. Nitrogen is especially valuable because soil does not generally contain enough, unless it is added intentionally in the form of fertiliser, as is the case in agriculture. Virtually every part of a plant must have nitrogen for vital tasks such as synthesising proteins and as a building block for nucleic acids. At the end of this process, all that is left in the leaves are yellow and red carotenoids, which give them their characteristic autumn colours.

Why do trees shed their leaves in autumn?
During the winter months, a tree's metabolism merely ticks over. To reduce the risk of them dying due to lack of water during the cold season, deciduous trees shed their leaves before winter sets in. This minimises or halts the loss of fluid from the leaves through evaporation. Although water is constantly evaporating from leaves in the summer, the arrival of ground frosts in winter has the potential to prevent roots from delivering water to the rest of the tree. This is because ground water freezes and cannot be carried from the roots to the trunk. By shedding their leaves, trees also rid themselves of metabolic waste products that have accumulated during the growing period.

Fruit growers spray orchards on a cold night as a thick layer of ice prevents fruit blossoms from being killed by frost.

How does ice protect fruit blossoms from frost damage?

It may seem paradoxical, but a layer of ice during a late winter frost can protect blossoms from damage. This is why fruit growers use sprinkler systems to spray their orchards on cold nights. As ice is formed, heat is generated – each litre of water that freezes produces around 335kJ of heat. This is known as 'heat of solidification' and is passed on directly to the blossoms. The layer of ice also insulates the flowers against evaporative heat loss. Some of the artificial rain evaporates and, instead of generating heat, produces a cooling effect in the same way as sweat cools our skin. Sprinkling has to be maintained throughout the night and interruptions must be kept to a minimum, otherwise the surrounding layer of ice will fall to the ambient temperature, with the result that fruit blossoms will be killed. Fruit growers therefore wait until two or three hours after sunrise to turn their sprinklers off, by which time the temperature should have risen to above 0°C.

This method only works at temperatures to -9°C, and then only if the frost periods are short. If sprinklers were used for longer than a few days, the protective layer of ice would become too heavy and would damage the trees.

How do snowdrops survive the frost?

Snowdrops bloom early because their bulbs contain many nutrients and they are able to protect themselves against frost damage. The plants do this by producing their own antifreeze – sugary compounds that build up inside cells in cold weather. This lowers the freezing point of sap in the plant's cells, so that no sharp-edged ice crystals with the potential to destroy tissues can form, even when temperatures fall as low as -10°C. The snowdrop's metabolism generates some heat, which thaws surrounding snow and, early in the plant's life, a special outer layer protects its new shoots from the cold wind.

Snowdrops have an arsenal of tricks to protect themselves from the cold.

How do animals cope with ice and snow?

Animals struggle when there is snow and ice, because food becomes scarcer and the cold causes them physical discomfort. If they can't escape the cold by migrating south or hibernating, they are forced either to make changes to their diet or rely on stocks of food. Beavers eat the bark off branches and tree trunks when there is no fresh foliage around in winter, while squirrels build up stores of nuts, fir cones, rosehips and other fruit.

How does global warming threaten the mountain pygmy possum?

A small marsupial, it is so perfectly adapted to life above altitudes of 1400m in the Australian Alps that it needs snow cover to survive. For seven months a year, it endures the harshest conditions by hibernating in rocky crevices insulated beneath a thick blanket of snow. The snow must be at least a metre thick to provide enough insulation, but there are signs that global warming may be reducing snowfall in the Australian Alps.

Why doesn't the lynx like powdery snow?

Soft, powdery snow slows a lynx down as it hunts for its chief winter prey: the snowshoe hare. The lynx can move faster over firm snow, while the hare stands a better chance of escaping predators in fluffy snow as its lighter weight and large paws give it a better chance of gaining a foothold.

How are fish able to survive the winter in a frozen pond?

When any body of water freezes, it does so from the surface downwards. As long as the stretch of water is deep enough, there will be sufficient quantities of liquid water in the lower reaches to allow fish to survive. The temperature of liquid water at the bottom of a frozen lake may only be around 4°C, but fish can adapt to that. Their body temperature drops and they keep their movements to a minimum.

The lynx depends on a covering of firm snow for success in winter hunts.

How do animals survive hibernation?

The vital functions of hibernating animals slow right down when they are dormant during the cold winter months. Their body temperature drops below 10°C, and breathing, heartbeat and blood pressure are all reduced. These changes enable animals to save up to 88 per cent of their energy. The small amount of energy that their bodies do require during this period is drawn from fat deposits, which means that hibernators have to eat enormous quantities of food before they can retire to the safety of their burrows or dens to sleep. Some animals never wake from their winter sleep. If a hibernating creature is woken up too frequently, it will be forced to consume its reserves. This means that it will not have enough energy available to get it through the waking up process in spring, when all vital functions have to be returned to normal levels.

How do insects protect themselves from the cold?

Flies and many other insects don't generally bother us in winter. Like all cold-blooded creatures, insects are unable to maintain a constant body temperature. When the air temperature drops, they lapse into a cold-induced torpor and their vital functions are reduced to zero. Some survive in this state in a sheltered spot, but most can only ensure that their offspring – in the form of eggs or larvae – can withstand the winter cold and live to see the spring. Insects in the form of eggs or larvae find it easier to endure low temperatures. Eggs are not especially sensitive to the cold and some larvae spend winter in deep water, which does not freeze.

What do marmots get up to when they wake in the spring?

When male marmots wake from their winter sleep in late spring, they set off to look for prospective mates, visiting any females that live in their territory.

During the period when they were hibernating, the animals woke up every now and then, but stayed in their burrows and went back to sleep. If a marmot wakes up in early spring it will venture no more than a few steps away from its burrow's entrance. However, towards the end of the hibernation period, males leave their burrows for periods of time that are long enough to enable them to visit their immediate neighbourhood and to call on any local females before returning home to rest. In this way, male marmots not only establish territories but also ties with prospective mates, which significantly improves their chances of mating soon after spring begins.

Some animals in tropical countries hibernate as well

Even regions where warm, summery conditions last all year round there are some mammals that hibernate. An example is the fat-tailed dwarf lemur, which lives in the tropics and sleeps for seven months of the year, even when temperatures rise above 30°C. Between the end of February/early March and the end of August/early September, this primate, indigenous to the island of Madagascar, retires to a hollow tree and sleeps its way through the dry season. During this period of dormancy, the lemur's body temperature adapts to the ambient temperature inside the tree. If daily temperature fluctuations cause the ambient temperature to rise repeatedly above 30°C, the animal's metabolism has sufficient energy to enable it to sleep through. On the other hand, like all hibernating animals, lemurs will wake from time to time if their body temperature repeatedly sinks below 28°C.

Nearly all of the world's inhospitable places are populated by animals and plants – from polar bears in the Arctic to cacti in desert regions.

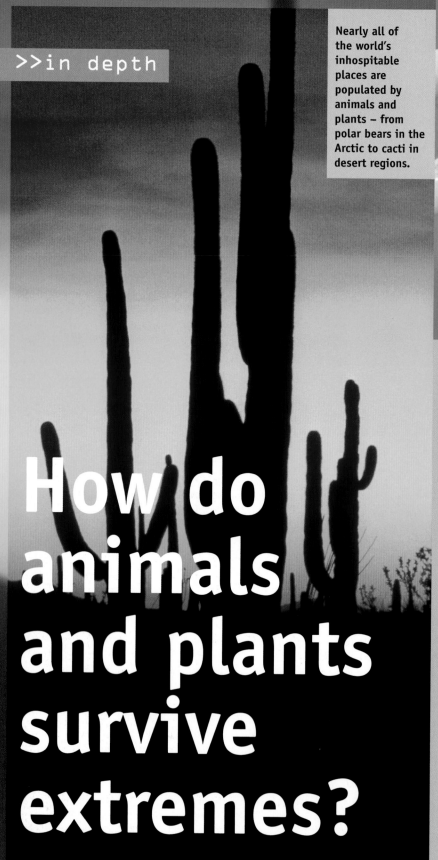

How do animals and plants survive extremes?

Planet Earth contains a range of habitats, some of which look very uninviting. But few habitats – however extreme – seem to be sufficiently hostile to prevent some plant or animal from living in them. Fish and other creatures live in the inky blackness of the deepest seas, penguins and polar bears have conquered the polar wastes, cacti bloom in the desert and salt-loving halophytes flourish in soils that would kill other plants.

How is it possible for plants to grow in soil with an extremely high salt content?

Most ordinary plants cannot survive in really salty soil because the salts seriously interfere with their ability to absorb nutrients and water. But there are plants – known as halophytes – that have specialised in surviving in precisely these kinds of habitats, where they enjoy the advantage of not having to deal with competition. With the exception of the polar regions, halophytes are found in many different habitats around the world – from the seaside to salt steppes, the margins of salt lakes and springs and even tropical rainforests. Some halophytes actually need salt to survive and would fail to flourish in ordinary soils, while others would probably fare better in conventional soil, but are also able to grow in a salty medium. Halophytes adopt various strategies to deal with the salt. Some species prevent salt from entering their roots, letting through only those salts required by their metabolism and blocking out the rest. Others actively excrete potentially damaging salts through special glands or hairs, while some species have mechanisms in their cells which prevent the salt from reaching concentrations that might result in damage to the plant.

Are there any plants growing on the Antarctic continent?

Plants have been discovered growing in the extreme climate to be found in Antarctica. The vegetation is not exactly rich in species, but some lichens, mosses and grasses have gained a foothold there. The local soil is low in nutrients, so most Antarctic vegetation grows in the vicinity of bird colonies. There is one more highly developed plant that has succeeded in overcoming the cold – the Kerguelen cabbage, which grows on several islands close to Antarctica. Most Antarctic plants grow in clumps just above the ground, where they present the smallest possible target for wind and weather. In addition, they are very robust and have an amazing capacity for regeneration.

Why is it that there are plants which can flourish in the desert?

Although deserts are characterised in part by a lack of vegetation, there are a number of cacti, grasses, bushes and even trees that have adapted to the inhospitable conditions. These plants – known as xerophytes – are experts in survival. They need very little water and have developed a variety of ways of collecting what little there is and surviving on it. Cacti, for example, can store water for several months, while some small herbaceous plants have developed extensive root systems that branch out over a wide area just below the surface of the ground, helping them maximise their water intake. Most types of desert trees adopt the opposite strategy, with very deep roots that reach down to the water table. In very long droughts, some trees shed their leaves to avoid losing what little water they have through evaporation.

How do animals cope with the extremes of heat and cold?

Not only plants, but animals as well have managed to adapt to extreme conditions. Desert reptiles and small mammals like jirds and jerboas can manage with just the fluid contained in their food. They can also retain water for very long periods of time and use it sparingly, with barely any water being wasted in their urine and faeces.

Some animals have also tamed the icy wastes. Mammals native to these regions grow a layer of insulating fur to protect themselves from the cold. In addition, sheer size can also offer protection against freezing temperatures. It is basic physics that as a sphere becomes bigger, the surface area/volume ratio decreases. The same holds true for animals, so that as a creature increases in overall size the surface area of its body – across which it can lose heat – increases at a smaller rate than its internal volume, within which it can retain heat. As a result, bulky creatures lose heat more slowly than smaller ones. According to the principle Bergmann's Rule, this is why many animal species tend to be larger the closer they live to the poles. Penguins are an example: the biggest survive closest to Antarctica.

Animals that live in cold regions also tend to have relatively small extremities and appendages, such as ears and tails. Too much heat would be lost through large extremities and they could also easily become frostbitten. This is why the Arctic fox has very small ears. At the other end of the scale, animals in the hotter parts of the world have large appendages that tend to stick out. This increases the surface area across which they can lose heat and helps prevent the body from overheating. Unlike its relative in the Arctic, the desert fox has ears that may look almost absurdly large to us, but they are perfectly adapted for the task of radiating excess heat

Why does the polar bear have black skin under its white fur?

A thick layer of insulating fur is one of the most important ways in which mammals keep warm in freezing temperatures, and the polar bear has developed a particularly sophisticated system. The hairs of its coat are hollow and provide extremely good heat insulation, since air is a particularly good insulator. The polar bear's fur may appear white or yellowish, but in fact each individual hair is transparent, which allows the Sun's rays to reach the bear's black skin where the heat is absorbed very efficiently due to the skin's dark colour. At one time, it was believed that the hairs worked like glass fibres, helping to direct light and heat to the animal's skin. Since then, however, the hairs have been found to be less efficient light conductors than was believed. Nevertheless, the polar bear's coat is extremely effective, and the hollow hairs also provide buoyancy.

The main function of the desert fox's large ears is to keep its body cool. Animals such as the fox are extremely well adapted to their natural habitat.

What exactly is a mammal?

Mammals are a class of advanced vertebrates. All possess the ability to suckle their young with milk and a capacity to maintain a constant body temperature. Palaeontologists believe that the first mammals were small rodent-like creatures that scurried around in the twilight of the Jurassic forests millions of years ago. Today's mammals are probably descended from those tiny creatures.

African elephants at a waterhole.

Why do camels have humps?

Contrary to popular belief, camels use their humps to store fat and not water. On long journeys across dry and barren desert regions, these cloven-hoofed animals draw sustenance from these reserves of fat. This is why the humps – dromedaries have one and Bactrian camels two – gradually shrink during an enforced diet until they eventually hang down loosely.

These mammals have adapted well to life in the desert, having developed mechanisms that enable them to go without water for as long as two to three weeks at a time. Their urine is highly concentrated and their faeces also contain little fluid. They can draw moisture from the air with the aid of fine nasal hairs, and are able to reduce the amount of fluid lost through sweating to a minimum thanks to an ability they have to vary their body temperature by as much as 10°C.

When camels reach water after a long trek, they can drink more than 100 litres in minutes. They are not in danger from hyperhydration – or water poisoning – because their red blood cells are oval-shaped, which prevents them from bursting following such a huge intake of water.

How long do elephants live?

These magnificent pachyderms can live to an age of 70 years. They live in herds of about ten cows and their young, led by an older matriarch. Males are excluded from the herd as soon as they reach sexual maturity. Young bull elephants will often get together to form bachelor herds, but will only approach females for breeding purposes. gradually worn down, the root is reabsorbed and the whole tooth is eventually replaced by a new one developing behind it. If an elephant is still alive when its last tooth wears out, it can no longer chew its food and will die of starvation. This rarely happens in the wild, since the animals don't usually live to such a great age.

How do elephants drink?

Elephants have short necks, which is why they use their trunks – formed from the animal's nose and its upper lip – for drinking. They suck up several litres of water, which they spray into their mouths.

How do elephants chew?

Elephants eat by grinding tough plant matter back and forth between four massive molars – one on top and one on the bottom at each side of the mouth. These are rasped away over many years and replaced six times throughout the elephant's life. As each tooth is

How do elephants communicate?

Elephants use trunk, ear and tail movements, as well as stamping and trumpeting, to communicate. An elephant's trunk can be positioned in a defensive or threatening attitude, although it is also used to pat fellow herd members. Elephants signal pleasure or excitement by flapping their ears. The tail also serves as an organ of touch, and is the means by which an elephant can tell whether another member of the herd is behind it. Stamping and low-frequency rumbling calls are used over longer distances.

Which animal should you be more frightened of – a hippopotamus or a lion?

Hippos can be more dangerous to humans than lions. In Africa, these friendly looking vegetarians are responsible for more deaths than the predatory big cats. Hippopotamuses are extremely territorial, quick to perceive a threat and able to defend themselves fiercely. They ram their heads into an opponent and inflict serious, often fatal, wounds with their canine teeth, which can grow to a length of up to half a metre. Hippos are also prone to panic and turn vicious quickly, especially on land. If a person is blocking a hippo's escape route back to the water, it will often react angrily and is likely to attack. Hippos should not be underestimated when they're in the water either, since they can easily capsize small boats.

How do bats see in the dark?

The ability of bats to 'see' in the dark is mainly thanks to their ears. These animals have developed a sophisticated echolocation system, which enables them to locate their prey with a high level of precision and avoid obstacles. When airborne, these skilled fliers produce sounds with an ultrasonic frequency of more than 20 000Hz, which are mostly inaudible to the human ear. Echoes of these loud cries are reflected from the bats' surroundings, giving the flying mammals a precise 'sound picture' of their environment. Bats are not completely blind, and can distinguish between light and darkness so are able to recognise contours. Bats also have a good memory for places, which they constantly update during the course of their flights.

How did bats learn to fly in the first place?

Bats are thought to have developed wings as a result of the mutation of a single gene that occurred some 50 million years ago. This genetic variation meant that the knuckle bones – between which the wing membranes stretch today – became greatly elongated, allowing the development of wings. This development may also explain why the transition from walking to flying animals occurred so rapidly. It is thought that only a few transitional forms bridged the gap between the small, flightless mammals and the bats. These transitional forms probably only lived for a relatively short space of time, which would help to explain the fact that no fossilised remains of any such transitional creatures have been found so far.

Why do cats' eyes glow in the dark?

The eyes of cats and other animals that are active after dark are equipped with a special reflective layer of cells – the tapetum lucidum – that lies behind the retina. This layer reflects light rays that are not caught by the eye's photosensitive cells and thus have no immediate effect on the retina. The light is not immediately lost, but has a second chance to stimulate the eye's photosensitive cells, considerably increasing the overall amount of light received. This is why cats can see so well in the dark. The tapetum lucidum is also the reason why cats' eyes sometimes seem to glow so eerily at night. The glow is a result of light rays passing back through the retina after being reflected in the light-sensitive cells so that it leaves the eyes and can be seen by an observer.

What do we mean by camouflage?

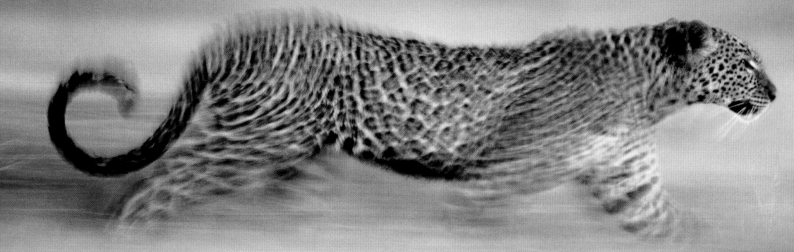

Thanks to the markings on its coat, the leopard blends with its surroundings, even when it is on the move.

The word 'camouflage' comes from the French and means misinformation or deception. Biologists talk about camouflage when an animal's shape, markings or colouring mean that it can easily be mistaken for another creature or object. The animal becomes harder to spot and blends into its background.

In the wild, these deception strategies are complex. Camouflage plays a major part in the survival strategies of both predators and prey, making potential victims more difficult for predators to spot while disguising predators as they attempt to approach prey unnoticed.

How do big cats hide from view when moving across open country?

Big cats prefer to be less conspicuous on grassy steppes, but have little to fear – other than humans – once they reach adulthood. It is to their advantage to remain invisible to their prey for as long as possible. The tawny fur of a prowling lion is hard to detect against a background of golden and brown dry-season grass. Striped coats are hard to spot in sun-dappled forest, and spots ,like those of leopards or cheetahs, allow the shape of a large predator to merge in with its background, even when it is on the move.

How do some animals maintain their camouflage as the seasons change?

Polar bears don't have a problem. They spend the entire year amid the snow and ice of the Arctic wastes, where their white fur ensures that they are always well camouflaged – and therefore virtually invisible to the seals on which they feed. But in regions where the environment changes markedly with the seasons, a brown coat in the winter snow would be just as fatal as white fur against the browns and greens of a jungle in summer. Some small mammals have developed a simple but effective strategy – their coats change colour with the seasons. Mountain hares and stoats have a greyish-brown coat in summer, which turns snowy white in winter. They are therefore relatively difficult to spot, whatever the season.

Why don't whales and seals suffer from decompression sickness?

During a deep dive, the pressure of the water forces nitrogen from the air that is breathed into body tissue. If a diver returns to the surface too quickly, the nitrogen in the tissue forms bubbles that can damage tissue and blood vessels, leading to mental confusion, severe pain, internal bleeding, paralysis and, in extreme cases, death from decompression sickness (known as 'the bends').

Unlike divers, whales, dolphins and seals encounter few problems when they surface from great depths because they don't continue to breathe underwater. They have a flexible thorax that is gradually compressed as the pressure increases, causing the contents of the lungs to be squeezed out gradually. The apparent collapse of the lungs does no damage to these aquatic mammals but provides an effective protection against decompression sickness.

No two whales have identical tail fins, a fact that helps researchers to distinguish one animal from another.

How do dolphins sleep?

Biologists observing both wild and captive dolphins have noticed that they swim around in circles when they are asleep. This is because only one half of their brain sleeps at a time. Researchers suspect that this behaviour serves to keep individual members of a group together. When dolphins are awake, they communicate by making whistling sounds that enable them to find one another easily. When they are asleep, such sounds could draw unwanted attention from predators. If all the members of a group swim steadily and silently around in circles, they can stay together without attracting uninvited visitors.

Do whales and dolphins have 'fingerprints'?

In principle, the answer has to be yes, although whales do not have fingers and therefore do not have individual line patterns on their skin. The equivalent of a fingerprint in the case of these aquatic mammals is the shape of the fins, which are almost as unique and recognisable as human fingerprints. Most distinctive of all are the notches along the edge of the fins, which vary significantly from one individual to another. These patterns seem to change little over time – barring accidents and misadventure – and can therefore assist biologists in identifying particular animals in the wild. Whales are most recognisable by their tail fins, dolphins by their dorsal fins.

...that the sea doesn't look blue to whales and seal?

The retinas in the eyes of whales and seals contain green cones but lack blue cones – the cells that react to blue light – so are colour-blind. Biologists suspect that the ancestors of today's whales and seals lost their blue cones when they changed their habitat, moving back into the sea. Initially, they would have remained in coastal regions where impurities made the water a greenish colour. The loss of the blue cones would have been an advantage, simplifying the processing of information in their brains and freeing up capacity for other functions.

Why do whales sing?

Biologists have not been able to explain why whales 'sing'. It is thought that the sounds they make are primarily for communication, but they also seem to contribute to a whale's development and wellbeing.

Researchers speak of whale 'song' because they make sound patterns that are regular and predictable. Male humpback whales only produce certain sequences of sounds during the mating season, presumably because this special song plays a central role in finding a mate. It is unclear whether this display is aimed at females or potential male rivals. Humpback whales also use individual sounds that are not part of complex songs to mark out their territory or when hunting in groups.

How are reptiles and amphibians different?

Amphibians grow up in water and adults can survive either on land or in water, although they need to keep their skin moist to avoid drying out. Reptiles are much better adapted to living on land, since they are equipped with a hard, horny, waterproof skin. They also lay waterproof eggs, which give the embryo greater protection and can be laid almost anywhere.

Vocal sacks pumped up, a frog adds its voice to the pond chorus.

Why are frogs slippery?

The layer of slime that covers frogs and other amphibians protects them from drying out. Most amphibians would not survive a fluid loss of 25 or 30 per cent, and because their skin is not horny, it provides little protection against evaporation. For this reason, frog skin is well equipped with glands that constantly secrete mucus and other substances. These glandular secretions enable frogs to control their water balance and also contain poisonous compounds that protect the amphibians from predators and discourage bacteria and fungi. Frogs can even produce sunscreen from glands in their skin.

Can frogs slip on wet surfaces?

Tree frogs have round suction pads under their toes that are designed to allow their feet to keep close contact with surfaces through static friction, despite the mucus secretions. They are able to maintain a grip even on smooth surfaces because the layer of slime around the suction pads is runny and less viscous than on the rest of their bodies. This improves contact with slippery surfaces. Sometimes the layer of slime disappears completely, presumably because the microstructure of the suction pads enables the liquid to drain away quickly. Because the slime on their feet is so runny, frogs can easily detach themselves from a surface to hop away in pursuit of prey.

Is breathing and drinking optional for frogs?

At the tadpole stage, frogs breathe through their gills. Adult frogs have simple lungs but also breathe through their skin and through the mucous membranes in their oral cavity. The skin is thin and has a plentiful supply of blood, which means that oxygen travels straight to the blood vessels. Depending on the species, frogs can take in between 50 per cent and 70 per cent of the oxygen they need through their skin. In very cold weather, frogs can even switch over to breathing through their skin alone. This is how they slow their metabolism in winter and are able to settle at the bottom of a lake and wait for spring. Their skin is also permeable to water, which means frogs do not need to drink. They take in water through their skin and store it in their lymph sacks and bladder.

How do snakes climb tree trunks?

Snakes move by using their muscles, sinews and scales. Land snakes have wide, flat scales along their stomachs, which are connected to their ribs via several muscles. When a snake moves along the ground these scales are splayed out so that they can hook onto uneven surfaces. The snake can then push and pull itself forward, sometimes achieving speeds of 15km/h. The same mechanism is used when snakes climb trees, even vertical ones.

How do geckos run up smooth surfaces?

Thanks to the special pads on the undersides of their toes, geckos can run up smooth glass surfaces or dangle upside down by one leg from the ceiling. These pads are covered with millions of microscopic hairs, and each one branches out into thousands of tiny bulges. The hairs can be brought into close contact with a surface so that they stick due to the weak electrostatic force, known as the Van der Waals force. The adhesive strength of an individual bulge is slight, but with several billion bulges a gecko is able to achieve its extraordinary clinging power. Using all four feet, a gecko can support a weight of about 140kg.

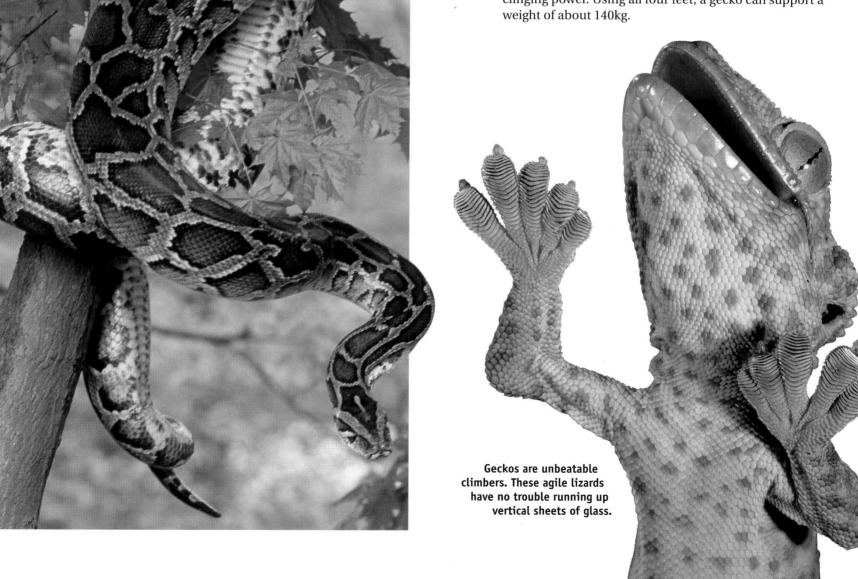

Geckos are unbeatable climbers. These agile lizards have no trouble running up vertical sheets of glass.

How did the dinosaurs live?

Dinosaurs lived on land, in water and in the air. The numerous land species ranged from creatures the size of chickens to sauropods more than 30m long – the largest land animals ever to have lived. Most dinosaurs were probably cold-blooded, although there is evidence that some were warm-blooded or developed hybrid temperature regulating systems. Nearly all were fast, energy-efficient runners, since their legs were positioned beneath their bodies rather than at the sides. The largest of the water-dwelling dinosaurs, known as ichthyosaurs, grew to lengths greater than 15m.

Do any flower species exist now that may have been sniffed by a passing dinosaur?

The first flowers bloomed 125 million years ago, at a time when dinosaurs walked the Earth. Dinosaurs, which lived between 230 and 65 million years ago, would probably have been familiar with magnolias, water lilies and peonies. These groups began to appear more than 100 million years ago and were among the earliest flowering plants. The conically shaped flower clusters of the magnolia suggests that the tree evolved from the more ancient conifers.

Why did the dinosaurs die out?

Was it volcanic eruptions, meteors, long-term climate change or is there some other explanation? The most likely cause was either a massive meteorite or unusually powerful volcanic activity 65 million years ago. It now seems certain that vast quantities of volcanic ash or meteorite dust injected into the atmosphere led to drastic climate change and the large-scale destruction of dinosaur populations. The sky remained dark for years, so that temperatures across the world dropped. Plant growth declined and some plants were affected by a drop in sea levels. The number of dinosaur offspring fell so low that they eventually died out.

There is evidence to support both the vulcanism and meteorite theories, but it seems likely that a combination of factors drove the dinosaurs to extinction. Sixty-five million years ago, intense volcanic activity in India meant that vast quantities of lava, ash and toxic gasses were released for about 700 000 years. On the edge of the Yucatan Peninsula in Mexico, a huge meteorite crater, 180km in diameter, was formed. The colossal amounts of rock dust blasted into the air by the impact must have had global effects. This can be seen from layers of rock that were formed at the time. All over the world – and especially around the crater itself – these contain high levels of the element iridium. This metal is fairly rare in the Earth's crust, most of what is there having arrived in meteorites. The even, worldwide distribution of iridium in rocks formed at the time of the Yucatan impact indicates that particles of the element were transported through the air.

Remains of the fearsome predatory dinosaur, *Tyrannosaurus rex*, were first discovered in 1902.

True or False?

Birds are the only surviving dinosaurs

Birds are the only genuine descendants of the dinosaurs. Many of today's reptiles, such as the crocodile, monitor lizard, iguana and some small lizards, share common characteristics with the prehistoric beasts, none are directly descended from any species of dinosaur. Today's reptiles have separate ancestries that developed independently of the dinosaurs. The prehistoric bird *Archaeopteryx* was descended either from therapods – two-footed, meat-eating dinosaurs – or from an ancestor shared with the therapods. The flying dinosaurs (pterosaurs) form a group of their own. Like bats, their wings were formed from membranes rather than feathers.

How do we know...

... that crocodiles would have known the dinosaurs?

Crocodiles are among the most ancient of reptiles. They developed at the same time as dinosaurs and – together with the flying dinosaurs and modern birds – are descended from the same forebears. This was a group of archosaurs called thecodonts, a reference to the fact that their teeth were set in sockets. The crocodile and dinosaur lines of descent diverged more than 250 million years ago and continued to develop independently. The first modern crocodiles already existed in the early Cretaceous period, almost 100 million years ago, and they have barely changed since. While the dinosaurs suffered a mass extinction 65 million years ago, the crocodiles managed to survive – perhaps because they needed less food and were able to slow their metabolism down to extremely low levels. Large crocodiles can live for more than a year without food.

A reconstruction of a prehistoric crocodile that lived 110 million years ago in North Africa, in the region that is now the Ténéré Desert.

Evolution – theories and facts

I
n the mid-19th century, the naturalist Charles Darwin shook the world with his theory of evolution. He suggested that all living things are descended from common ancestors and that over the course of time they develop by a process of natural selection which brings about constant change. His ideas challenged the idea of a God as the creator of every single species. While some greeted Darwin's theory enthusiastically, it also met with widespread scepticism and remains controversial. Although Darwin produced plenty of evidence to support his theory, it did not amount to proof in the eyes of many of his critics. Since Darwin's time, scientists have amassed a lot of evidence in support of the evolutionary mechanisms that he proposed. Findings in the field of molecular biology – Darwin knew nothing about genes – clearly show how evolutionary processes are based on genotype mutation, selection and heredity.

Was Darwin the first to develop a theory of evolution?

Charles Darwin was not the first to wonder about the diversity of organisms. The French naturalist Jean Baptiste de Lamarck had suggested that species changed over time 50 years before Darwin published his *Origin of Species* in 1859. Lamarck also gave thought to the mechanisms behind these processes, suggesting that features and characteristics survived solely through use. He maintained, for example, that giraffes had long necks because they were constantly reaching up for leaves, and this acquired characteristic was passed on to succeeding generations. Unlike Darwin, Lamarck did not hit upon the principle of random variability and natural selection, by means of which only the better adapted survive. Some characteristics that result from chance alterations to the genome prove useful and are retained, while others which confer no advantage are lost.

Darwin was influenced during his early career by

Thanks to fossil finds we now have a reasonably clear picture of how modern humans evolved from their ape-like ancestors.

The axolotl, found only in central America, is an extraordinary creature. It spends its entire life in the larval stage of development and its limbs can regenerate if they are removed in an accident.

Lamarck's work, but by the time he began writing the seminal *Origin of Species* he had disregarded virtually all of the French naturalist's theories. Working at around the same time as Darwin, the naturalist Alfred Russel Wallace also formulated a theory of evolution, which in many respects coincided with Darwin's. The two scientists kept up a correspondence, but each developed his own theory independently. Darwin was able to support his theoretical observations with the extensive evidence collected on his travels, which was something Wallace could not do.

Are mammals former reptiles?

Darwin's theory turned the contemporary view of the world upside down. Suddenly, human beings were no longer God's supreme creation, but instead shared a common ancestry with apes. To many at the time, the idea was impossible to accept, but today most people are more comfortable with the idea.

We now know that all mammals are descended from reptiles, their direct forebears being the vertebrate group known as the synapsids. Later forms of these reptiles displayed typically mammalian characteristics. Their metabolism was faster than that of reptiles and so they were warm rather than cold-blooded. Their skin was already partly covered by hair and their teeth included incisors, canines and molars. Moreover, unlike reptiles, their legs did not grow out of the sides of their bodies but were situated beneath their bodies. Of the synapsids, it was the therapsids that became more and more similar to mammals and it is thought that all of today's mammals are descended from them.

Do some animals never grow up?

Over the course of millions of years, evolution has been responsible for producing the huge diversity of life – both plant and animal – we see around us. It has also produced some very strange results, such as the animals that never grow up. Some salamanders spend their entire lives as larvae, yet still reach full sexual maturity. They retain their gills and stay in the water, rather than venturing onto land as most adult amphibians usually do. Among this group are the European olm and the axolotl.

Can single-celled organisms see?

The eye is an organ that has been used by both the advocates and opponents of evolutionary theory to support their case. Darwin himself had doubts about his theory when he considered the eye's extremely complex structure. But the eye can be extremely helpful in understanding how evolution works. Even single-celled organisms like euglenas and many of the dinoflagellates have the sense of sight, and this can

Euglenas are among the single-celled organisms which have a primitive sense of sight – evidence in favour of the theory of evolution.

sometimes be highly complex. For example, the dinoflagellate *Erythropsis pavillardi* contains most of the components that make up a vertebrate eye. These include a lens, a retina-like structure and an opaque spot of pigment that shields against light from one side. A similar structure also enables euglenas to move towards a light source. The first evidence of eyes has been found in fossils about 540 million years old. This is when the evolution of the eye began, and all the eyes we know today appear to have developed within a period of a mere 100 million years.

Can apes see in colour?

Not just sight itself, but also the ability to see colours is an example of continuing development during the process of evolution. The first reptile-like mammals probably lived in semi-darkness and so had little need of an ability to recognise colours. As a result, their visual sense would have deteriorated and most of their descendants have at most two different types of sensory cells – known as cones – which are responsible for colour vision. But certain primates – including humans – are among the few mammals with three types of cones in their retinas, a feature that results in much better colour vision. Creatures with the three types of cones can distinguish between red and green, which is a distinct advantage when most of life is lived mostly in bright daylight, a fact that helps to explain why this ability has been retained by humans.

Why is the underwater world so colourful?

Colours are used by the inhabitants of coral reefs in much the same way as by land dwellers. They act as signals, assist in helping to find a mate, provide camouflage and send out warning messages. At depths of greater than 200m everything looks black to us, but not to reef fish. Many have colour receptor cells in their eyes and are able to detect ultraviolet light, which penetrates further into the depths than visible light.

How does colour affect the mating behaviour of fish?

Like many other creatures, some female fish react to colours displayed by males during the mating season. There is a female sand goby, for example, that has a preference for large, bright-blue males. Interestingly, females in cloudy water are less choosy, with the result that when visibility is poor, even the smaller, paler males stand a chance of mating. Research has shown that small males living in murky water are just as successful as their larger rivals. In the long term, this pattern of behaviour could result in size differences between the sexes disappearing altogether in muddy waters.

Corals and other reef creatures appear even more vibrant to fish, which have a more highly developed colour sense than we do.

In the mating season male sticklebacks develop a distictive red colouration on their undersides.

Are fish deaf and dumb?

Fish are anything but dumb. Squirrelfish communicate with clicking sounds, wrasse grunt, gurnards growl and male drumfish have special muscles around their air bladder which produce a drumming sound when contracted rapidly. Most remarkable of all, when herrings fart it has nothing to do with their digestion, but everything to do with communication. In order to have a conversation they swallow air which they expel through the anal orifice in their gas bladder. In this way the fish can produce notes extending over more than three octaves at frequencies of between 1.7 and 22kHz. They can hold each note for almost eight seconds.

Since fish can make such a range of noises, it seems only logical to assume that they can also hear. They do so through their inner ear which is connected to the air bladder by what is known as the Weberian apparatus, which functions as the eardrum in the process of hearing. The Weberian apparatus then transmits the sound to the inner ear – in a similar fashion to the way the auditory ossicles function in mammals.

Wrasse, native to the western Pacific Ocean, communicate by making grunting noises.

Why do female sticklebacks see red during the mating season?

During the mating season, the throats, breasts and bellies of male sticklebacks turn bright red. To make it even easier for female sticklebacks to recognise their suitors, their colour vision adjusts to adapt to prevailing external conditions. Scientists have found that the ability of the female fish to recognise red light depends on the character of the light that is available in the stretch of water in which they live. Female sticklebacks living in murky waters – where most of the red light is absorbed by particles in the water– are much more sensitive to the colour red than those living in clear water.

Not surprisingly, the males also adapt their behaviour to suit prevailing light conditions. In cloudy water, where red objects are harder to distinguish, the red body colouration of the male sticklebacks tends to be much brighter.

How does a fish's sixth sense work?

Fish have sensory equipment that can detect impulses caused by movements in the water. This is often referred to as a fish's sixth sense. Because the sense organ responsible for this ability runs along the body, from eye to tailfin, it is known as the lateral line. The lateral line consists of small pores – which often appear as a pale line along a fish's body – that lead to canals beneath the skin lined with sensory cells called neuromasts. These cells enable fish to sense even the smallest vibrations and movements in the water, which is why they never swim into the wall of a fish tank, and why they can always manage to find their way around, even in the murkiest water. This sense also gives fish some idea of the size of other creatures that may be lurking in their vicinity, important for predator and prey alike. Sharks, for example, use this ability to locate nearby prey.

Can coral reef fish change their sex whenever the fancy takes them?

The sex lives of coral reef fish are really quite complicated. Gobies, for example, are fully mature before they become definitively male or female. The decisive factor is the gender of the members of the same species that they happen to meet during the mating season. If a goby in search of a mate has a chance encounter with a male, it will turn into a female, and vice versa.

While gobies can undergo multiple sex changes in both directions, most reef fish can transform in only one direction on a single occasion. This is the case with clownfish, where all females are initially male. If a male loses his mate and then finds a new, smaller male, he can change into a female. The larger of a clownfish pair is always female and the smaller, male. But with clownfish, only males turn into females, never the other way round.

Coral reef fish have to make the best use of every possible opportunity to produce young for the very good reason that reefs are often a long way apart, and it can be difficult to find a suitable mate.

The clownfish's ability to change its sex greatly improves its chances of finding a suitable mate.

What do fish need to survive?

Although fish live in water, like all animals they need one thing above all else – oxygen. Without it, they would suffocate in a matter of minutes. This is why the quality of water is so important. A fish needs water that contains sufficient oxygen, is low in toxins and harmful substances and with about the same level of acidity as drinking water suitable for humans.

Some fish are herbivores, but many need animal matter as food, which is why zooplankton – comprised of tiny creatures such as water fleas and minute crustaceans that feed on microscopic plants – forms the basic diet of many types of fish. Because aquatic life is just as varied as life on land, there are fish with widely differing needs, depending on their local ecosystem and the kind of waters they live in.

Are some fish cowards?

Fish behaviour can be complex and unpredictable. In some cases, fish even appear to learn through experience and modify their behaviour accordingly. So timid fish can become daring and vice versa by responding to their own experiences and those observed in other fish. Some rainbow trout lose their courage after a few defeats in fights against others of their kind, while others may increase in confidence after a triumph. Just watching how other members of the same species behave is enough to make a fish re-evaluate itself. The timid become braver when they are confronted with shy fish, but not when they meet bold ones.

One of the most timid fish of all is the coral goby. It will not leave the shelter of its cave even under the most extreme conditions. Even if the cave is briefly drained of water, it will stay put. The coral goby needs very little oxygen to survive and will survive for up to four hours in the open air.

Gobies hide from their predators in small hollows.

Most sharks have to keep moving to breathe, although leopard sharks can remain stationary on the ocean floor for a while.

Do fish need to drink?

The answer is yes and no – saltwater fish drink, freshwater fish do not. The reasons are all down to physics and a process known as osmosis. When two fluids of differing concentrations are separated by a membrane – in this case the skin of the fish – water always flows towards the more highly concentrated solution. The body fluid of freshwater fish is more highly concentrated than the surrounding water. So water is absorbed into the body through the skin and gills and the fish must actively excrete water, otherwise it will eventually explode. The opposite is true for saltwater fish. In their case, the surrounding water is more highly concentrated and draws fluid out of the fish. This means that saltwater fish must actively take in water to avoid drying out. They take in water through their mouth, mucous membranes and gills. The salt is excreted by the gills before reaching the body.

Do fish sleep?

Although fish do not have eyelids, they regularly take short naps, just like the higher vertebrates. Some sleep during the day, others at night. Sleeping fish hide in caves, crevices or coral canyons, lying on their sides or surrounding themselves with a coat of slime, which also helps to protect them from predators. As they sleep, fish slow their metabolism down and they become less alert. But they are never completely unconscious. Electroencephalograms (EEGs) taken of fish brains have shown that resting fish do not have phases of deep sleep. Some fish can even manage to continue swimming slowly while they are asleep.

Can any fish shoot their prey?

Archer fish can shoot insects off leaves or straight out of the air with carefully aimed jets of water. Every jet of water is adjusted to suit the size of the prey – the bigger the target, the more powerful the jet. The fish works as efficiently as possible. It balances the force required to release the jet against the quantity of water. With double the amount of water it will also double the force of the jet. If, instead of adopting this strategy, it were to double the initial pressure and speed of the water jet it would expend four times as much energy.

Can we distinguish individual fish by the way they swim?

Observations of goldfish have shown that each individual has its own swimming style. The likelihood that two goldfish will go through the same sequences of movement is almost zero, so each will execute turns or changes of speed in its own characteristic way.

Why do flying fish fly?

When flying fish leap out of the water it is usually because they are fleeing from an enemy. Rapid movements of the tail fin make them accelerate so rapidly that they are catapulted out of the water. Once airborne, they spread their large, fan-shaped pectoral fins and can glide along above the surface of the water for distances of up to 200m. When a fish falls back onto the water's surface, it can briefly beat its tail fin and take off again if necessary. Unlike birds' wings, the fins are rigid and so cannot be moved up and down.

Fish, such as these yellow-striped fusiliers, don't bump into each other when swimming in schools because they can detect very precisely the position of others in the group around them.

Why do dead fish float upside down?

Like all living creatures, gases build up inside a fish's body when it dies, and this keeps it afloat in the water. The gases collect mainly in the abdominal cavity, on the underside of the fish. When this happens, the lighter, gas-filled abdomen becomes buoyant and the more compact back forms the centre of gravity, with the result that the fish is turned upside down. It is only when decomposition is further advanced that the gases are released and the fish sinks.

How do animals look for food?

Herbivores don't need to find and chase food in the same way as animals that eat other animals. For them, the biggest challenge is often to avoid being eaten. But carnivores have developed a range of methods to locate and catch their food. They range from hunting to constructing traps made of sticky threads and often involve clever deception and camouflage.

How do glow-worms catch their prey?

Glow-worms are not actually worms. They are the carnivorous larvae of tiny flies called fungus gnats. The larvae grow to about 30mm in length and live on sheltered rock ledges in humid environments such as caves, cliff overhangs or forest margins. These tiny creatures have developed an extraordinary method of catching their prey. They position themselves on a ceiling or on a sheltered rock face, from which they suspend long, sticky threads, similar to spiders' silk. The glow-worms then use bioluminescence to produce a bluish light which attracts insects that get tangled up in the threads. The larva then simply reels in its catch.

Insects are lured to their doom by the glow-worms' deadly light show.

Do all crocodiles eat meat?

There are no vegetarians among the crocodiles. As well as fish and birds, crocodiles will try to catch larger prey such as antelope and buffalo and will even attack other predators such as lions and hyenas. Lying half-submerged in the water a crocodile waits for animals that come to drink. It will then rush onto the river bank or lake shore, grab the victim with its sharp teeth and pull it into the water. Once submerged, the prey will drown if it has not already been crushed to death by the reptile's powerful jaws.

How do crocodiles recognise their prey at night?

Crocodiles don't have to rely on their eyes and ears when it is dark. Instead, pressure sensitive organs called dermal pressure receptors allow them to detect even the tiniest movements in the water and to locate their prey. The pressure receptors cover almost every scale on a crocodile's body, but in alligators and caimans are only found around the jaw line. As long as the jaw is held partly above and partly below the surface of the water, the crocodile can use these sensory organs to detect vibrations transmitted along the water surface. The vibrations tell it where its next potential meal is having a drink.

More people are killed by wasps and bees than by sharks

From a statistical point of view, the likelihood of being killed by a shark attack is far less than that of dying as a result of a wasp or bee sting. Every year these insects are to blame for many more deaths than are marine predators. Our chances of encountering bees and wasps are obviously much greater, so stings are more likely. This is not really a cause for concern for most people since stings are usually only fatal if the victim has a serious intolerance to the venom.

Human and animal – which is the greater threat to the other?

People do get killed by animals, but humans represent a far greater threat to other members of the animal kingdom than the reverse. Animals usually appear on lists of endangered species because their numbers have been reduced by hunting and poaching, or because their habitat is so endangered that entire populations are at risk. Compared to the damage inflicted on humans by animals – from bee stings to shark attacks – the damage done to fauna worldwide by humans has had far more serious consequences for the planet.

The mighty rhinoceros has no predator except humans.

How much of a danger to humans is Jaws?

Sharks can undoubtedly pose a threat to humans. But although the number of shark-related incidents is on the rise, this can be explained simply by the increase in human populations and not by any theory that sharks are becoming more aggressive.

Currently we can expect to see an average of about 60 shark attacks each year worldwide, although only very few of these will be fatal. Compared to the number of people who swim in the sea each day, or who ride the waves on surfboards, these events are incredibly rare. Statistically, the great white is the most dangerous of all sharks. But even so, in the period from 1876 to 2006 there were 232 incidents worldwide involving great whites, 63 of which were fatal. But since researchers suspect that in many attacks the shark has mistaken a human for a seal or other prey, it is not accurate to conclude that humans are deliberately being targeted.

...that masks provide protection from tiger attack?

Although humans do not form part of a typical tiger's diet, individual animals have been known to hunt humans on occasions. The big cats rarely enter human settlements in search of prey, but woodcutters and honey gatherers who leave their villages can become victims. Where a threat of this kind exists, people have discovered that wearing a mask on the back of their head seems to offer some protection. Tigers usually approach their intended prey from the rear, and it seems to be the case that eye contact with the victim acts as a deterrent.

Are non-human animals right- or left-handed?

Chimpanzees in the wild tend to be left-handed. Unlike chimpanzees in captivity – which are usually right-handed – they prefer to use their left hand for certain tasks, such as delving for termites, which requires a great deal of dexterity in the use of a stick. It seems to be the case that mothers pass this preference on to their young, but we still don't know whether it is inherited or learnt behaviour.

Do octopuses have a 'favourite' tentacle?

We might assume that all eight of an octopus' tentacles are used more or less equally. It is certainly the case that when hunting, many of these creatures will settle on a rock and search for prey beneath it using all eight tentacles simultaneously.

But octopuses actually do tend to favour one tentacle for certain tasks. When exploring a new hiding place or touching an unfamiliar object, for example, they use just one tentacle or a particular combination of tentacles. The favoured tentacles are usually on the creature's front, close to their preferred eye. The rear tentacles are generally used by the octopus in the same way as legs, for moving around.

Do walruses favour a particular flipper?

These large marine mammals have developed several techniques for finding molluscs, their favourite food, which usually live buried a few centimetres deep in sediment on the sea floor. Both flippers and muzzles are used to burrow into the sand, and they will even occasionally use jets of water to help to blow sediment out of the way. But most frequently walruses use their right flippers to dig with. Studies have shown that, for two-thirds of the time, walruses will use this referred flipper in the search for food. It seems to be the case that, just like humans, walruses tend to be right-handed.

An octopus may look symmetrical, but not all its tentacles are equal.

Why do some animals try to look like something they are not?

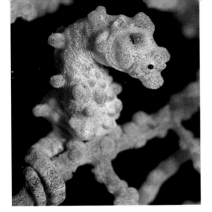

This pygmy seahorse is perfectly camouflaged.

In the eternal struggle for food and survival, animals and plants have developed many ways to outwit and deceive rivals and enemies. One of the simplest methods is camouflage, where a creature changes its appearance so that it becomes extremely difficult to distinguish from its surroundings. The result can be a certain amount of protection from enemies or a chance to catch prey which has been lulled into a false sense of security. Other animals pretend to be something they are not by imitating particular objects or even other living creatures.

How does mimicry work?

Mimicry is when one creature imitates the characteristics of another in order to deceive. The imitator will generally copy some obvious distinguishing marks on the creature they are imitating, hoping to send out the same message in order to confuse predator and prey alike. The wasp fly, for example, imitates the conspicuous black and yellow markings on real wasps, thereby signalling to possible predators that it is able to defend itself and wouldn't make a good meal – even though the flies are in fact harmless and perfectly edible. This type of protective mimicry – known as Batesian mimicry – is widely used in nature. Peckhamian mimicry is another, more aggressive form of the strategy of deceit. This occurs when a predator disguises itself as something else in order to deceive its prey. The angler fish, for example, has a small flap on its anterior dorsal fin which it uses to imitate a worm to attract other fish.

What is the difference between mimicry and mimesis?

When an animal sends out a false signal which results in a reaction from the signal's recipient, this is known as mimicry. Mimesis, on the other hand, is when a creature imitates a generally immobile, living or inanimate object which is of no interest to the deceived observer. Unlike camouflage, the deceiver is perfectly visible, it is just that it is not seen for what it really is. Stones, branches, leaves and flowers provide perfect objects for imitation. It isn't always easy to precisely differentiate mimesis from camouflage. When an animal imitates its surroundings the result is also that others have difficulty spotting the creature doing the imitating.

Why does a chameleon change its colour?

The claim that chameleons can adapt their colouring to suit a particular background is more of a myth than a fact. They can certainly change their skin colour, but this comes about primarily when the animal's skin reacts to fluctuations in temperature or light. Chameleons also use skin colour to communicate with one other and also to express their mood. The lizard's use of its skin colouring for camouflage is of secondary importance.

Several mechanisms inside the skin cells of a chameleon work together to bring about the colour changes. Within specialised skin cells known as chromatophores, pigments are able to mass together at the centre or spread across the whole of the cell, thereby causing the change in colour. There are different types of chromatophores which contain different shades of pigment in the form of shimmering crystals which allow a range of colour graduations and effects.

What do butterflies signal with their colourful patterns?

The colourful patterns on butterflies' wings strike us as attractive, but in fact they often serve to repulse and deter predators. When a butterfly comes to rest on a blade of grass or a flower, we usually see only the unspectacular, grey-brown undersides of its wings. When the butterfly suddenly spreads its wings, the bright colours on the wings' upper surface become visible. In some cases – such as with the peacock butterfly – the markings are remarkably similar in appearance to mammalian eyes. The hope is that a potential predator will be startled, or at least made to hesitate momentarily, giving the butterfly sufficient time to escape.

By turning a bright green, a chameleon signals that it is ready to find a mate.

Faster, higher, farther – records from the animal kingdom

Many of the physical things human beings are able to do only with the help of technology, certain animals can achieve unaided. A number of animals can run faster than an Olympic sprinter, while others can maintain speeds through the water that human world-record-breaking swimmers can only dream about. In any contest between humans and the animal kingdom, humans would be the losers.

The two-toed sloth is considered to be marginally faster than its easy-going three-toed cousins.

Which is the slowest moving mammal?

Representatives of the three-toed sloth family – which are natives of South America – move at a speed of 0.16km/h across the ground. What's more, they have great difficulty in achieving even that leisurely speed. But when travelling in the branches of the trees where they have their actual habitat, these creatures become a bit more mobile and have been known to achieve speeds as high as 0.27km/h.

Which is the fastest animal on land?

The cheetah holds the sprint record. This agile predator is able to achieve speeds of almost 120km/h over short distances. Over long distances, however, the cheetah must defer to the pronghorn antelope, which can sustain speeds of about 88km/h.

Which is the fastest animal of all?

On average, a diving peregrine falcon will reach speeds of around 350km/h. The fastest dive recorded so far reached a stunning 389km/h.

Which animals are the fastest swimmers?

With a top speed of 111km/h, the sailfish is the fastest swimming fish. The fastest swimming marine mammal is the orca or killer whale, which can reach speeds of up to 55km/h. The Gentoo penguin holds the record for aquatic birds of 27km/h.

Which animal jumps the farthest?

The record in this category is held by one of the smallest contenders – the flea. World champions at the long jump, these tiny insects can cover 200 times their own body length with just one leap.

Which is the heaviest land mammal?

An average male African elephant grows to a height of about 3m to 3.7m and weighs four to seven tonnes, which makes it the heaviest land mammal. Some fully mature elephant bulls can reach a height of 4.9m and a weight in excess of seven tonnes.

Which is the tallest land mammal?

This record is held by the giraffe, with males growing to a height of about 5.5m. Like most mammals, these long-necked residents of the African savannah have just seven cervical vertebrae – in their case, however, each vertebra is much enlarged.

Which is the longest and heaviest snake?

Two species – the green anaconda and the reticulated python – compete for the world record in this category. The anaconda, which reaches a maximum length of between 8m and 9m, does not grow to be any longer than the reticulated python, which in rare cases can grow up to 10m. With a weight of more than 200kg, the anaconda is considerably bulkier than its rival, making it the world's largest and heaviest snake.

Which is the largest marine animal?

The blue whale grows to an average length of 27m and can weigh up to 150 tonnes. This makes it the largest mammal ever to have lived on the planet. The record holder is a specimen that was caught in waters around Antarctica in 1947. This massive female was 33m long and weighed 190 tonnes.

Which was the largest of the dinosaurs?

For a long time, the vegetarian *Brachiosaurus*, which weighed about 70 tonnes and grew to a height of 13m and a length of 25m, was thought to have been the largest dinosaur. However, recent fossil finds seem to indicate that much larger dinosaurs may once have existed. The *Argentinosaurus*, the *Seismosaurus* and the *Supersaurus* may have grown to lengths of about 40m, which means they would have been considerably larger than the former record holder. However, so far no complete skeleton of these giants has been found.

Which were the largest insects?

With a wingspan that could reach 700mm, the largest insects ever to have lived are thought to have been the *Meganeura*, which resembled huge dragonflies. These giant insects first appeared on Earth around 320 million years ago and became extinct 70 million years later.

Which is the smallest mammal?

The bumblebee bat – also known as Kitti's hog-nosed bat – and the Etruscan shrew jointly hold the record for being the world's tiniest mammals. The bat measures between 29mm and 33mm from head to rump, has a wingspan of about 130mm to 150mm and weighs no more than 2g. The Etruscan shrew is 30mm to 35mm long from head to rump, has a 25mm- to 30mm-long tail and weighs about 2g.

How large are the largest molluscs?

Capable of achieving a length of 17m and a weight of around 900kg, the giant squid *Architeuthis dux* is both the largest known mollusc and the largest invertebrate. The biggest bivalve mollusc is the giant clam *Tridacna gigas*, which can grow to a weight of about 200kg, with a 1.5m-long shell.

Which animals live the longest?

Sponges of the species *Scolymastra joubini* may have been alive for 10 000 years or more. Scientists have calculated the age of these sedentary sea creatures based on their present size, given their extremely low oxygen consumption and slow metabolism. Specimens of giant sponge – which live in Antarctic waters – have been measured at 2m in height with a diameter of up to 1.7m.

Which is the strongest of the animals?

The record holder in this category is the rhinoceros beetle. These sturdy insects can carry loads of up to 850 times their own weight.

What insects form the largest swarms?

The largest swarms are formed by the desert locust. In 1954 a locust swarm in Kenya covered an area of about 200sq km. The enormous cloud of insects was thought to contain 50 million individuals per square kilometre, which means that the swarm was probably made up of about 10 billion insects.

Which is the noisiest insect?

Of all the insects, the African cicadas produce the loudest noise. At a distance of 500mm, their chirping can reach a level of 106.7 decibels – louder than a chainsaw. The song of a cicada plays an important role in communication, helping it to find a mate and possibly defend its territory against intruders.

Which animal is the fussiest eater?

The most pernickety animal when it comes to food has to be the koala. It eats almost nothing but eucalyptus leaves, as well as the tree's bark and fruits. Of the 700 species of eucalyptus that grow in Australia, it will only snack on about 70 – but in some cases only five or so. Koalas will search through several kilograms of leaves every day in order to eat just 500g. They look specifically for older leaves, since these contain fewer toxins than young leaves.

Which birds fly the highest?

The highest altitude at which a bird has been observed flying is 11 300m. In late November 1973, a Rüppell's vulture collided with a commercial aircraft at that altitude over Côte d'Ivoire.

Which are the loudest animals?

The loudest animal noises of all are emitted by blue whales and finback whales. The low frequency sounds they emit reach a sound level of 188 decibels.

What makes birds so special?

Birds developed from reptiles and are closely related to the extinct dinosaurs. They are the only class of vertebrates of which almost all of the members have conquered the skies. Most birds are able to fly and their bodies are adapted to this way of life. Their frontal appendages have developed into wings, and their bones are hollow – which keeps them light while at the same time strong enough to withstand the stresses of flight. Birds also have a respiratory system that is proportionally larger than that of the average mammal.

Why do parrots speak?

Parrots don't actually speak, they merely imitate the sounds they hear. This could be human speech, an alarm clock, other birds or the telephone. In their natural habitat, parrots generally live in large groups. In captivity and deprived of their usual company, this social instinct drives parrots to mimic human language in order to forge a relationship with the human beings who have become their replacement companions. In birds, the air flow required for speaking or singing is produced in the syrinx. The bird uses its tongue to form various sounds and even the smallest change of position is enough to produce a different sound.

American coots count their eggs to give their chicks the best possible chances of survival.

Can birds count?

Primates are not alone in being able to count or even to add up – there are birds that can do that as well. The American coot has learnt to count its eggs in order to defend itself against brood-parasitism. This is when birds try to palm their eggs off onto other females of the same species. They are also adept at recognising alien eggs, which are pushed to the edge of the nest where their chances of incubating are considerably reduced. The hen then continues to lay eggs until she has reached a particular number – eight on average. By keeping count, she ensures that her offspring are not neglected in favour of any parasitic chicks.

Can birds recognise their relatives?

The American north-western crow can distinguish between its relatives and other crows that are not related to it. The birds are friendlier towards relatives than they are towards unrelated crows.

It isn't unusual for these birds to steal food from each other. But the thieves are far less aggressive when the food-owning crow is a relative. They come close, but are generally passive as they loiter in the vicinity of the other crow, which will often let them have the food anyway. Where thieves and victims are not related, the thieves approach aggressively and will touch their victim. If the victim attemps to flee, the thieves will pursue it.

Why don't woodpeckers get headaches?

Woodpeckers have several physical features that minimise the effect of the blows caused by pecking and protect their brain from shock. Their skulls are around twice as thick as those of other birds, and their brains are surrounded by twice as much fluid as is ordinarily the case. This prevents the brain from striking against the inside of the skull.

The most important of the anti-shock measures is the structure and construction of the bird's beak. It is not centrally arranged on the cranium, as it is with other birds, but is positioned lower down. This ensures that the force of a blow with the beak is not transferred directly to the brain. Instead, the force is deflected along the skull bone. Secondly, cartilage on the lower half of the beak absorbs the shock produced by the pecking. The interplay of these factors protects the woodpecker from damage to its skull and brain, while allowing it to seek food and a place to nest inside the wood of a tree.

The pecking of the great spotted woodpecker can often be heard in the woodlands of central Europe.

Although the winter wren is only 9.5cm long, its lusty song can be heard from several hundred metres away.

Do songbirds have perfect pitch?

Songbirds do indeed have perfect pitch. A scientific study has shown that various songbirds, including zebra finches and white-throated sparrows, are much better at determining, distinguishing and remembering isolated pitches than human beings. When human subjects were provided with a second sound for comparison, they were still far less skilled at determining its pitch than songbirds.

The company of member of their own species seems to play a role in the development of hearing in birds, since those reared in isolation seem to do less well in experiments of this kind. Early experiences with music play a role in human beings. Most babies have perfect pitch, and this is usually preserved if they are given music lessons at an early age. Speaking a tonal language such as Mandarin – where words can have different meanings depending on tone and pitch – also helps to develop perfect pitch.

How do birds know when to start their song in the morning?

Every species of bird wakes up at a particular time in the morning and begins to sing. This is because each species has its own specific waking stimulus, which is linked to the brightness of the dawn light. These waking times are so precise that we can tell the time by them – a phenomenon that has given rise to phrases such as 'up with the lark' and 'to rise before cockcrow'.

In central Europe, the first to sing is the common redstart, which starts precisely one hour and 30 minutes before sunrise. This bird is followed at intervals of between five and 10 minutes by the robin, the blackbird, the cuckoo, coal-tit, chiffchaff and chaffinch, and then at 20-minute intervals by the sparrow and the starling.

Is birdsong learnt or innate?

Just as human beings have to learn to speak, so birds also have to learn how to sing. They do this in several stages. First, they practise tones and sounds, which is comparable to the babbling stage in human language development. During the second stage the birds practise their song for eight to nine months, until memory and practice match up. The singing is strengthened and polished during the final phase.

Humans and birds both appear to go through a phase when the brain is particularly receptive to learning language or song. This is why birds have to hear other birds of their own species sing while they are still young, otherwise they won't be able to produce much more than a whistle later on.

Do nightingales sing only at night?

Nightingales got their name as a result of their nocturnal singing but both males and females sing during the day. The nocturnal song is used mainly for mating purposes, which is why only bachelors sing at night. After he has succeeded in attracting a female, the male nightingale falls silent when darkness falls. He will strike up again at dawn, to warn off any possible competitors for his female's attentions. This is vital, because male nightingales tend to be in the majority and each season about half the males are left without mates.

Do birds sing in dialect?

Birds don't all sing the same songs, not even within a species. The song of a robin in London will be quite different to that of a robin in Paris or Berlin. Although the members of each bird species shares a vocabulary of sounds, dialect differences are common. Among yellowhammers, a distance of a few hundred metres is sufficient for the birds' song to change. The individual dialects are not innate, but are learnt while the birds are still chicks in the nest – just as children adopt the dialect of their parents. The young birds always sing as well or as badly as their teachers, and because the adults on which they model themselves vary in style and talent from region to region it is possible for different dialects to develop. Some bird species

have more than 60 dialects, and many singers can be described as multilingual. Males that master several dialects have a better chance of finding a mate, since females prefer mates from the same dialect family.

Bird dialects also help to drive evolution, because different songs lead to the formation of groups. This can cause the formation of new sub-species and eventually even new species, as happened with the marsh warbler and the reed warbler. These two birds look the same, but have different songs.

In which seasons do we hear the greatest amount of birdsong?

Birds are seasonal singers. We only hear intensive, many-voiced bird concerts between spring and high summer. This is the time when birds are looking for mates, and when the males aim to impress the females with their song, which simultaneously serves to mark out their territory for the benefit of rival males.

After they have found a mate and the chicks have hatched, male birds continue to chirp and twitter because this is the time when they are showing their offspring how to sing. The young males have to learn how to attract females when they are mature, and the young females must discover what song it is they have to pay attention to later on. When the chicks have left the nest, the male's sex hormone levels fall and the birds cease to sing. Moulting then takes place, a process which will drain them of energy.

These three young tits are still practising the songs they have heard their parents sing.

The bald eagle can even see fish swimming under water.

True or False?

Eagles have the best eyesight of all living creatures

An eagle's vision is among the best in the animal kingdom, but the eyesight of the falcon, the buzzard and the hawk – other birds of prey that hunt in daylight – is at least as good. All have large, elongated eyeballs, allowing plenty of room for the lens and retina. There are many more sensory cells in the retinas of birds of prey than in a human eye, and they are more evenly distributed across the retina. This allows them to see sharply across a larger proportion of their field of vision. The more tightly the sensory cells are packed in, the sharper the image. In those areas where vision is sharpest, birds of prey have two to eight times more sensory cells per square millimetre than human beings. They are able to focus on objects faster and over greater distances. Their temporal angular resolution is also higher: while people perceive 25 images per second as a sequence of fluid movement, birds of prey can recognise 150 images per second as individual images. They are able to see ultraviolet light, so can follow the urine trail of a mouse when hunting.

How does bird vision differ from that of humans?

On most birds, the eyes are arranged on the sides of the head, which is why they can see a large portion of their surroundings at any one time. The woodcock's eyes are so far apart that they have 360° vision. The drawback with having eyes positioned far apart is that the area that is seen by both eyes simultaneously – something that is necessary if a creature is to have binocular vision – ends up being relatively small. The birds have to compensate for this by moving their heads from side to side continuously. This results in enough separate visual information being received by each eye for the brain to be able to use it to construct an integrated, three-dimensional image.

Many birds also have large eyes, which gives them good vision even in dim light, and there are a large number of birds that are able to see ultraviolet light in addition to the usual colours. This can be of benefit both for finding food and a partner.

With its eyes on the side of its head, the black-tailed godwit can see enemies approaching from the rear.

How do migrating birds find their way?

A fascinating sight, the V-formations of migrating wild geese conserve the birds' energy.

Migratory birds seem to know instinctively when it is time to move on and where it is they need to go. They are guided on their journey by the position of the Sun during the day and the stars by night, and they can also make use of the Earth's magnetic field for navigation. The exact details of the process, the precise nature of the sensory organs used, how the mechanism works and where it is located in the body continues to be a puzzle.

It seems certain that the sense organ that allows the birds to orient themselves magnetically is located somewhere in their heads. Before they fly off they can be seen determining where north is by moving their heads to and fro many times. This is probably how they find the position of the magnetic field, and it is thought that special receptors or molecules that undergo chemical changes when in contact with magnetic fields are probably involved. These receptors may be located in the retinas of birds' eyes, and among possible candidates are the chryptochromes, which are found in particularly high concentrations there.

Why do wild geese fly in a V-formation, while flocks of sparrows do not?

Wild geese are migratory birds, and they can travel for thousands of kilometres to get to their northern breeding grounds. They fly in a V-formation in order to save as much energy as possible during long flights. The energy is saved by reducing air resistance, since each bird benefits from having another bird flying in front of it. With every wing movement, a flying bird produces an up-current that can be used by the bird behind. Wild geese that fly in a V-formation are on average over 70 per cent faster than birds flying alone. When the lead goose begins to tire it falls back and another goose moves to the front. Sparrows, on the other hand, are sedentary birds. They live in the same region whatever the time of year. Since they do not have to travel long distances, they have no need to organise themselves to fly in any particular pattern. Most birds that migrate – including many of the songbirds – don't fly in a V-formation because they are more likely to fly alone and tend to travel at night.

How do pigeons find their way home?

There are indications that, like migratory birds, pigeons are guided by the Earth's magnetic field and the position of the Sun and stars. Pigeons have small magnetic particles in their heads that they use to locate the Earth's magnetic field. Studies have shown that pigeons will also rely on visual clues when they are flying home, using prominent traffic arteries such as motorways, railway tracks and main roads to orient themselves once they have settled on the correct direction. They will continue to follow the course taken by an artery even when it is not the most direct route.

How do we know...

...that albatrosses don't move their wings when they fly?

Albatrosses spend most of their lives in the air. They have narrow, straight, long wings that are perfectly adapted to their way of life. They don't even need thermal up-currents to soar, since the ordinary movements of air above the surface of the sea are sufficient for them. The wandering albatross, with a wingspan of up to 3.4m, is the largest of the albatrosses and it can glide through the skies for days, covering vast distances without once flapping its wings. Flapping its long wings would rapidly bring it to the limit of its physical strength, as a single wing beat uses up more of the bird's energy reserves than many hours of almost effortless gliding.

Why do albatrosses have problems at take-off and landing?

An albatross taking off and landing is not an elegant sight. Although ideal for gliding through the air, its long wings tend to get in the way when the bird is closer to the ground. Flapping uses up energy, and the bird's great weight – up to 12kg – adds to its difficulties. An albatross therefore needs a long, quick run-up and a headwind of at least 12km/h before it is able to take off. When landing, it tries to reduce its momentum by running. Since it tends to fly faster than it is able to run, landings frequently end with a crash and possibly even a somersault.

Homing pigeons are placed in a special transporter and taken up to 1000km from their loft before being released as part of a competition.

Konrad Lorenz – mother to the grey geese

In 1935 Konrad Lorenz published an essay about the behaviour of grey geese, which proved to be a breakthrough in animal psychology.

Grey geese do not have an innate image of their mothers. This is something they have to learn. Konrad Lorenz observed that newly hatched chicks that have been separated from their mothers go through a highly sensitive phase in which they tend to follow any moving object whilst making peeping sounds – just as they would if they were following their mothers. Lorenz called this behaviour imprinting, and discovered that any object the chicks see after their birth can replace their biological mother. He even managed to imprint the chicks on himself, and discovered that once the birds are imprinted on an object it is impossible to erase the learnt behaviour, even if the chicks' real mother is returned to them later. Lorenz shared the 1973 Nobel prize for Physiology and Medicine for his work.

Do wind turbines pose a danger to birds?

The question is still being assessed, although previous studies have suggested that these installations pose much less of a danger than had previously been assumed. An organisation in Germany calculated that each year an average of 0.5 birds die for each turbine in the country, which makes an annual total of approximately 8000 dead birds in Germany alone. Other hazards – such as windowpanes, electricity lines and road traffic – are responsible for far more bird deaths. Some calculations have shown that as many as 600 times as many birds are killed each year as a result of collisions with road vehicles.

Migrating birds will spot wind farms from a distance of about 3km, and can therefore fly around them. It has been suggested that in avoiding wind farms, birds are missing out on major breeding sites, but this has not been confirmed.

In order to to protect native wildlife and minimise the risk, it is important that all proposed wind farm sites are investigated thoroughly before being given approval. Wetlands and areas close to stretches of water are not suitable locations for wind turbines because birds of prey, such as eagles, red kites and others, are regular visitors. Areas adjoining woods or forests should also be avoided, since wind turbines pose a danger to bats and other animals.

Do birds make flying errors that cause them to crash?

Fledglings make flying errors, since they have yet to learn how to assess distances correctly. Healthy, mature birds only make mistakes in exceptional circumstances. When a bird first comes across electricity lines, or windowpanes in which the sky is reflected, it is possible for a crash to occur. If it survives such an incident, a bird is unlikely to make the same mistake again, since it memorises all the obstacles in its environment.

Migratory birds following a course in late evening can make flying errors. They can be deflected from their route by the presence of strong artificial light and small migratory birds can be driven off route by strong winds. Plumage is important for accurate flight because the different types of feathers help the birds to follow a precise course. Flying wears feathers down quickly, so birds moult regularly to replace the worn feathers with new.

Why don't penguins stick to the ice?

Although it has a body temperature of 39°C, a penguin's feet are always cold. This adaptation enables the birds to stand on an icy surface without it thawing – and since no liquid water is formed, their feet do not stick to the ice. To ensure that the feet always stay cold, the flow of blood is restricted. Veins and arteries in the legs are surrounded by muscles that are able to contract so that the blood supply is kept to the minimum necessary to maintain the temperature of the feet at just above 0°C. Since this also cools the blood, it is necessary to ensure that it is not too cold when it returns to the body. It is therefore directed through numerous small blood vessels and past warm arteries, becoming warmer as a result.

Why do penguins seem to be wearing evening dress?

Penguins' black markings act as camouflage when the birds are in the water, protecting them from birds of prey and marine predators. Seen from above, the penguin's black 'dinner jacket' merges with the surface of the sea, making it barely visible from the air. Seen from below, the bird's white stomach merges with the bright, sunlit surface of the sea. The result is that penguins are more or less invisible to predators both in and out of the sea.

How do penguins keep their nests clean?

Penguins avoid soiling their nests by launching their droppings like projectiles. They can shoot their excreta for distances of up to 500mm out of their nests. To achieve this feat, penguins must produce up to 60kPa of pressure in their intestines. These calculations have been made on the basis of the trajectory of the droppings, their consistency and the distance travelled. The birds manage to produce an internal pressure which is more than half the atmospheric pressure. This is about four times as high as that produced by human beings.

What do king penguins feed their young when their mates are off looking for food?

King penguin parents take it in turns to look after their young. One parent stays with the chicks while the other goes hunting. In order to ensure that it has enough to feed itself and its chicks, the parent that remains behind stops its digestive processes and is therefore able to store food in its stomach for up to three weeks. It then regurgitates the food to feed the chicks. King penguins are somehow able to kill off their stomach bacteria, possibly with the release of an antibacterial substance from the gastric wall.

Why do emperor penguins keep their eggs on their feet?

Most penguins build nests for their eggs and chicks, but emperor and king penguins use their bodies to protect the eggs and keep them warm, placing them on their feet to do so and using a fold of skin as a covering. They have adopted this strategy because there isn't sufficient nest-building material in the areas where these birds breed. Emperor penguins breed on the pack-ice so the best and only protection for the eggs is balanced on their parents' feet. Emperor penguins are even able to hatch their chicks in the polar winter, when temperatures fall as low as -40°C. Once hatched, the chick is transported around on its parents' feet, still protected by the fold of skin.

How do insects see the world?

Insects have a different view of the world to ours. As they are so much smaller, their surroundings appear greatly enlarged. Many insects are able to fly, and this gives them a broader overview. And while we see our environment as a single image, insects see it through compound eyes, therefore their view is likely to be a mosaic made up of numerous individual images.

Close-up views of insects – such as the compound eyes on this housefly – reveal an alien world that exists alongside our own.

Why are insects so small?

The amount of oxygen in the air determines how large an insect can be. If the air contained more oxygen, the invertebrate trachea would be more effective and insects would be bigger. This labyrinthine system of tubes runs through an insect's body and supplies it with oxygen. The larger the insect, the larger and more complicated its oxygen supply system has to be. This places a limit on the size of insects, but an increase in the oxygen content of the atmosphere would allow the system to function more effectively.

During the Palaeozoic era, when the oxygen content of the air was 35 per cent compared to today's 21 per cent, dragonflies with a wingspan of 760mm could be found. But there is another factor that places an upper limit on the size of insects. At the point where limbs and body meet, the trachea can only be as thick as the delicate structure of the joints permits. Once a certain size is reached the extremities can no longer be supplied with sufficient oxygen.

How long do mayflies live?

There are more than 2800 species of mayflies in the world.

The life of an adult mayfly is very short. It may even – as in the case of the dayfly – be a single day. There are about 2800 species of mayflies and, depending on the species, a mature insect may live for anything from just a few hours to a few days. These insects, which are between 2mm and 3mm long, have neither functioning mouthparts nor a working digestive system. They draw their nourishment solely from their fat deposits. Once they have hatched from their pupae, mature flies exist only to reproduce, after which they die – the males shortly after mating and the females after they have laid their eggs. But prior to reaching their final form, the insects will have spent a lot of time – sometimes even a few years – as larvae.

How do flies use their feet for tasting?

A fly's sense of taste isn't restricted to the mouthparts through which it takes in food. Insect taste receptors are located on hairs that are distributed all over their bodies – including on their wings, ovipositor and even legs and feet. In this way a fly can always tell whether something is tasty or not, regardless of how it happened to come into contact with the potential food source.

Flies, like humans, are able to distinguish between bitter and sweet, and have similar receptors for this.

Taste receptor hairs on the leg of a fly.

Why and how do glow-worms glow?

Most glow-worms, fireflies and lightning bugs are actually beetles belonging to the family *Lampyridae*. They are found in temperate and tropical areas around the world. The larvae of all species glow, which may be a warning to deter predators. Most adults glow when looking for a mate and, depending on the species, both sexes may glow, or only the females. Every species produces its own characteristic glow, which avoids mistakes. The signals vary in terms of colour, light impulse frequency and the arrangement of the body's light-emitting organs. The females are often unable to fly and so wait on the ground or in foliage for passing males, which they try to attract. Some female fireflies maintain a constant glow throughout the hours of darkness, while others don't begin to glow until they have spotted a glowing male of their own species.

The typical glow – usually greenish-yellow – is produced in the firefly's light-emitting organs, called photophores, by means of a process known as bioluminescence. This is the ability of some living creatures to produce light with the help of biochemical processes. Bioluminescence involves the production of energy by the oxidation of special luminous substances – known as luciferins – which are then transformed into light. This process takes place with the help of an enzyme known as luciferase. Hardly any heat is produced along with the light, which is why it is known as cold luminescence.

The female glow-worm is wingless and therefore unable to fly. It sits in the grass and attracts males.

How far are butterflies able to travel?

Some butterflies are able to fly across huge distances. One of the most accomplished travellers is the American monarch, which leaves North America in the autumn and flies to Mexico, where it spends the winter. The first three or four generations that hatch during the summer stay in the north where they mate, but the last generation to emerge before the autumn does not become sexually mature, although it will have well-developed flying muscles. These butterflies then embark on a long journey, flying more than 3000km in eight to 12 weeks. The insects travel on average a distance of 70km per day, but with favourable winds they can cover as much as 300km or more. They use the position of the Sun for navigation. In the spring, these creatures embark on the return journey north, during which they reach sexual maturity and lay their first eggs. Once they have laid their eggs a lot of females die, as do many males. Their descendants continue a journey that is completed in many stages and often takes several generations.

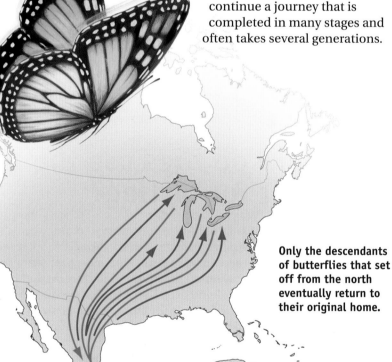

Only the descendants of butterflies that set off from the north eventually return to their original home.

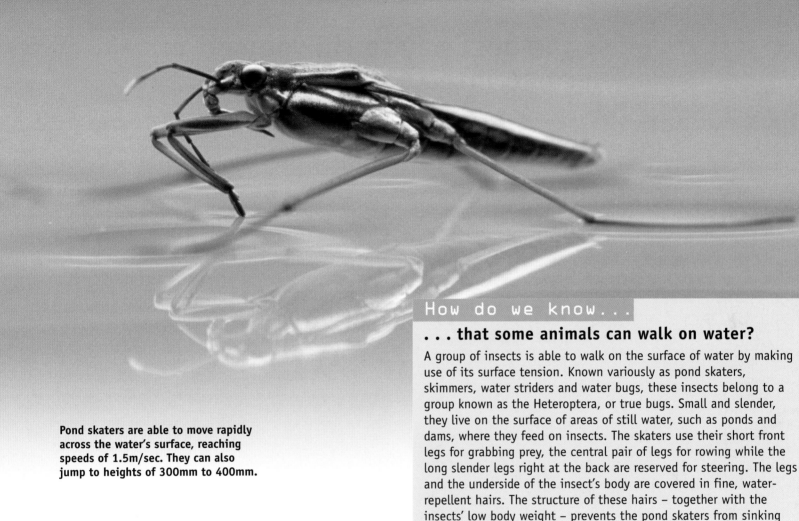

Pond skaters are able to move rapidly across the water's surface, reaching speeds of 1.5m/sec. They can also jump to heights of 300mm to 400mm.

How do we know...

...that some animals can walk on water?

A group of insects is able to walk on the surface of water by making use of its surface tension. Known variously as pond skaters, skimmers, water striders and water bugs, these insects belong to a group known as the Heteroptera, or true bugs. Small and slender, they live on the surface of areas of still water, such as ponds and dams, where they feed on insects. The skaters use their short front legs for grabbing prey, the central pair of legs for rowing while the long slender legs right at the back are reserved for steering. The legs and the underside of the insect's body are covered in fine, water-repellent hairs. The structure of these hairs – together with the insects' low body weight – prevents the pond skaters from sinking below the water's surface.

Can insects help to solve murder mysteries?

Forensic entomologists are specialists who analyse the insects that colonise corpses. On the basis of the species that are found and their stage of development, it is sometimes possible to pinpoint the time of death andthe cause or circumstances of a fatality. This is possible because many organisms are involved in the decomposition of dead organic matter, among them fungi, bacteria and scavenging insects like the maggots of certain flies and beetles. For many flies, a dead body provides more than just a source of food – it is also an ideal breeding place as it provides offspring with plentiful nourishment as soon as they have hatched.

Factors such as the location, temperature, humidity and the stage of decomposition can all, to varying degrees, determine which insects will colonise a corpse. This information can be used to give an indication of whether the death occurred where the body was found or whether it had been transported there from somewhere else. Whether eggs, larvae or empty pupa shells – perhaps from several generations of insects – are found can be used to give a fairly precise indication of the actual time of death.

What is the best way to rid yourself of troublesome insects?

A fly swat is not the only way to deal with annoying insect guests. The pest control industry has provided a variety of options for combating irritating pests such as flies, ants, moths and mosquitoes. We can choose anything from sticky flypaper and elaborate traps to natural pest control substances and the insect equivalent of chemical warfare agents. Not all methods are equally effective.

It may not be a pretty sight, but flies are better on a sticky trap than on food.

How do mosquito repellents work?

Mosquitoes rely mainly on their sense of smell to locate potential victims. It is thought that they are attracted by carbon dioxide emissions and particularly the smell of sweat, although it is possible that they react to body heat as well. Some people are particularly prone to attack due to their individual body smell.

Most mosquito repellents evaporate on the skin and work by blocking a mosquito's sense of smell, preventing it from finding its target. This is how substances such as icaridine or diethyltoluamide (DEET) work. DEET is the most reliable insect repellent available at present – it remains active for up to four hours after it has been applied – but is not considered safe in high concentrations. Natural essential oils such as citronella or tea-tree oil seem to cause fewer health problems, but they are less reliable. A physical barrier, such as flyscreens on windows, mosquito nets over beds or appropriate protective clothing, is the safest protection.

How do pheromone traps work?

These lure pests by exploiting their instinct to mate. The traps contain female sexual attractants – pheromones – but instead of finding a willing female to mate with, the males encounter a sticky mat or a trap from which they cannot escape. Pheromone traps are species specific, and will only work with the particular pest they are designed to attract, such as flour moths, clothes moths or bark beetles. The traps are non-toxic, so harmless to humans, and can reduce the insect population considerably. Since it is possible to count the number of insects trapped, these devices are also useful for estimating the severity of an infestation.

Why do yellow glass containers work as cockroach traps?

Not all animals perceive colour in the same way as humans. Cockroaches cannot see yellow because they don't have cells in their eyes that are sensitive to light of that frequency. What they see instead is black, and this is why a yellow glass container looks like a safe hiding place to a cockroach. It is then a simple matter to gather up and deal with any insects that have entered the supposed hiding place.

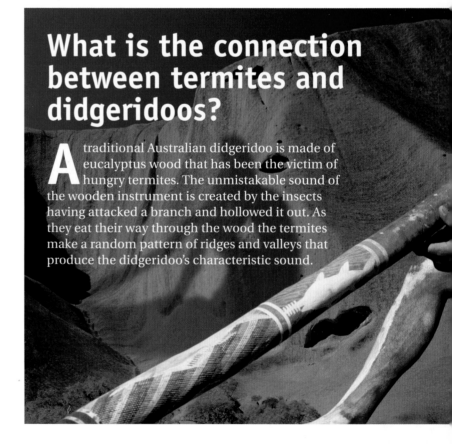

What is the connection between termites and didgeridoos?

A traditional Australian didgeridoo is made of eucalyptus wood that has been the victim of hungry termites. The unmistakable sound of the wooden instrument is created by the insects having attacked a branch and hollowed it out. As they eat their way through the wood the termites make a random pattern of ridges and valleys that produce the didgeridoo's characteristic sound.

...that termite mounds have airconditioning?

Termite mounds are large nests that can reach several metres in height, with up to three million individual insects in residence. The insects build their mounds in tree trunks or directly onto the ground. A complex structure of chambers, shafts and insulating systems ensures that the temperature and humidity of the air supply is carefully regulated. Warm, used air from the mound's lower levels rises slowly through a central chimney and flows out sideways through many corridors at the top. When the air strikes the external walls, it cools and sinks back down. In addition, fine pores in the walls ensure that gases are exchanged – with methane and carbon dioxide being expelled, while fresh oxygen enters from outside. A constant draught transports cooler air through chambers and corridors, keeping the nest at an ideal temperature.

The life of social insects like bees is strictly organised by a division of labour.

How do social insects organise their colonies?

The central principle around which communities of social insects – ants, bees, wasps and termites – are organised is the division of labour. Supplying and protecting the nest, care of the young and reproduction are organised in such a way that no individual performs all the tasks that need to be done if the community is to thrive. Among bees, the workers are responsible for the everyday jobs like providing food and looking after the young, while among ants the soldiers take care of defence, never considering their own lives in a fight to protect the nest. Since they are infertile, workers and soldiers must rely on others to replenish the population. In most communities it is the queen that is responsible for producing the offspring, and she only mates with a few select males.

Ants have particularly complex societies, and will even 'farm crops' and 'keep livestock'. Some species feed and protect aphids and caterpillars, which they milk for the sweet secretions, while others grow a special fungus that is used to feed the colony.

How do bees tell each other where to find sources of food?

When a bee on a foraging trip discovers a good source of food she performs an intricate series of movements, popularly known as a 'dance', when she returns to the hive in order to tell the other bees about her find. By means of this special dance the insect encodes information about whether the food source is plentiful, how far it is from the hive and – if the distance is great – in which direction it is to be found.

If the food source is less than 100m away and easily found, it is sufficient for the bee to dance in a circular pattern to draw the attention of other bees to it. Where the distances are greater, the information is transmitted by means of a tail-wagging dance, which allows the other bees to work out in which direction the food source can be located, relative to the position of the Sun. To communicate whether they have found nectar, pollen, honeydew or water, the insects simply provide the aroma or a sample of the food.

How do bees defend themselves against enemy attack?

If a bee stings an enemy the bee will die. Some bee species have other ways of defending themselves, such as using heat to destroy intruders in a process known as thermo-balling. Bees are able to withstand temperatures of almost 50°C for short periods, whereas most of their enemies cannot. A cluster of bees will surround an enemy and raise their own body temperatures by making their muscles tremble. Within a short space of time the temperature at the centre of the cluster has risen to 47°C and the intruder dies. Japanese honey bees can kill hostile hornets in this way. This ability to produce heat also comes in useful when the hive is attacked by fungi. Then the bees heat up the entire hive to combat the fungal infection.

How do bees survey a new home?

After the young queens have hatched in the summer, around half the bees leave the hive; approximately 30 000 bees leaving with each new queen. A new nesting place for such a large swarm will need a volume of at least 40 litres. When scouts are looking for a new home they therefore need to be able to make a rough assessment of the size of any potential new quarters – and they need to be able to do it in the dark. They pace out the walls of any promising location and complete several short flights from one wall to the opposite side. The information they gain from this helps them to make a rough estimate of the size of the potential hive.

Are ants more aggressive when they are closer to their own nest?

Experiments have shown that it is an insect's sense of the distance between itself and its nest – and not landmarks or the smell of the nest – that determines just how aggressive it will be when reacting to possibly hostile intruders. Desert ants are are relatively peaceful while out on reconnaissance, yet get more aggressive towards potential enemies the closer they are to their own nest. An internal distance gauge tells the insect how far it is from home.

Two wood ants fighting – a dramatic scene captured in Claude Nuridsany and Marie Pérennou's 1995 film *Microcosmos*.

Our blue
planet

The Earth is a planet of transformations. Its land masses and oceans are in a state of constant change. Earthquakes and volcanoes reveal the titanic forces still active within the planet, while the oceans and a protective atmosphere help to shape the land and provide an environment within which life can flourish. During a period of climate change it is even more important that we understand the processes at work in and on our planet to make sure it remains a good place to live for generations to come.

What controls the weather?

Confronted with many different climatic conditions, human beings have tried to discover the laws that govern the weather for thousands of years. At first they relied on rural folklore, which has often proved to be accurate. Later on, scientists who kept regular records recognised some of its basic mechanisms – and that the force driving it all is the Sun.

Solar winds affect the Earth's magnetic field, and the flow of charged particles creates the polar lights. People used to believe that the polar lights heralded a cold, clear winter.

What part does the Sun play in the weather?

The Sun's light and heat are the source of life on Earth, and they are also responsible for the weather. The main factor driving changing weather patterns is the difference in temperature between the Earth's surface, which is warm, and the cooler layers of air above it. The effect of solar radiation also depends on additional factors such as geographical location, distribution of land and sea, vegetation and other forms of life found in a region.

Do sunspots affect the weather?

Sunspots – dark patches on the visible surface of the Sun – are a well-known phenomenon and occur in large numbers every 11 years. Sunspots are created by local disturbances in the Sun's massive magnetic field. At temperatures of about 4000°C, sunspots are cooler than other areas of the Sun, where temperatures average 6000°C. Scientists have tried to discover if large numbers of sunspots cause cooling on the Earth by using tree growth to monitor the effects of the 11-year cycle. This has been unsuccesful, and any direct link between sunspots and the weather on Earth seems unlikely.

Earth's surface, but the area that affects the weather is the troposphere, a part of the Earth's atmosphere that only reaches a height of about 15km.

Can fine weather be dangerous?

The negative effects of too much Sun are now well known. Overexposure can cause painful sunburn and lead to skin cancer in the long term. Energy-rich ultraviolet rays (UVA) are the main cause of sunburn. The atmosphere, and especially the ozone layer, would normally provide us with good protection from UVA radiation. In recent years, the concentration of ozone has reduced considerably, especially above the polar regions. This is due largely to chlorofluorocarbons – or CFCs – which are used mainly as propellants or cooling agents. Scientists disagree about whether the ozone layer will regenerate following the banning of CFCs, or whether it will continue to shrink.

The ozone layer is particularly thin over the Antarctic and the effects of this can be felt even in Australia. But this doesn't discourage sun-seekers on Sydney's Shelly Beach.

Can solar winds be felt on Earth?

Increased solar winds occur in conjunction with sunspots. These 'winds', which consist of electrical particles, have nothing to do with air currents but their effects are noticeable. Solar winds cause more electrically charged particles to enter the Earth's atmosphere, where they interfere with radio communications. They also cause an increase in the occurrence of aurorae – the polar lights. The polar lights are not meteorological phenomena, but in some countries it is thought that they affect the weather. The polar lights are formed more than 60km above the

Is the air subject to tides like the sea?

The gravitational pull of the Moon and the Sun are powerful forces, responsible for the ocean tides that can rise and fall by several metres under their influence. The atmosphere is affected by them, too, but as air is about 800 times lighter than water, these 'tides' in the air are fairly weak. Compared to the power of the winds, their effect is almost non-existent. But air pressure can rise by minute amounts when there is a 'high tide' in the atmosphere, and drop when the 'tide' is low.

Does the Moon have an effect on the weather?

The Moon reflects sunlight onto the Earth, but it does not give off heat that would influence weather patterns. Its gravitational pull helps to move the planet's oceans, causing tides to ebb and flow and coastal water levels to fluctuate, sometimes by many metres. The air does not have sufficient mass for there to be any noticeable effect on atmospheric pressure (see above), so there is no direct link between the Moon and weather on Earth.

Is there any change in the weather when there is a new Moon?

A new Moon occurs when the Earth shields the Moon from sunlight. For centuries, people have believed that the weather is more changeable than usual when there is a new Moon. Research has found nothing to connect the weather and the phases of the Moon. During its remaining phases, the Moon reflects only visible sunlight and not the warming radiation that would have a major effect on our weather.

Do we get better weather during a full Moon?

A full Moon is only visible when there are few, if any, clouds in the sky. This has lead to a full Moon being associated with fine, clear weather but there is no meteorological basis for the belief. The full Moon isn't visible when the sky is cloudy. When there is a full Moon, it is full on the same night the world over and there has never been a period of fine weather that has affected the entire planet at the same time.

Does the weather change when there is a ring around the Moon?

According to folklore a ring around the Moon signifies bad weather. There is a grain of truth in this, since the bright ring around the Moon is caused by millions of tiny ice crystals associated with high-level cloud. These disperse the moonlight on its way to an observer on Earth, and provide a clear indication of moisture in the air. The halo shows that the air, which is oversaturated with moisture, has already sunk to lower levels and will probably bring a warm front and precipitation.

What causes the air pressure to be high or low?

Solar radiation and the rotation of the Earth keep the air masses in the atmosphere in motion. This causes high- and low-pressure areas to be distributed across the planet. A low-pressure area – also known as a cyclone, and not to be confused with the tropical system of the same name – develops when the surrounding air converges towards an area of low pressure at the surface and then rises into the atmosphere. The formation of low-pressure areas – known as cyclogenesis – occurs when air moves away from an area of high pressure and the air above subsides.

Why doesn't the wind blow directly from high- to low-pressure areas?

Air flows from a high-pressure area into one where the pressure is lower, as natural forces try to balance out the differences. The wind created will be weak or strong, depending on how great the differences are. The ideal direction for the air to flow, and the shortest, would be in a straight line from a high- to a low-pressure area. The rotation of the Earth gives rise to the Coriolis Force, and this prevents the air from moving in a straight line. Instead, the air in low- and high-pressure systems moves in a spiral. In the Northern Hemisphere, the winds in a low-pressure system spiral upwards, inwards and anticlockwise; in a high-pressure system they blow downwards, outwards and clockwise. The directions of air movements are reversed in the Southern Hemisphere.

High- and low-pressure areas determine the weather over wide regions.

Is the weather always going to be fine when the air pressure rises?

When the air pressure rises, the result can be an area of high pressure, also known as an anticyclone. Anticyclones move slowly and are more stable than low-pressure areas. A high-pressure system indicates a period of dry, cloudless weather as large masses of air sink downwards. As the pressure steadily increases tthe air to warms up. Droplets of water in the air evaporate and clouds dissolve.

High-pressure systems extend over wider areas than low-pressure systems, and can form a barrier against approaching lows. Clear weather can cause problems, when high pressure leads to heat waves and drought in summer or extreme cold in winter.

Where would you find the most intense high-pressure cell in the world?

The 'Siberian High' forms an area of consistently high atmospheric pressure. Reaching its greatest extent and intensity during the Northern Hemisphere winter, it covers a vast area of Siberia for months on end. It affects the weather patterns across much of the Northern Hemisphere, including Europe and northern Canada. Surface-level pressures frequently exceed 1040 hectopascals, and the highest atmospheric pressure ever officially recorded was 1083.8 hectopascals at the Siberian town of Agata on 31 December 1968.

Did Goethe invent the barometer?

The poet and dramatist Johann Wolfgang von Goethe (1749–1832) was not only famous for his contribution to literature, but also for his scientific interests.

Goethe had a device for measuring air pressure named after him – the Goethe barometer. This consists of a pear-shaped vessel with a long, narrow neck, open at the top and filled with liquid. If the atmospheric pressure rises, the descending air presses the fluid in the neck downwards. If the pressure falls, the fluid rises. Goethe did not invent this barometer, he only owned one; we do not know who did invent it. It was Torricelli (1608–1647) who developed the principle on which the first functioning barometer was based.

The Italian physicist Evangelista Torricelli demonstrated that liquid in a narrow tube rises until the weight of the column of liquid corresponds to the ambient air pressure – so discovering the principle of the barometer.

How much water can be stored in air?

Air can absorb a certain amount of water vapour without it condensing to form clouds or separating from the air as dew or rain. The more water vapour there is in the air, the higher the humidity; the hotter it is, the higher are the levels of absolute humidity that can be reached before the water condenses. At a temperature of 20°C, one cubic metre of air can absorb about 17g of water vapour. The point at which the air becomes saturated with water vapour is known as the dew point.

How is atmospheric humidity measured?

The amount of water vapour in the air is expressed as humidity. Absolute humidity is the amount of water vapour that can be held by air at a particular temperature, while relative humidity is the amount of water vapour in the air expressed as a percentage of the amount needed to saturate it at the current temperature. Humidity is measured using a hygrometer, with different types used to measure relative or absolute humidity. Hair hygrometers measure relative humidity. They contain a bundle of hairs that stretch when they absorb moisture. A finely tuned mechanism transmits the changes in hair length to a pointer on a calibrated scale. Human hair – especially blonde female hair – is particularly suitable because it is highly sensitive to fluctuations in temperature. Capacitive hygrometers use electricity to measure atmospheric humidity. Greater tension is created between two conductive sheets –

which form a condenser – when humidity is low than when it is high. The difference in tension and hence the relative humidity can be measured and read.

Why is it impossible to see through fog?

Fog is a cloud – or condensed water vapour – close to the ground. It is formed when finely scattered droplets of water float in the air. Depending on the temperature, air absorbs water, with about $30g/m^3$ being the maximum. If more water evaporates into the air, or if the air temperature drops so sharply that it can no longer hold the moisture in the form of vapour, the vapour condenses into fog. The droplets of water then cast an opaque veil over everything and visibility becomes poor. Officially, fog occurs when visibility drops below 1000m. Mist, haze and reduced visibility are preliminary stages to fog.

Foggy for more than 120 days a year, the Grand Banks off Newfoundland, Canada, is arguably the 'world's foggiest place'.

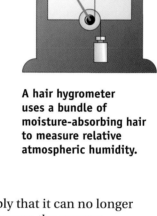

A hair hygrometer uses a bundle of moisture-absorbing hair to measure relative atmospheric humidity.

When visibility drops below 1000m, you are driving in fog - finely scattered droplets of water floating in the air.

How are vapour trails formed?

The long white strips seen criss-crossing the sky are caused by the exhausts of jet engines, and are made up mainly of water vapour. Aircraft engines also emit aerosol particles, and these act as nuclei around which the water vapour can condense. As the water vapour rapidly expands and cools, ice crystals are formed and it is these that make the vapour trails.

For vapour trails to form, the temperature of the air needs to be about -40°C, so the aircraft must be flying at altitudes of 10km or more. At this height, the atmosphere absorbs so little vapour that the water quickly condenses and freezes into small ice crystals.

Vapour trails are thought to contribute to global warming. They prevent some of the Sun's rays from reaching the Earth, and also block heat flowing from the Earth back into space.

Why are there so many different cloud forms?

Clouds occur mainly in the troposphere, the bottom layer of the Earth's atmosphere, at an altitude of up to 18km. Within this band there is a constant interplay of wind, solar radiation, air pressure, condensation and evaporation, which results in the great variety of cloud types. Clouds may form small piles or be scattered across wide areas by the wind. In most areas, certain types of cloud seem to predominate but all kinds of cloud formation can be seen almost anywhere on Earth. The study of clouds is known as nephology and forms part of meteorology.

Approximately 27 types of cloud can be found distributed across the layers of the atmosphere, ranging from small, fleecy white clouds to the threatening, dark mountains we see when storms loom. All clouds consist mostly of ice crystals and water droplets.

High clouds: 5–15km

Cirrocumulus Less common than cirrus clouds (see opposite), they are more fluffy and appear as thin, white patches, broad expanses or layers. They consist mostly of ice crystals, but also contain super-cooled droplets of water.

Cirrostratus Looking like a misty veil, these clouds cover wide stretches of the sky. Because of their great height they are almost entirely made up of ice crystals.

Low clouds: ground–2km

Stratus Drizzle often falls from this dense, thick type of cloud, which is made up almost entirely of small droplets of water. These clouds wrap themselves around skyscrapers and mountain tops and can persist for a long time in the lower layers of the atmosphere.

Stratocumulus Often referred to as fleecy clouds, they can vary considerably in size, shape and density. Made up of both small and large droplets of water, they form grey to whitish patches, broad expanses or layers that almost always include dark areas.

Cirrus Cirrus clouds are composed of ice and snow crystals. They appear as small spots, clusters or narrow, wispy bands. Cirrus often herald the approach of frontal systems.

Middle-layer clouds: 2–5km

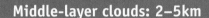

Altocumulus Drops of water predominate in this type of cloud. Many small clouds come together to form a regular pattern. Altocumuli also frequently appear as lenticular or almond-shaped banks. These stretch across wide expanses of sky and have clear outlines.

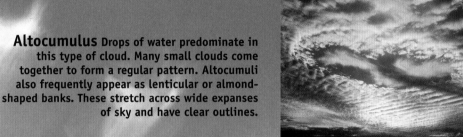

Altostratus Raindrops, small droplets of water, snowflakes and ice crystals can all be found mixed together inside these clouds. Altostratus clouds form grey or bluish expanses or thin, wispy layers that cover all or part of the sky. In places they are so thin that the Sun shines through, as if through frosted glass.

Vertical clouds

Cumulus Cumulus clouds can take a variety of forms, including hills, domes and towers. They consist mainly of water droplets and are always relatively dense and clearly defined. These clouds frequently produce showers.

Nimbostratus The Sun is rarely visible from beneath these clouds. They extend over several levels of the atmosphere, forming extensive, grey and often dark layers with frayed undersides. Persistent rain, snow or hail frequently falls from clouds of this type.

Cumulonimbus The lower part of this type of cloud is made up of water droplets and the upper part of ice crystals. They are the classic storm clouds, and their huge, mountainous, sometimes anvil-shaped formations penetrate through every layer of the atmosphere, bringing heavy showers. The usual name for cumulonimbus is thundercloud.

Is rain more than simply water?

Meteorologists distinguish between ordinary rain – with drops measuring between 0.6mm and 3mm in diameter – and drizzle – with drops measuring less than 0.6mm. When there is a shower, water is not all that falls to the ground. The rain takes dust and pollen from the air, as well as oxygen, nitrogen and acids, which erode stone and act as a fertiliser.

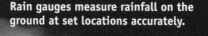
Rain gauges measure rainfall on the ground at set locations accurately.

Weather radar is a precise way of locating precipitation. It is used principally in aviation to help pilots locate areas of bad weather and avoid them. Meteorologists also use the radar to forecast potentially dangerous situations, such as floods, hail or snow. The radar works by emitting radio waves that travel up into the sky. If water is present, then some of the waves will encounter droplets and be reflected back. The greater the amount of radiation that returns, the more water droplets a cloud contains and the heavier the rain.

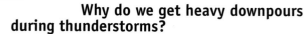

Why do we get heavy downpours during thunderstorms?

The torrential rain that often accompanies severe storms occurs as a result of updraughts inside clouds. These updraughts prevent small water drops and ice particles from falling, and instead they are carried back up into the cloud where they freeze and form more and more ice. When these ice crystals become too heavy for the updraughts to support, they fall as sleet or hail, or they melt and fall as large drops of water. At any one time, about 2000 thunderstorms are in progress across the surface of the planet.

What causes 'blood rain' to fall?

When red rain falls from the sky it leaves behind a reddish dust on the ground and can even give a pink tinge to snowy alpine peaks. In some areas, red rain is known as 'blood rain', a term that was used for the phenomenon during the Middle Ages. Scientists have found that sand from the Sahara Desert can be carried through the air for thousands of kilometres. Southern winds carry the desert sand across France and Italy all the way to Central Europe. If the dust combines with rain it pours down in the form of red rain.

How is precipitation measured?

A rain gauge is used to measure how much water falls from the sky – whether as rain, snow or ice. It consists of a container with a defined volume and no outlet. Rainfall is measured in millimetres. If 25mm of rain is measured in a rain gauge it means that the surrounding area would have been covered to a depth of 25mm if the water had not been able to disperse.

Huge hailstones like these occur when hot, humid air collides with cooler layers of cloud, causing unusually strong updraughts.

Did you know...

...that hailstones can travel at 90km/h?

In the summer of 1984, the inhabitants of Munich discovered just how dangerous hail can be. After a relatively cool start to the season, the city became very hot within just a few days. The temperature climbed to 37°C, but only a few days later a cold front moved across Germany, with catastrophic consequences. Hailstones the size of tennis balls beat down on Munich, causing damage worth over 1.5 billion euros.

	Speed	Size
Hailstones	25m/s (90km/h)	up to 100mm
Raindrops	8m/s (29km/h)	0.6–3mm
Snowflakes	0.8m/s (2.88km/h)	5–30mm

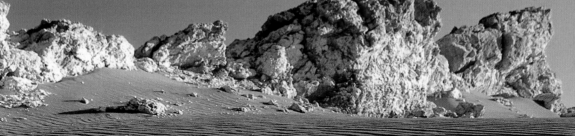

The dry Atacama Desert provides NASA scientists with excellent conditions for testing space probes designed to work on Mars.

Where are the driest regions on Earth?

No life can survive without at least some water. Although rain is rare in the Sahara Desert, the trade winds bring with them sufficient moisture to produce useful rain and therefore life.

The Atacama Desert in northern Chile is even drier. Some weather stations in the Atacama have not recorded even a single drop of rain. This is due to the desert's position in the rain shadow of the Andes. Easterly winds from the interior are dry and virtually no moist air blows in from the sea because the cold Humboldt Current that sweeps along the coast allows scarcely any water to evaporate. Despite the extreme dryness, some life forms – bacteria and lichens – have been discovered recently.

It is perhaps surprising that Antarctica is the driest continent. There is little precipitation and nearly all of it occurs as snow, although it sometimes rains in coastal areas. The average annual precipitation across the entire landmass is a mere 160mm.

What causes climate change?

The Earth's climate has undergone many changes. Ice ages have alternated with long periods of warm weather, while volcanic eruptions and meteorite strikes have also had an effect. Today, human beings have a major influence on the climate. The next few decades will see the planet warming by at least 2°C, caused by the burning of fossil fuels like coal, gas and oil and the resulting emission of greenhouse gases. In order to distinguish between climate change brought about by natural events and that caused by human behaviour, it is crucial to look back at the history of the Earth's climate.

Ice cores from the Antarctic and Greenland reveal a great deal about the history of the Earth's climate.

How is it possible to find out about the climate in the past?

Drilling into ancient layers of polar ice provides the most important insights into the history of the Earth's climate. The ice masses, which are thousands of metres thick, contain clues about the Earth's climate extending back for hundreds of thousands of years. Over this period, new layers of ice and snow have been deposited and in these are preserved evidence of the weather conditions at the time they were formed. Analysis of the cores began in the 1930s, and since then has proved to be the best source of evidence about the Earth's climate in the past.

Ancient coral reefs, bacterial cultures and fossils also provide insights into the history of the climate. Growth rings on trees and historical records of agricultural crop yields help climatologists analyse the more recent past.

Samples of coral provide information about changes in ocean currents over the past 100 years or more.

How does pollen help climate research?

Flower pollen contains a wealth of climate data. Because some plants need warmth, while others have adapted well to cold conditions, the types of plants that are able to flourish in a particular region depends on prevailing climatic conditions. Their pollen can show when, where and under what weather conditions plants were growing. Climatologists find traces of pollen in sediments, glaciers and the soil from ice-age bogs. They are also able to gather information about climate fluctuations that occurred over the past 150 000 years with the help of pollen that has sunk to the bottom of the sea.

What is the link between grape harvests and the Earth's climate?

It wasn't until the 19th century that people began to observe and accurately document the weather, but records of grain and grape harvests have been kept since antiquity. These reports have helped climatologists learn a lot about periods of drought, hailstorms and damaging frosts. The records kept during the grape harvest are precise because grapes are sensitive to temperature changes. French scientists recently discovered records of Burgundy's average summer temperatures dating back to the year 1370 while they were trying to establish when the traditional Pinot Noir grape variety was first harvested. With temperatures of 4.1°C above average, the summer of 1523 was found to be almost as hot as that of 2003.

Weather conditions dictate when grapes should be harvested, a fact that has revealed much about climate in the past.

The growth of a great worldwide network

The organisation that helps you to plan next weekend's picnic.

In 1657, the Grand Duke of Tuscany in Florence established the first organised network of weather stations. These were located in major cities across Europe, including Florence, Milan, Innsbruck, Paris and Warsaw. This was the forerunner of the meteorological networks that spread around the world over the next 200 years or so. The International Meteorological Organization (IMO) was formed in 1873, becoming the World Meteorological Organization (WMO)in 1950, an agency of the United Nations with its headquarters in Geneva, Switzerland. The WMO has over 180 member nations and is the umbrella organisation for all national weather services around the world. Through the patronage of the WMO, meteorological observations from hundreds of instrument networks are freely exchanged between nations. Often operating in hostile political environments, the WMO is a model of successful international cooperation.

Satellites and advanced modern weather stations, like this one in Germany, enable today's meteorologists to forecast the weather accurately several days ahead.

When did people begin to observe the weather systematically?

Polar-orbiting satellites – which operate close to the Earth – send back a constant stream of data.

Atmospheric pressure, precipitation, temperature, humidity, hours of sunlight, wind and cloud cover all need to be monitored and recorded for any serious study of the weather. This wasn't possible until suitable instruments were available, such as the thermometer (invented in 1592), the barometer (about 1643) and the hygrometer (1783). Comprehensive records of the weather began to be kept towards the end of the 19th century. Today, there are many thousands of professional weather stations all over the world, as well as countless simple weather stations run by amateurs. Modern meteorologists have access to a stream of weather satellite images, which enables them to make reliable predictions for several days ahead.

Why are weather forecasts sometimes inaccurate?

As highly complex interactions take place in the atmosphere, today's weather forecasts are remarkably accurate. Satellite pictures and data from a network of weather stations, coupled with information drawn from past experience and sophisticated computer simulations, make reliable forecasts about the coming three to five days possible. Forecasters are usually right about what and where, less so about when and how much. The further forecasters look into the future, the less precise their forecasts will be, since it is still only possible to see basic weather trends.

Where do weather balloons fly and why?

Tethered weather balloons are filled with helium, a lightweight, inert gas, which enables them to rise to altitudes of up to 30km. At heights greater than this the gas expands to such an extent that there is a risk the balloon will burst. Weather balloons carry numerous instruments aloft with them to measure temperature, atmospheric pressure, humidity and the composition of gases in the atmosphere at different altitudes. A radio module transmits the data back to meteorologists on the ground, who use it to draw up a vertical profile of the atmosphere. The data also allows scientists to better understand the atmosphere, and particularly the effects that humans may be having on factors such as ozone concentrations and the build-up of greenhouse gases. Once all the data has been collected, the weather balloon can be pulled back to Earth using its anchor line.

A weather balloon is prepared for its ascent at a weather station on the Norwegian island of Spitsbergen.

When Gunung Merapi on the island of Java – considered to be one of the world's most dangerous volcanoes – erupts, the effects can be felt all around the world.

How do volcanic eruptions affect the climate?

There are about 600 active volcanoes on Earth today. When they erupt, they endanger their immediate environment with streams of molten lava and red-hot rocks and also propel thousands of tonnes of ash and sulphur into the atmosphere. The volcanic particles are scattered across the globe by winds, and can block out the rays of the Sun for many years, leading to extended periods of cooler weather.

Can a volcanic eruption in South-East Asia affect the climate in Europe?

The effects of the eruption of Mount Pinatubo in the Philippines in June 1991 were felt all around the world. The 1500m-high volcano spewed out 17 million tonnes of ash and aerosols – suspended particles – into the atmosphere. The Sun was obscured by ash for several days afterwards, and the sulfurous aerosols lingered even longer in the atmosphere. About five per cent of the sunlight was reflected back into outer space, with the result that in 1991 global temperatures dropped by an average of 0.5°C. Climatologists also observed an increase in the reduction of ozone in the stratosphere, and the hole in the ozone layer above the South Pole reached record size. The ash cloud in the stratosphere that resulted from the eruption lasted for three years.

Which is more dangerous – a volcanic eruption or a meteorite strike?

This would depend on the size of the meteorite and the violence of the volcanic eruption. Both natural phenomena have the potential to fill the atmosphere with dust and aerosols and produce global cooling for years. The immediate damage caused by a large meteorite would certainly be greater. The meteorite that struck the Yucatan Peninsula in Central America 65 million years ago had far more catastrophic results than any single volcanic eruption. Large areas of the Earth were devastated and the climate was changed drastically. Because the event was followed by a cold period, the strike may have been responsible for the disappearance of the dinosaurs. However, the likelihood of such a huge meteorite bringing disaster to the planet is slight, whereas many volcanoes represent a constant danger.

How do oceans influence the weather and climate?

Without the oceans, the climate on Earth would be much drier and it is unlikely that any life could exist on the planet. While the larger connections between the climate and the water cycle are now understood, many questions remain unanswered about the interaction between the oceans and climate. We do know that it is not just the water itself that is important, but also the powerful ocean currents that supply life-giving warmth to entire continents.

What are El Niño and La Niña?

El Niño is a climatic phenomenon that occurs at intervals of between two and seven years in the Pacific Ocean, between the west coast of South America and the coasts of South-East Asia and Australia. The name El Niño is Spanish, and means 'the little boy', with the little boy in question being the infant Jesus. The phenomenon got its unusual name because its effects – which include heavy rainfall along the South American coast and severe droughts in South-East Asia and Australia – are particularly noticeable around Christmas time. The climatic effects are caused by changes in sea surface temperatures, such as when the Humboldt Current along the east coast of South America becomes increasingly weaker, allowing water that is warmer than usual to penetrate into these areas.

La Niña – Spanish for 'the little girl' – usually occurs immediately after El Niño. Here too there are changes in ocean currents, which cause the trade winds to drive warm surface water towards South-East Asia. This causes cold water from the ocean depths to well up along the Peruvian coast. La Niña can bring increased rain to South-East Asia and Australia and lead to extremely dry conditions in South America, although its effects are less severe than those caused by El Niño.

What happens when the seas become warmer?

One of the worst problems caused by global warming could be a rise in ocean temperatures. Warm ocean currents are beneficial, but warmer seas storing up vast amounts of heat could have a dramatic effect on the Earth's climate. Ice in the polar regions and Greenland would melt, causing sea levels to rise by several metres. The interchange of warm and cold water masses would decrease and some ocean currents could come to a standstill. At the same time, more water would evaporate and, in some parts of the world, this could lead to increased rainfall, with disastrous effects. Other areas would suffer water shortages, causing droughts in what are now fertile regions.

A loss of its ice mass through melting in areas such as Sermilik Fjord in eastern Greenland would have far-reaching effects on the global climate.

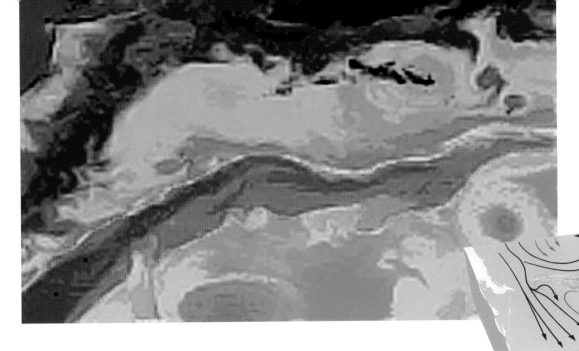

A satellite-based thermal imaging camera provides an impressive picture of varying water temperatures in the Gulf Stream (left). The diagram below shows warm and cold currents along the coasts of North and South America.

Why is the Gulf Stream called Europe's central-heating system?

For Europe, the most important ocean current is the Gulf Stream. American statesman Benjamin Franklin gave it its name because water temperature increases considerably in the Gulf of Mexico before flowing northwards along the east coast of the United States. As one of the world's most powerful surface currents, the Gulf Stream drives warm water across the Atlantic Ocean to Europe. At this point it divides: westwards towards Iceland and northwards towards the British Isles and Norway, before turning south-eastwards along the French and Iberian Atlantic coasts. The Gulf Stream is responsible for the mild climate in western and northern Europe because it transports about 1 billion megawatts of energy on its journey – equivalent to the output of a million or so nuclear power plants.

True or False?

The next ice age is on its way

Global warming could lead to cooler weather in Europe, and some climate experts believe that a new ice age could soon engulf the continent. The event that could trigger such a catastrophe would be the Gulf Stream coming to a standstill, thereby depriving Europe of the warmth provided by its waters. The Gulf Stream might cease to flow because the north-flowing water would no longer be cooled sufficiently, as a result of the seas becoming warmer. Without heavy, cold water, the force that drives the Gulf Stream would cease to operate.

What drives the ocean currents?

The Gulf Stream is only one of a network of surface ocean currents that flow around the world, all of which have a profound effect on temperature and rainfall patterns. These currents are generated by the prevailing global wind patterns as well as the spin of the Earth, but are also heavily influenced by the distribution of continental land masses. The main ocean currents in the Southern Hemisphere rotate in an anticlockwise direction and draw cool water northwards along continental west coasts and warmer water from topical latitudes southwards along eastern coastal areas. Cool surface currents flowing northwards along the tropical west coasts of southern Africa and South America suppress the development of tropical cyclone activity in those areas. In the Northern Hemisphere, the situation is more complex because of larger continental land areas, but several clockwise rotating ocean currents are still identifiable.

When the climate becomes threatening

In 2005, Hurricane Wilma left large areas of Havana under water. Many scientists warn that events like this will become more common as the planet heats up.

Global warming may also cause droughts, increasing the risk of dangerous forest fires.

We can already feel it. Summers are becoming hotter and the winters milder. Few experts now doubt that climate change is a fact and that we are beginning to feel its effects. The fourth and most recent report of the IPCC (Intergovernmental Panel on Climate Change), published in 2007, explains the serious consequences of the greenhouse effect. In the 21st century, average temperatures are set to rise by up to five per cent, or at least 2°C. Ice in the polar regions and mountain glaciers will melt, raising sea levels around the world. Over the next few decades, rivers in Central Europe will flood more often and coastal defences will more frequently be put to the test by storm tides. Rising sea levels are a threat to millions of people in low-lying countries, such as Bangladesh. Island states such as Tuvalu, Kiribati and the Maldives could be destroyed completely. Global warming will also lead to water shortages and an expansion of the deserts – even southern Europe is threatened by drought and crop failure.

The effect of climate change on the animal and plant world will have a creeping but critical impact. Germany's Federal Nature Conservation Agency considers between 15 and 40 per cent of all species to be in acute danger. While the effects of global warming are currently felt mostly by the world's poorer nations, the global economy is threatened by an enormous crisis. British scientists have calculated that climate change will consume between five and 20 per cent of gross worldwide domestic production. This is why they are calling for about one per cent of the money being earned in the world today – some 270 billion euros – to be spent on climate protection in order to avoid the worst consequences.

Will hurricanes become more common in the future?

The year 2005 was a record year for tropical cyclones – including Hurricane Katrina, which devastated the city of New Orleans. Climatologists confirm that the number of tropical cyclones has doubled since 1970. It is impossible to say whether this trend is linked to global warming – there is insufficient long-term data available – but a connection does seem likely.

Tropical storms are becoming increasingly violent and dangerous. Because the surface of the oceans is heating up, hurricanes, typhoons and cyclones are charged with ever greater amounts of energy. Sea temperatures in the tropics have already increased by about 0.5°C – and storm duration as well as wind speeds appear to have increased over the past 50 years.

What are some of the global effects of climate change?

Climate modelling has indicated that global warming will produce massive effects across all continents during the next 100 years. Predictions from the IPCC include increased water shortages across Africa by 2020, increased flooding for coastal

Asian regions by the 2050s and more heatwaves across North America throughout the 21st century. North America is also expected to experience a decline in snow-pack depth, increased flooding and a decrease in summer flows in many inland rivers. Massive vegetation changes are expected across South America, with large areas of the Amazon jungle giving way to grassland as rising temperatures lead to lower soil moisture. Australia is among the most vulnerable continents, with a lack of water affecting agricultural areas in eastern Australia and a simultaneous increase in the frequency of droughts and bushfires.

What is likely to happen when the polar ice caps start melting?

The melting of the ice caps at the North Pole and South Pole is already having a major impact on the Earth's climate. In Greenland, glaciers are shrinking at a rate each year that causes the sea level to rise by 0.1mm. According to NASA estimates, a complete meltdown of the ice in Greenland and Antarctica would raise sea levels by 70m, although such an event would only take place over hundreds of thousands of years. The melting of ice at the North Pole does not affect sea levels, however, because all its ice floats on the Arctic Ocean.

It would take hundreds of years to submerge half of Europe by floods, but climate experts believe that an increase in sea levels of between 40cm and 80cm in this century is

possible. This, and the simultaneous warming of the oceans, could severely disturb the global system of ocean currents. The Gulf Stream, Europe's central-heating system, could cease to flow. As a result, Europe could become considerably colder despite the general warming trend worldwide.

are already referring to the thousands of West Africans streaming into Europe via the Canary Islands as climate refugees. The UN estimates that 18 million people are currently on the move across the African continent, heading for the more prosperous south or for Europe. They are not escaping persecution or oppression,

Indonesian farmers are being forced to turn from growing rice to growing peanuts as their fields start to dry up.

Many people are already fleeing from the consequences of climate change, among them thousands of former New Orleans residents who will not return to their home town. Overall, it is difficult if not impossible to distinguish between 'climate refugees' and 'economic refugees'. Environmental organisations

but poverty and starvation caused by drought. These migrations will increase as global warming continues.

The 100 million people who now live in areas 1m above sea level could also become climate refugees. Many of them will be forced to seek a new home this century.

What do we know about the effects of the weather?

With lightning, dazzling displays of colour and devastating hurricanes, the weather can put on quite a show. Many phenomena remained a puzzle for a long time, and were often attributed to divine intervention. The actual causes involve a complex interplay of Sun, wind, the atmosphere and the nature of the Earth's surface. Explanations for some rare phenomena, such as ball lightning, have only recently been discovered.

Why does the sky sometimes change colour?

Without sunlight, we would not only be in darkness but the world would be devoid of colour. Sunlight appears to us as a pale whitish-yellow, but it is actually made up of the whole spectrum of colours visible to the human eye – the colours of the rainbow. The coloured components of the Sun's light becomes visible when it is broken up as a result of refraction and scattering in the atmosphere. During the day, the sky is predominantly blue because many different solid and gaseous particles in the atmosphere scatter the sunlight in a particular way. The British physicist Lord Rayleigh (1842–1919) was the first to come up with a conclusive explanation of the sky's blue colour – which is why the phenomenon is known as Rayleigh scattering.

Why are dawn and sunset skies often red?

The reddish tints seen in the sky at either end of the day also lies in Rayleigh scattering. As it passes through layers in the atmosphere, light with shorter wavelengths is scattered more than light with longer wavelengths. Blue light, with its short wavelength, is scattered more than red light. When the Sun is high in the sky, and the journey that sunlight takes through the atmosphere is relatively short, blue is predominantly scattered and the result is a blue sky. When the Sun is low, the light travels much further through the atmosphere and scattering reduces the blue content to such a degree that red predominates. Blue is literally scattered away, leaving a red sky at sunrise and sunset. Sunlit clouds and droplets of water in the air further intensify the red tint.

True or False?

A red morning sky is a portent of bad weather

'Red sky in the morning – shepherd's warning!' This old proverb tends to be accurate in mid-latitudes, because bad weather usually arrives from the west. A rosy dawn means that the morning Sun in the east is casting its light on rain clouds that are approaching from the west and will arrive within the next few hours. 'Red sky at night – shepherd's delight' also contains a grain of truth. The setting Sun in the west shines onto clouds in the east, an indication that a bad weather front has moved from west to east and there is a good chance of fine weather to come. While these proverbs are often accurate in Europe, North America and southern Australia, they do not apply in the tropics because weather systems in low latitudes travel from east to west.

Do yellow clouds spell approaching danger?

Clouds of sand or dust pose a threat to people with respiratory problems, and also adversely affect visibility for drivers.

When the clouds turn yellow, the cause is solid particles floating in the air. Huge clouds of flower pollen blown into the atmosphere or grains of sand or dust carried by strong winds blowing from arid areas give everything a yellowish tinge. Sand and dust storms can carry up to 100 million tonnes of suspended particles across wide areas. The pollen-impregnated clouds only pose a threat to people with allergies, and the sand or dust is harmless to anyone who doesn't suffer from respiratory problems – it is merely a nuisance when it coats cars and garden furniture with fine layers of dust. In some areas it is believed that yellow clouds herald hail. Thinly distributed grains of sand could promote the formation of hailstones, but no direct link has been established. Orange skies are also associated with bushfire smoke.

The colours of secondary rainbows appear in reverse order.

How are the colours in a rainbow formed?

The ideal conditions for a rainbow occur when a brief morning or evening shower is followed by a rapidly brightening sky. Large numbers of water droplets are still present in the air, and the Sun shines onto them from a relatively low position in the sky. This is why morning rainbows only appear in the west and evening rainbows only in the east. Rainbows are only ever seen on the side of the sky opposite to the Sun.

The technicolour display is the result of white sunlight being broken down by refraction into its component colours. This happens on the curved surfaces of the almost spherical droplets of rain, which is why rainbows have their characteristic arched shape. As is the case with white light broken down by a prism, blue light, with its shorter wavelength, is refracted at a greater angle than red light, which has a longer wavelength. This is why the colours appear in their familiar sequence: red, orange, yellow, green, blue, indigo and violet.

Sometimes it is possible to see another, weaker, secondary rainbow. This is created by the light waves being refracted a second time in the water droplets. With the secondary rainbow, the colours appear in reverse order to those of the primary rainbow.

Thunderstorms occur when warm air rises and collides with layers of colder air, forming dark thunderclouds.

What causes thunderstorms?

Rumbling thunder and flashes of lightning are the central features of thunderstorms. About 2000 of these violent weather events are occurring at any one time around the world. A thunderstorm develops when there are major temperature variations in the layers of the atmosphere. Warm moist air close to the ground rises vertically into colder atmospheric layers, especially after long periods of fine weather. This causes first droplets of water and then ice crystals to form. Friction develops between the ice crystals in the turbulent air currents, and water molecules bearing different electrical charges become separated. If this charge separation is great enough, it can create huge potential differences between areas that can amount to several tens of millions of volts. When a discharge occurs, the result is a flash of lightning. The thunder is caused by the sudden rise in temperature and the explosive expansion of air along the path taken by the lightning.

Why are thunderclouds dark?

The colour and brightness of clouds is mostly determined by sunlight and water. Finely distributed water droplets or ice crystals absorb, diffuse and reflect light. If a cloud is impregnated with a large amount of moisture that it is about to be released as rain, little sunlight can penetrate it and the cloud appears to be darker than usual. This is the case with thunderclouds, which are known to meteorologists as cumulonimbus clouds. These clouds often take on an anvil shape, formed by strong updraughts in the storm zone. Unlike many other types of cloud, thunderstorm cells break up quickly, usually after about an hour. Large cumulonimbus clouds, or supercells, are an exception. They can produce tornadoes, large hail and flash flooding and affect the overall weather patterns for much longer.

Where and when do most thunderstorms occur?

The days when observers at a weather station record at least one roll of thunder are defined as thunderstorm days. In Central European latitudes, meteorologists record up to 30 thunderstorm days per year, with the summer months of July and August seeing on average five times as many thunderstorms as during winter months.

Across the world, the frequency of thunderstorms varies widely. In the Midwest of the USA – an area known as Tornado Alley – thunderstorms occur frequently, and in tropical rainforests they occur almost daily. In the Polar regions, a peal of thunder is a rare event.

What effects do thunderstorms have?

When hot, sultry air gathers, the barometer falls and friction between clouds creates high voltage. This discharges as lightning, and lightning strikes are the most frequent natural cause of forest fires. Thunderbugs benefit from storms, because these insects make use of the favourable thermals created and let the air currents carry them over long distances. Micro-organisms also enjoy warm, humid air since they can breed in even greater numbers.

Does milk really turn during a thunderstorm?

Milk really can go sour after a thunderstorm. The culprit is not lightning but the warm and humid air that occurs during thunderstorms. This encourages bacteria in the milk to breed and make the milk curdle.

However, the souring process appears to be much more complicated as it can occur even behind the closed door of a cold refrigerator. Some experts think that electromagnetic impulses, or 'sferics' (short for atmospherics), can penetrate closed fridges and cause trouble. However, they still don't know exactly how this affects the bacteria in milk.

Why do insects fly low before a thunderstorm?

Many small insects, such as mosquitoes, hover in the air. When atmospheric pressure is high, the air is relatively heavy, giving insects sufficient upward impetus to allow them to venture into higher layers of air. Before a thunderstorm, however, air pressure falls and the insects drop to the lower layers. In hot, sultry weather, more small black bugs can be seen. Although they have wings, they hardly move at all under their own steam but use the wind and favourable thermals before thunderstorms to push them along. Because they are concentrated in the lower layers of air before a storm, they are known as thunderbugs. Their fringed wings gave them their scientific name: *Thysanoptera*, from the Greek *thysanos* (fringe) and *pteron* (wing). There are some 5000 species worldwide, variously known as storm flies, corn lice or thrips. They cause problems for gardeners, as they puncture plants and suck out the juice. Flowers and leaves then die more quickly and plants cease to flourish.

Thanks to the fine hairs on their wings, thunderbugs rise easily on an updraught.

Does lightning always travel downwards?

The Etruscans studied lightning bolts as far back as 700 BC because they used them to predict the future. It has only been in recent times, with the availability of scientific instruments, that the true nature of these massive electrical discharges has become apparent. One surprising discovery is that lightning doesn't usually travel from the sky to the Earth, but in the opposite direction. Discharges also often occur between clouds. At the end of the last century pilots and astronauts were able to confirm that lightning also occurs above the clouds, at heights of between 40km and 90km from the Earth's surface. Known as 'blue jets', 'sprites' and 'elves', these discharges produce impressive blue and red light effects, which are directed upwards as well as downwards.

How large is a bolt of lightning?

In a fraction of a millisecond, a bolt of lightning can discharge an electrical current of about 20 000 amps, with a field intensity of 200 000 volts per metre. Initial discharges, known as leaders, create a channel along which air molecules are ionised. Lightning channels can change direction several times, which is how the characteristic zigzag pattern of the main discharge is formed. Between thunderclouds and the Earth's surface, bolts of lightning reach an average length of between 1km and 2km. Between clouds, the discharge can extend for distances of up to 7km. In some cases, lightning has been known to travel up to 140km.

Does ball lightning really exist?

For centuries, many scientists refused to believe that there was such a thing as ball lightning. However, we now know that it exists. Like a luminous football, it flies through the air for a few seconds. Eyewitnesses have reported ball lightning lasting up to 30 seconds, but between two and eight seconds is more common. Ball lightning is not always round, it can also be oval or rod-shaped. The largest example ever seen measured between 15cm and 40cm in diameter. In nature, ball lightning is thought to occur after an ordinary bolt of lightning strikes the ground. This causes matter to melt and be propelled upwards in the form of a ball. In 1991, Japanese researchers were able to produce ball lightning in the laboratory for the first time. Recently, Brazilian scientists were able to keep a ball of lightning 'alive' for eight seconds.

This ball lightning was produced by a high-voltage discharge in a tank of water.

Can lightning be used in technical applications?

The amount of energy produced by the dazzling displays of light may seem huge but because lightning lasts for such a short time, a single bolt doesn't produce a large amount of usable energy. Scientists are looking for ways of producing cheap, environmentally friendly power from the 300 000 or so bolts of lightning that strike the Earth every hour – so far without success. However, the plasma from the hot charged particles that makes up each lightning flash can be used to develop new kinds of surface coatings. This type of controlled plasma – a quasi-lightning bolt produced and captured in the laboratory – may also form the basis for future fusion reactors.

Why is lightning attracted to metal?

A bolt of lightning is not attracted to metal, but is trying to find the easiest way to balance a voltage difference. It seeks out materials that offer the least electrical resistance. This is why metal – which is a good conductor of electricity – is struck more often than insulators, such as dry wood.

People in open spaces run the same risk of being struck by lightning as a metal rod does in the same place. This is because the shape and position of a lightning conductor is more important than the material it is made of. Anything that protrudes from the ground is more likely to be struck by lightning because of the shorter distance required to reach the discharge point.

Can the energy from lightning be harnessed?

Ideas have been put forward for lightning power plants. The energy could be used to heat water and the steam produced would then drive a generator, just as in a conventional power station. Alternatively, the flash could be used to charge a capacitor where the energy could then be stored. However, none of these suggestions has been put into practice. Although a bolt of lightning produces a few hundred kilowatts – enough to power a 100-watt light bulb for a few months – most of the energy is converted into heat and light in the lightning channel and so is no longer usable.

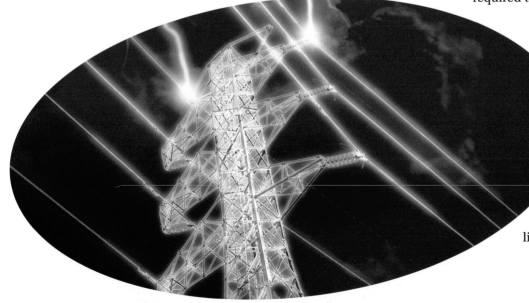

Are we wise to be afraid of lightning?

Few people are killed by lightning each year in developed countries. A century ago, the annual figure was much higher because many more people worked unprotected in the countryside. Open spaces, swimming pools and lakes should all be avoided during a thunderstorm. Most incidents involving lightning occur during outdoor pursuits, and current statistics show that men run almost twice as much risk of being hit by lightning as women.

Can fish be killed by lightning?

Fish are relatively safe since, unlike humans, they do not swim with parts of their bodies protruding above the water. The reason why human swimmers are at risk is because their heads provide a direct target. If a bolt of lightning strikes water, the lethal current density declines rapidly, and after a short distance can no longer do any damage. Fresh water conducts the electrical current about as efficiently as the body of a fish, so the current is distributed equally between fish and water. Salt water has even less electrical resistance, so most of the current will flow straight past any fish. But it is possible for fish in the immediate vicinity of a lightning strike to be killed. The electrical impulse causes the fishes' muscles to contract so violently that it breaks their spines, and it can also cause internal bleeding. For this reason, fish farmers protect their tanks with lightning conductors.

Can we really be safe from lightning?

According to an old saying, during a thunderstorm people should avoid oak trees and seek shelter under beech trees. However, because of their height, trees of any variety are among the objects most likely to be struck by lightning. Trees are also dangerous during thunderstorms because the current from the lightning can travel along roots beneath the ground, possibly causing serious burns and paralysis. Oak trees show more obvious signs of damage after they have been struck by lightning than beeches, which may have been the source of the old saying.

The best thing to do in a storm is to seek refuge in a building or vehicle. A closed car provides perfect protection because the current can be diverted around the outer shell. If neither a house nor a car is within reach, squat on the ground keeping both feet together. Do not touch any metal object, such as an umbrella. Cyclists and motorcyclists should dismount and move away from their bikes until the storm has passed.

How did Franklin invent the lightning conductor?

In 1747, American scientist Benjamin Franklin began experiments in the new and fascinating field of electricity

Franklin attracted attention with his theory that a flash of lightning was no more than a very large spark. To test his thesis, he constructed a kite. He believed that the electric charge from a bolt of lightning would travel down the kite's wet string to a metal key attached to its end. If Franklin's theory was right, it would then be possible to draw sparks from the key. The experiment was carried out successfully in 1752, although possibly not by Franklin himself and not in the manner described, which would almost certainly have had fatal results. Franklin concluded that the damaging effects of lightning could be diverted from a building by placing an iron rod on or adjacent to it and connecting it with the ground.

How is a tornado formed?

A tornado is a compact and often devastating whirlwind. In the build-up to a tornado, warm, humid air spirals upwards beneath a large thundercloud. The rotation of air about the tornado's central axis becomes faster and faster, like that of a figure skater. Finally, a funnel appears on the underside of the cloud, which gradually descends towards the Earth's surface. This creates a vortex, and as soon as this column of rotating air touches the ground everything in its path is hurled skyward. Scientists still don't know exactly what the mechanisms are that lead to the creation of a tornado. While their effects are usually limited to relatively small areas, tornadoes can form in almost any part of the world where heavy rain and thunderstorms occur. In the USA they occur most frequently in an area of the Midwest known as Tornado Alley, and it is here that they have been most intensively researched. In Tornado Alley, scientists can observe between 500 and 600 tornadoes each year. Because of the publicity given to American storms, the term tornado has gradually replaced the European term whirlwind.

How long does a tornado last?

A tornado is almost invisible at first because it has yet to gather up any particles of dust or water. Only when the pressure inside the vortex drops, causing it to cool, does water vapour condense and the storm start to churn up dust, debris, water and anything else in its path.

The appearance of a tornado varies, ranging from thin, tubular forms to funnels of varying widths. The lifespan of a tornado can be anything from a few seconds to more than an hour, with the average being about ten minutes. As the vertical winds ease, the tornado gradually dissipates and the column of air draws back up into the clouds.

How fast does a tornado rotate?

A tornado spout can measure from a few metres to as much as a kilometre in diameter. Its rotational movement is based on an increase in wind speed and changes in wind direction at different levels above the ground. Thanks to these differences, a strong, rotating updraught is created in the storm cell. At the centre of the lower section, air continuously streams in towards the axis of rotation. Thanks to the spin effect, the air gathers speed as it spirals inwards.

Meteorologists use a special radar system to measure the rotational velocity of air at the centre of tornadoes, and have discovered that speeds can approach 500km/h. These are among the highest wind speeds ever measured on the surface of the Earth. Only in the upper layers of the atmosphere are greater wind speeds to be found. Here, jet stream winds can reach speeds of 540km/h.

The speed at which a tornado moves across the Earth's surface is much slower. Depending on the prevailing winds, the average is about 50km/h. However, when the parent thunderstorm is embedded in a fast-moving air mass, tornadoes can race across the ground at speeds of up to 100km/h.

The arrows show the horizontal progess of a tornado (in yellow) and the rotation of air as it spirals upwards (in pink). Tornadoes tend to be accompanied by rain, hail and thunder. The world's deadliest tornado was probably during the storm that swept across Bangladesh in April 1989. As many as 1300 people were reported killed, with 12 000 injured.

Why are tornadoes so dangerous?

The high wind speeds at the centre of the storm are the reason for the destructive power of tornadoes. Surface winds inside intense low pressure cells and hurricanes can reach speeds of about 250km/h. Inside a tornado, wind velocities can be nearly double this. No motor vehicle or tree, and few buildings, can withstand such a force. Even a fragment of straw can become a dangerous missile when it moves fast enough to pierce a sheet of steel. Constant improvements in tornado warning systems have reduced the average number of victims in the USA to fewer than 100 a year. Half a century ago, tornadoes claimed an average of 250 American lives annually.

Can a tornado suck people out of buildings?

In principle, the intense low pressure in a tornado spout should suck people out of buildings. However, when a tornado moves over a house, the building is more likely to be blown to smithereens. When all the doors and windows are closed, the pressure inside the building cannot compensate for the rapidly falling pressure inside the tornado. Meteorologists don't know the precise pressure levels inside a tornado. Although Hollywood movies such as *Twister* have described the phenomenon, it has not been possible to gather data directly from the interior of a tornado.

What was the deadliest tornado to strike the US?

On 18 March 1925 the most destructive twister in US history rampaged across the three American states of Missouri, Illinois and Indiana generating winds estimated to have been in excess of 480km/h. The twister produced a path of destruction 350km long, smashing through several townships and killing 695 people. The death toll included 50 schoolchildren who were killed when the tornado tore apart two separate schoolhouses. The Tri-State Tornado, as it was later called, remains the benchmark of severity against which US tornadoes are measured.

A large tornado may be a fascinating sight, but few people want to see one at close quarters.

Where does wind come from?

Without wind there would be no weather, because it is wind that transports heat, moisture and energy through the atmosphere. Wind blows constantly around the planet – sometimes strong enough to cause damage, sometimes so gentle that it is almost imperceptible. Without wind, the continents would not shaped as they are today.

The main causes of wind are variations in atmospheric pressure between different air masses. Air from high-pressure areas streams into low-pressure areas, giving rise to a wind that does not cease until the air pressure is balanced. The greater the difference in atmospheric pressure, the more powerful the flow of air into the area of low pressure, and the stronger the wind created by this movement of air. Apart from local gusts, the direction of the wind is governed by the relative positions of low- and high-pressure areas, which generally travel from west to east around the globe.

Why are evenings so often windless?

An evening calm is a particular worry to sailors, since it occurs mainly in regions where areas of water are closely bordered by land. The cause is the force responsible for driving the wind – the Sun. As radiation reaches the Earth, land and water surfaces, as well as the atmosphere above them, are heated at different rates. During the day, the temperature over land is higher than it is over water. Warm air rises and moves over the sea at higher altitudes while cooler air above the sea flows towards the coast – in what is known as a sea breeze – creating a cycle that constantly seeks to establish equilibrium. After sunset, this cycle comes to a halt. However, because the water takes more time to cool than land, the process goes into reverse as cooler air flows from the land to the sea – known as a land breeze – warms up and rises to higher altitudes.

How fast is the jet stream?

The strong winds that streams constantly around the Earth at altitudes of between 8km and 16km are known as jet streams. This system occurs in the area between the upper troposphere and the stratosphere. Jet streams are driven by strong temperature differences in the upper atmosphere, together with the rotation of the Earth. They can reach speeds of 540km/h, even stronger than the wind speeds measured inside tornadoes.

Because jet streams blow in the same regions, aircraft on intercontinental flights make use of them. Although this means deviating from shorter routes, it reduces flying time and cuts fuel costs.

The regularity of jet streams was first recognised during high-altitude Second World War bombing raids.

A jet stream is made visible by this band of cloud streaming over Egypt and the Red Sea.

What is the Beaufort Scale?

As the Richter Scale is used to classify earthquakes, the Beaufort Scale is used to measure wind strength

It was while he was in command of a man-of-war that the British admiral Francis Beaufort (1774–1857) came up with the idea of using a simple, universal system to express the force of the wind. In 1806, he created what later became known as the Beaufort Scale, based entirely on observations of the effects of wind on land and sea. The scale ranges from Force 0 – calm, with flat seas – to Force 12 – hurricane, with huge waves and spray. The British Royal Navy officially adopted the Beaufort Scale in 1838 and its inventor was rewarded with a knighthood. On its centenary, the Beaufort Scale was adopted by the International Meteorological Committee. In 1946 it was extended to include a total of 17 categories, although the additional categories are only used in areas affected by typhoons.

How do meteorologists measure wind speed?

Wind is measured by an anemometer, a word derived from the Greek *anemos*, meaning wind. Anemometers record wind speed in either metres per second or knots (nautical miles per hour). Today, fixed velocity levels are categorised in accordance with the Beaufort Scale.

The simplest measuring system is based on a rotating fan wheel. The stronger the wind, the faster the fan blades spin. Wind speed can also be measured using dynamic pressure in a small tube. The faster the air stream that enters the tube, the more pressure it exerts on a gas or liquid. The same principle underlies the Pitot tube (named after its inventor, the French engineer Henri Pitot, 1695–1771), which is used for measuring the speed of aircraft.

Other measuring devices use ultrasound or the cooling effect of the wind on a heated wire to measure the speed of air currents. According to an internationally agreed standard, wind speed is measured at a height of 10m, because speeds on the ground can be subject to local variations.

Are the seasons gradually becoming more uniform or more extreme?

Summer, autumn, winter and spring define the time of the year for many areas of the world. Closer to the equator, these seasons become less well defined and are replaced by a more uniform temperature pattern, upon which is imposed an emerging wet-season/dry-season cycle. This tropical regime seems to be spreading further from the equator and moving into mid-latitudes. This is reflected in rising temperatures across many continents, together with decreasing summer snow cover in some alpine areas. Some climatologists predict that the traditional seasons will become less distinct as time goes on, with only high latitude areas still experiencing the familiar four-season cycle by the end of the 21st century.

How do the seasons come about?

The tilt of the Earth's axis on its orbit around the Sun is responsible for the seasons. Our planet's rotational axis is not perpendicular to its orbit but tilts at an angle of about 23.5°. When the North Pole is tilted towards the Sun, it is summer in the Northern Hemisphere and winter in the south. When the South Pole is tilted towards the Sun, the opposite is the case. However, the Earth's orbit and the tilt of its axis are subject to constant change. This can alter the amount of sunlight reaching the Earth and affect its climate.

A group of French astronomers have predicted that the Earth's axis will shift by 0.4° over the next ten million years. Their forecast is based on sophisticated computer simulations of the movement of the Earth over a period of 500 million years. Such a change in the angle of the Earth's axis could lead to a noticeable shift in the seasons – in about ten million years' time.

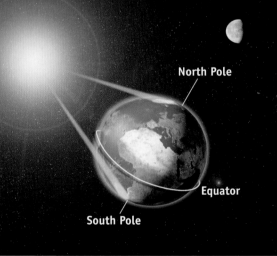

Every cloud has a silver lining. Sunbathers in northern Germany make the best of an unseasonably warm winter in January 2002.

When the North Pole is tilted away from the Sun it is winter in the Northern Hemisphere and summer in the south.

Which areas have the greatest temperature fluctuations?

In the course of a day, the greatest temperature fluctuations often occur in the Earth's great deserts, such as the Atacama in South America and the Sahara in Africa. The ground is heated near the surface only because it is covered in loosely packed quartz sand, which does not conduct heat efficiently. In addition, dry desert soil can store far less heat than damp soil.

The lack of cloud during the day means that the rays of the Sun are rarely interrupted before they strike the ground, with the result that temperatures can rise as high as 70°C. At night, the small amount of stored heat can radiate away unhindered back into the atmosphere. Because of this, temperature differences of 30°C in a day are common, with extreme fluctuations of up to 50°C possible.

The greatest recorded temperature fluctuation over an extended period of time was on the plains of Siberia, where the town of Verkhoyansk has recorded a temperature fluctuation of 104°C. The highest temperature measured there reached 36°C, and the lowest –68°C. Verkhoyansk also holds the record for the lowest temperature ever recorded in Siberia.

Are there seasons on the equator?

The closer you are to the Earth's equator – a 40 075km-long imaginary line that stretches around the centre of the planet – the fewer differences there are between the seasons. The tilt of the Earth's axis has little effect on the climate in this area and there are few significant temperature variations during the year. There are rainfall variations within equatorial regions, with a wet-season/dry-season cycle emerging as you move away from the equator.

Places along the equator are hotter because the amount of solar radiation that reaches the ground is greater. Atmospheric pressure is low over warm areas and, as a result, low pressure areas and violent thunderstorms in the middle of the day are common occurrences close to the equator.

The meteorological equator is not located at precisely 0° latitude but at about 5° north. The reason for this shift northwards is the uneven distribution of land and sea on the planet. The larger land masses in the Northern Hemisphere store more heat.

True or False?

The Earth is closest to the Sun in summer

On 22 June each year the Sun is directly overhead the Tropic of Cancer, and people living in the Northern Hemisphere experience the summer solstice. This corresponds with the winter solstice in the Southern Hemisphere. The situation is reversed on 22 December, when the Sun is directly overhead the Tropic of Capricorn. During the summer solstice, that hemisphere is closer to the Sun than the hemisphere experiencing its winter solstice.

How does the weather affect our health?

From early times, people have suspected that there is a link between the weather and human health. The Greek physician Hippocrates wrote a treatise on the subject in about 400 BC called *On Air, Waters and Places*. It examined the effect of wind and temperature on human reactions and the effect of seasons on disease pandemics. Since then, evidence has accumulated that connects the weather with several diseases, such as arthritis, rheumatism, bleeding ulcers and coronary thrombosis. While the weather is not a cause, it may act as a trigger in susceptible individuals. There is a statistical connection that links the onset of spontaneous labour in childbirth with falling atmospheric pressure – more women go into labour when a cold front or low-pressure cell is approaching. No universally accepted theory has been proposed to explain this.

Is seaside air really as healthy as many claim?

Sea air really is good for the health. The bracing climate by the sea is especially beneficial to the respiratory organs and the skin, but also improves circulation and strengthens the body's defences. Doctors recommend seaside climate therapy for asthmatics and people with allergies or skin problems. The healing powers of a bracing climate are particularly effective in autumn and winter, because it is at these times of year that people with respiratory problems are most affected by the dry air caused by central heating, damp cold and air pollution.

Sea air contains tiny droplets of sea water enriched with salt, iodine, magnesium and trace elements, which are scattered into the air by wind and waves. Known as surf-generated aerosols, they stimulate immune reactions by the skin and respiratory organs. A period of inhalation will result in the loosening of the mucus in the respiratory tracts, which makes coughing easier and so helps clear airways. Sea air is almost completely free from harmful vapours, exhaust fumes and soot particles, especially on offshore islands.

Who is sensitive to the weather?

Weather sensitivity involves changes in personal feelings of wellbeing that are highly subjective. Experts in medical meteorology see increased weather sensitivity in people who are affected by changes in the weather. The autonomic nervous systems of these individuals have a lower stimulus threshold. As a result, they experience a stronger physical reaction to weather and climate. Some scientists believe that this is an ancient protective reflex that once forewarned people and animals of changes in the weather. Depending on its severity, a disproportionate reaction can lead to tiredness and lack of concentration. Complaints linked to the weather can resemble symptoms of disease, although a physician is unlikely to find any pathological changes to the affected organs. Weather-related illness must therefore be considered a symptom of a weakened organism that cannot adjust to atmospheric changes. In some European countries, 30 per cent of the population is thought to be sensitive to the weather. Training and lifestyle adjustments to can reduce this.

Is it possible to issue 'health forecasts'?

As the statistical connections between human health issues and the weather became better known, will it be possible to forecast human health trends by using these statistical connections? This area of research, known as biometeorology, has been found useful in general health planning. By processing weather forecasts in various ways, 'health forecasts' can be issued for periods of up to four days ahead, identifying peak danger periods for a variety of human conditions

believed to be weather sensitive. These include mood swings, sleep patterns, migraines, blood pressure, physical performance and respiratory diseases. As well as alerting individuals to the development of these danger periods so that optimum medication can be planned, health authorities can prepare themselves to provide improved quality of care and reaction time to incoming patients.

Weather forecasts may soon be accompanied by 'health forecasts', especially in large cities where poor air quality from pollution can be a serious concern for the vulnerable.

True or False?

Old wounds hurt when the weather changes

Weather changes, especially when it turns colder or damper, cause old wounds to hurt more. Many people with rheumatism and osteoarthritis are familiar with this phenomenon, and a US study has furnished proof of it. A comparison of patient data with weather data showed that when air pressure fell, many patients had more severe pain in their knees. A drop in temperature can also cause pain, although it is usually less serious.

Record-breaking weather

Statistics prove that extreme weather events have become more frequent in recent years. There is heated international debate as to whether or not these changes in the climate are due to human activity. However, we need to be prepared for challenging times ahead as it seems likely that some of the records listed below will be broken.

Which place holds the record for the highest temperature ever recorded?

The highest air temperature ever recorded (at 2m above the ground) was 57.3°C at El Azizia in Libya.

What are the hottest temperatures recorded on other continents?

North America: 56.7°C in Death Valley, California, on 10 July 1913.
Asia: 53.9°C at Tirat Tsivi, Israel, on 22 June 1942.
Australia: 53.3°C at Cloncurry, Queensland, on 16 January 1889.
Europe: 50.0°C at Seville, Spain, on 4 August 1881.
South America: 48.9°C at Rivadavia, Argentina, on 11 December 1905.
Oceania: 42.2°C at Tuguegarao, Philippines, on 29 April 1912.
Antarctica: 15°C at Vanda Station, Scott Coast, on 5 January 1974.

Where is the coldest place on earth?

The lowest temperature of -89.2°C was recorded at the Russian Vostok Station in Antarctica on 21 July 1983.

What are the coldest temperatures recorded on other continents?

Asia: -67.8°C at Oimekon, Russia, on 6 February 1933.
North America: -63.0°C at Snag, Canada, on 3 February 1947.
Europe: -55°C at Ust'Shchugor, Russia, date unknown.
South America: -32.8 at Sarmiento, Argentina on 1 June 1907.

Africa: -23.9°C at Ifrane, Morocco, on 11 February 1935.
Australia: -22.2°C at Charlotte Pass, New South Wales, on 29 June 1994.
Oceania: -11.1°C at Mauna Kea, Hawaii, on 17 May 1979.

How much rain can fall in a day?

Nearly 2m of rain (1870mm) fell between 15 and 16 March 1952 on Reunion Island in the Indian Ocean to the east of Madagascar.

Which region has the greatest number of rainy days?

There is a reason why tropical rainforests are given that name – it really does rain on just about every day of the year. The highest average annual rainfall occurs on Mount Waialeale on the Hawaiian island of Kauai, where meteorologists have recorded an impressive 11 684 mm/year.

Where does the least rain fall?

The regions with the lowest rainfall are the world's deserts. At Calama, in Chile's Atacama Desert, meteorologists estimate that not a single drop of rain has fallen for more than 400 years. While not totally devoid of water in the form of windblown snow and ice, some parts of Antarctica's remarkable Dry Valleys – situated on the edge of the continent beside McMurdo Sound – have received no rain for an estimated two million years.

Which place has received the most rain in a single year?

The highest annual rainfall – 26 461mm – was recorded in Cherrapunji, northern India, between August 1860 and July 1861.

How long was Europe's longest period of fog?

For more than 10 days (242 hours) in May 1996, the fog did not lift at the Neuhaus weather station in eastern Germany's Thuringian Forest.

The summit of the 1569m-high Mount Waialeale on the Hawaiian island of Kauai is the rainiest spot on Earth.

What is the record speed for a wind gust?

On 12 April 1934 a gust of 416km/h – faster than a Formula 1 racing car – swept over Mount Washington in the American state of New Hampshire.

What is the strongest sustained wind ever recorded?

Although gusts of wind are stronger in the short term, a strong, sustained wind has much greater effect. On 12 April 1934, a wind continued to blow at 372km/h for 10 minutes above Mount Washington in the state of New Hampshire, USA

When was the highest atmospheric air pressure ever recorded?

A record atmospheric air pressure of 1083.8hPa was recorded on New Year's Eve 1968 at Agata in north-western Siberia.

How low can the air pressure fall at mean sea level?

The atmospheric pressure at the centre of Typhoon Tip fell to 870hPa on 12 October 1979, as the storm raced across the Pacific Ocean to the west of Guam.

How big were the biggest hailstones?

The largest and heaviest hailstone ever recorded weighed 10kg and landed during a thunderstorm in a Swiss suburb. Meteorologists believe it impossible for a naturally formed hailstone to be that large and heavy, so it seems more likely that the giant stone was formed from water falling from an aircraft flying at high altitude.

How thick was the thickest blanket of snow?

The record for the greatest amount of snow in one day is held by Silver Lake, Colorado, USA, where 1930mm of fresh snow fell in a single day in April 1921.

Where did the greatest amount of snow fall in one year?

In the winter of 1971–72, Paradise Ranger Station on the slopes of Mount Rainier in Washington State, USA, recorded 26.1m of fresh snow. The same region also holds the record for the greatest amount of snow to fall over a whole year, with 31.1m falling between February 1971 and February 1972.

Which place enjoys the longest period of sunshine in a single day?

The Arctic and Antarctic are the leaders when it comes to this record. During the polar summers, there can be sunshine for 24 hours a day for about six months (depending on how far you are from the pole), provided there are no clouds. The downside to perpetual summer Sun is permanent winter darkness.

In Browning, at the foot of the Rocky Mountains, the temperature fell overnight by 55°C.

What is the record for the highest average annual temperature?

Even in the winter months it is hot around the Ethiopian city of Dallol. In the 1960s, a world record-breaking annual average temperature of 34.6° C was recorded.

Where were the fewest hours of sunshine recorded in a single month?

This record is held by London, where the Sun shone for a mere six minutes during the whole of December 1890. The lowest annual average hours of sunshine was recorded in the sub-Antarctic South Orkney Islands, where the Sun shines for 478 hours a year. Most of the time it is obscured by cloud.

Which place can boast the biggest temperature variation in a single day?

Browning, in the US state of Montana, holds the record for the greatest temperature variation in a 24-hour period. On 27 January 1916, the temperature stood at 6°C until very cold air swept down from the Arctic, causing the temperature the next morning to drop to –49°C, a difference of 55°C.

Is the face of the Earth changing?

Planet Earth has gone through many changes in the course of its history, and the constant process of change continues, albeit slowly. Mountains are growing because of the movement of tectonic plates, and about 600 active volcanoes have the capacity to devastate their surroundings. However, it is not only nature but also climate change brought about by human activity that now affects the growth of deserts and the migration of vegetation zones.

How quickly do mountains grow?

The Earth's mountain ranges are growing – although they are shrinking at the same time. When tectonic plates collide, mountains are forced upwards by as much as a few centimetres each year. This applies particularly in the case of 'young' mountains like the Alps and the Himalayas. At the same time, wind and rain erode soil and rocks from mountain summits.

How long would it take for the natural elements to erode a mountain chain?

Several hundred million years will have to pass before wind and weather erosion cause mountains to vanish completely. Research has shown that the Ural Mountains – which divide Europe from Asia – were more than 4km high 350 million years ago, and even now they are still more than 2km high. Some geologists estimate that, were it not for erosion, mountains would rise to a height of more than 30km, nearly four times that of Everest.

Although no one alive today will live to see the end of a mountain range, some spectacular changes can be observed in a single lifetime. There has been an increase in rockfalls, and only recently large sections of the famous Eiger in Switzerland collapsed. The cause of such dramatic change is probably global warming, which is pushing back the height at which ice cements together fragile, shattered rocks.

How were the world's highest mountains created?

Colliding continental plates are responsible for the birth of the world's highest mountain range, the Himalayas. After the break-up of Pangaea some 200 million years ago, the Indian, Arabian and African continents began travelling northwards, slowly closing the Tethys Sea until they slammed into the Eurasian continent. The titanic forces gradually pushed rocks skywards along the line of collision, resulting in a mountain belt stretching for some 10 000km, including the Himalayas, Hindu Kush, Zagros Mountains, Alps and the Atlas Mountains.

The rocks making up these mountains have recorded this cataclysmic event. The hard, metamorphosed sandstones, shales, limestones and dolomites were once sea floor sediments abounding in the fossils of shellfish and other marine organisms. The collision and mountain-building process is still going on today, which is why the mountains remain so high.

The world's mountain ranges – such as the European Alps seen here – are evidence of the titanic forces shaping the face of our planet.

Why don't all mountains look the same?

Two geological processes – the collision of tectonic plates and volcanism – are responsible for the huge variations in the formation of mountains. Mountains in the 'crumple zones' between colliding tectonic plates are usually sheer, towering masses of rock. Those that are formed as a result of volcanic eruptions have conical peaks. Over time, wind and weather erode the different types of rocks from which mountains are formed. It takes millions of years for erosion to smooth down once jagged mountains to form gently rounded peaks, like those of the Urals and Appalachians. Geologically younger ranges, such as the Andes, Himalayas and Alps, still retain steep slopes and jagged rock formations.

Are we about to be engulfed by desert?

The world's deserts currently cover around one-fifth of the Earth's land surface – nearly 30 million square kilometres. Add to this other dry regions known as semi-deserts, and the total area adds up to one-third of all the land on the planet. Despite measures to halt the spread of deserts, erosion and global warming are causing these dry regions to expand ever more rapidly – by up to six million hectares each year.

Why are the Earth's deserts growing?

The fact that the deserts are expanding is almost entirely due to human activity. Overgrazing, unsustainable agricultural practices and deforestation are turning once fertile land into steppe, and then gradually into desert. Without a protective network of plants, fertile soil is easily attacked by the erosive effects of wind and rain. Even the water that gathers after a heavy downpour of rain can no longer be stored, and instead it simply washes away a further layer of fertile topsoil. The natural causes of desertification are long periods of drought.

The growth of deserts is depriving millions of people of their livelihoods, which is why the United Nations has now put in place a large number of programmes to fight the further expansion of these hostile landscapes.

Can desert be transformed into a landscape of trees and flowers?

With sufficient water, new sources of energy and clever planting techniques, even the desert can be made green again. However, calls for major financial investment. The oil-rich states of the Arabian Peninsula are currently putting their money into projects aimed at reclaiming the desert. The water used to make the desert fertile again comes mainly from the sea, and has to be

How do we know...

... that sand dunes can travel long distances?

Dunes are often crescent-shaped. The wind carries sand particles a short distance and then drops them along the edges of the dune. This creates a crescent-shaped dune, which is also known as a barchan dune. The speed at which wandering dunes can travel depends on the speed and direction of the prevailing winds and the size of the grains of sand. They usually move between 8m and 10m per year. Even in Europe, wandering dunes in Lithuania and Poland have covered large swathes of land, smothering all life in their path.

Built in 1900, Rubjerg Knude lighthouse in Jutland, Denmark, is gradually being engulfed by shifting sand dunes.

In many parts of Africa, huge investments are being made to make the desert bloom. Here water is being pumped out of the ground into an oasis.

Deserts are regions with extremely low rainfall, inhospitable to all except specially adapted forms of life

Size	Desert	Continent
13 200 000km²	Antarctic	Antarctica
8 700 000km²	Sahara	Africa
1 560 000km²	Australian Deserts	Australia
1 300 000km²	Arabian Desert	Asia
1 040 000km²	Gobi	Asia
900 000km²	Kalahari	Africa
330 000km²	Taklamakan	Asia
320 000km²	Sonora	North America
273 000km²	Karakum	Asia
273 000km²	Thar and Cholistan	Asia

processed in desalination plants before it is used. Libya is putting its money into pumping out ground water, huge quantities of which are hidden under the Sahara's oases. When the 'Great Man-Made River' is completed, Libya hopes it will supply five million cubic metres of water each day to the fields, towns and cities on its Mediterranean coast. This means that the water table will sink, and there is a possibility that the desert's few surviving springs will dry up. In Iran, they have been slowing down the rate of evaporation from sand dunes by sprinkling them with oil, which prevents water vapour from escaping. One day plants will be able to grow on the dunes for the first time. More than 60sq km of so-called petro-forest has already been grown near the town of Ahwaz.

Is it possible to stop the growth of deserts?

Human beings are mostly to blame for the expansion of the deserts, and it is humans who can stop it – but only if the root causes are tackled. Measures have been taken to stop deforestation, and many reforestation projects are now underway to try and reverse the damage. Efforts are being made to use water and agricultural techniques more effectively and more sustainably.

Will Europe turn into a desert one day?

In southern Europe, the risk of desertification has increased considerably. According to a study conducted by the UN, about 12 per cent of Europe's land mass is acutely threatened with desertification. Mediterranean countries are witnessing the most rapid desertification. In Spain, climate change has led to increasingly frequent water shortages during the summer months.

The process of desertification in Europe is a result of excessive cultivation of the soil, massive deforestation and inadequate irrigation systems. Climate change is accelerating the process by making these areas more arid. In the USA, about one-third of the land area is threatened with desertification, and worldwide there are already some 250 million people affected by encroaching deserts that are gradually consuming what was once fertile land.

Do islands always stay the same?

Erosion is constantly changing the appearance of islands, while rising sea levels mean that flat islands, especially in the South Seas, are in danger of disappearing beneath the waves. Volcanic eruptions can also destroy island landscapes, but on rare occasions volcanism can lead to the formation of new islands.

Are new islands still being formed today?

Although the Earth has become a less dynamic place over the millennia, new islands can spring up even today. They are created by active volcanoes, like the one that is forcing fresh masses of rock upwards in the Hawaiian Islands. A new island will rise out of the waters of the Pacific there one day, but not for several thousand years.

In 2006, the yachtsman Frederik Fransson spotted plumes of steam in the Pacific Ocean near the island of Tonga. They were rising from an island that was not marked on any map. This volcano had surfaced twice before – in 1984 and 2004 – but on both occasions it had been washed away within a few months. Perhaps this time the island – named Home Reef – will manage to stay permanently above the surface.

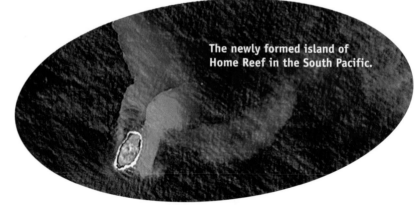

The newly formed island of Home Reef in the South Pacific.

Most new islands are created by artificial means, such as the three huge archipelagos now being constructed off the cost of Dubai. Commissioned by the Crown Prince, the project involves shifting several hundred million cubic metres of rock and sand at a cost that runs into billions of dollars. One archipelago – in the form of a palm tree – is nearing completion, while the other two, one of which is in the form of the continents of the Earth, are not due to be completed until 2015.

Celebrities such as David Beckham and Boris Becker have bought homes on The Palm, Dubai's vast, artificially constructed archipelago.

It would be a great pity if beautiful islands, such as this one in French Polynesia, were to vanish beneath the waves.

Are coral reefs doomed to disappear?

Corals belong to the class of animals known as anthozoans, which are famous for the colourful colonies they create. However, scientists are witnessing the worldwide destruction of coral reefs. More intense ultraviolet radiation, an increased concentration of carbon dioxide in the seas and the warming of the oceans are causing corals gradually to bleach and die. More than two-thirds of the world's coral reefs are threatened by climate change. Some are already severely damaged and one-fifth has already been destroyed.

In order to protect these habitats, which are home to an extraordinary variety of species, factors that are having a harmful effect on the oceans must be reduced. There are even proposals for an awning to be erected along large stretches of Australia's Great Barrier Reef to cool the waters surrounding the world's largest coral reef. However, experts are sceptical about the idea because it could disrupt the flow of warm currents.

What is the difference between an atoll and an island?

Atolls are islands, but of a particular kind. They take the form of a circular coral reef – a string of narrow islands – surrounding a lagoon. Atolls almost always begin life as a volcano. When molten rock rises from the sea as a volcanic island, corals settle on it and gradually their skeletons form a reef. As the volcano slowly sinks beneath the sea under its own weight, the coral continues to grow upwards towards the sunlight, forming an atoll.

Atolls are generally found in tropical regions of the Pacific and Indian oceans. The Maldive Islands are composed entirely of atolls, and the largest atoll is Christmas Island (388sq km) in the Pacific island state of Kiribati. Rising sea levels and declining coral reef growth may result in atolls disappearing altogether.

The Morning Glory Pool in Yellowstone National Park is a hot spring created by volcanic activity.

Volcanoes – spectacular and dangerous

Molten lava often streams from Mount Mayon in the Philippines. The last major eruption was in 1993.

Since the Earth was formed around 4.6 billion years ago, it has gradually cooled down and its rocks have hardened – but only on the surface. Internally, a vast mass of molten rock – or magma – is still churning away. Frequently magma finds its way to the surface in the form of red-hot lava via the world's approximately 600 active volcanoes.

Why doesn't the magma stay in the Earth's interior?

Most volcanoes are located where oceanic and continental plates meet. As these plates float on the molten rock of the Earth's interior they collide with one another (see p. 256), releasing huge amounts of energy as one tectonic plate is pushed under another. The rocks of the lower plate melt in the Earth's hot interior, creating even more magma. Mixed with gases, it accumulates until the pressure becomes so great that it takes the only available way out – through fissures in the Earth's surface.

Less frequently, new volcanoes are created by so-called 'hot spots'. These occur when molten rock in the Earth's interior is particularly active at one point under an apparently stable plate, and must find an escape route to the surface. The Hawaiian Islands are an arc of volcanoes created in this way. All the islands originate from a single hot spot, but because the Pacific Ocean's tectonic plate has been moving gradually across the spot for millions of years, the magma has had to keep finding new ways to reach the surface. This is why the islands are strung together like the stones on a necklace. As long as the hot spot continues to supply enough magma, new volcanoes and even new islands will be created.

How dangerous are super volcanoes?

Super volcanoes emit up to 100 times more magma than ordinary active volcanoes. When they erupt, they can devastate entire continents and change the Earth's climate for decades afterwards. Super volcanoes are also deceptive because they are barely discernible from the surface and don't form the familiar cone-shaped mountain usually associated with a volcano. Instead, after an eruption, the Earth's crust simply subsides.

The world has experienced the eruption of at least two super volcanoes in recent times. Taupo, in New Zealand, exploded 26 500 years ago, while the eruption of Toba, on the island of Sumatra 74 000 years

ago plunged the world into darkness for six years. The Earth's climate became much cooler, and genetic studies have shown that as few as 5000 or 10 000 human beings survived the disaster.

Today, there are still some super volcanoes that lie dormant beneath the Earth's surface, and geoscientists cannot rule out the possibility that they may come back to life one day. The gigantic magma chamber beneath the super volcano in Yellowstone National Park in the USA is 60km long, 40km wide and 10km deep. It contains about 24 000m³ of molten rock and experts agree that an eruption is overdue. According to British geoscientists, there are two massive volcanoes in Europe that are also due to erupt. One is situated in the Phlegraean Fields near Naples, the other in the eastern Mediterranean close to the island of Kos. Scientists are calling for an international monitoring system equipped with many sensors to give sufficient warning of an impending catastrophic eruption.

Can eruptions be predicted?

Every volcano is different, which makes their behaviour difficult to predict. Geoscientists are steadily improving their ability to assess the risk of an eruption. It is helpful to look at a volcano's history, since eruptions tend to occur in a more or less regular pattern. Sensors are also used to analyse small earth tremors around a volcano to look for

signs of an impending eruption, while detectors are placed on a volcano's slopes to measure the composition of the gases emerging from fissures. Mount Etna on the island of Sicily is the subject of a permanent study into this.

Despite the work being done, precise predictions are still a long way off. Fortunately, some volcanoes announce an impending eruption by emitting increased amounts of smoke and ash, and this allows people to leave the danger zone in time.

Why is volcanic soil so fertile?

A volcanic eruption leaves the landscape looking scorched and devoid of life. However, it will not take long for new life to emerge. The ash emitted from a volcano contains valuable plant nutrients and lava is rich in phosphorus, potassium and calcium. Also, volcanic ash retains water and releases it into the soil effectively. On the slopes of Mount Etna in Sicily, lemons,

oranges, grapes, olives and figs thrive. Even 1000m above sea level on the mountain, wheat, cherry and apple trees flourish.

Which cities are at greatest risk from volcanoes?

Scientists and city planners study many of Earth's most dangerous volcanoes in a bid to understand their behaviour and improve the survival chances of residents living in nearby towns and cities, or farming their fertile slopes. As a result of this, evacuation plans can be prepared. The volcanoes being studied were chosen because of their history of large, destructive – albeit infrequent – eruptions. Large cities can develop rapidly in densely populated areas, even spreading up the flanks of the volcano. Cities at greatest risk include: Naples, build around the base of Mt Vesuvius; Seattle, which is endangered by Mt Rainier; Tokyo near Mt Fuji; and, Mexico City near Popocatepetl.

French vulcanologist Charles Rivière – seen here taking samples from lava flowing down Mount Etna in the summer of 2001 – faces enormous risks.

What lies beneath the Earth's surface?

Masses of liquid rock are in constant movement beneath the solid crust of our planet. The heat of the planet's core, which exceeds 5000°C, is the force that keeps the interior in constant motion. Researchers have only been able to bore about 12km into the Earth's outermost crust, but earthquakes and seismic waves provide a precise picture of the Earth's interior – which is layered like an onion.

How thick is the Earth's crust?

Our planet is covered by a relatively thin crust, rather like the skin of an apple. It is made up of a mosaic of many tectonic plates that form the continents and oceans. The continental crust, which is between 30km and 60km thick, is composed of more than two-thirds granite. It is less dense than the rocks that make up the Earth's viscous mantle immediately below, and is therefore able to float on it. Beneath the oceans, the Earth's crust is only 5km to 7km thick and consists mainly of basalt. The oceanic crust presses deeper into the Earth's mantle because its rocks are denser.

Is the Earth's interior liquid?

At a depth of 5100km, the Earth's mantle gives way to its outer core. This consists mainly of the metals iron and nickel, which, at a temperature of 2900°C, form a liquid alloy. If it were possible to bore 1300km deeper, the drill would strike against the Earth's solid inner core. Here, at the centre of the Earth, the pressure is enormous – around 3.5 million times that at the surface – and the temperatures can reach between 4000 and 5000°C. Because of the extreme pressure, the iron at the inner core has crystallised into a solid, heavy block. This turns around within the liquid metal alloy of the outer core, creating the Earth's magnetic field in the process. With a circumference of approximately 3500km, the solid inner core is slightly smaller than the Moon. What we know about the interior structure of our planet has been obtained by measuring the Earth's magnetic field and probing its interior with long radio waves. Additional clues about the core's characteristics can be deduced from a study of the behaviour of seismic waves caused by earthquakes and explosions.

All the layers inside the Earth shown in red or yellow are viscous or liquid. Only the inner core (white) and the crust (blue) are solid.

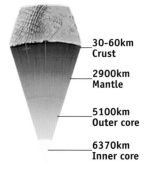

30-60km
Crust

2900km
Mantle

5100km
Outer core

6370km
Inner core

The 83m-high derrick at the German Continental Scientific Deep Drilling Programme.

How deep can we bore into the Earth?

In 1970, geoscientists drilled to a depth of 12 262m on Russia's Kola Peninsula

The Russian drilling record still stands to this day. The core samples that were gathered provided valuable information about the Earth's history, because the scientists struck 2.5-billion-year-old rock formations. With temperatures reaching much higher than the expected 180°C at this depth, drilling was forced to cease in 1992.

Between 1987 and 1994, scientists involved with the Continental Scientific Deep Drilling Programme at Windisch-Eschenbach in the Upper Palatinate of Germany penetrated to a depth of 9101m. The location of the borehole was chosen because it is at this point that two former continents collided, and nowhere else in Europe is the Earth's crust as varied in its structure. Data gathered here continues to provide the foundations for many new geoscientific studies.

How do scientists examine the Earth's interior?

Boreholes, even the deepest ones, are little more than pinpricks in the Earth's outermost surface – its crust or lithosphere. There isn't a drill that can penetrate to the depths that Jules Verne had his intrepid explorers reach in the famous 1873 novel *Journey to the Centre of the Earth*. However, the interior of the Earth is not completely hidden from scientists. Studies into the way in which shock waves travel through the Earth following an earthquake can help to reveal facts about its internal structure.

The technique involves measuring shock waves that reach the Earth's surface, and a similar method is used to locate oil and gas deposits at depths of several kilometres beneath the surface. Scientists do not have to wait for an earthquake to do this, since the vibrations caused by minor explosions or heavy blows to the Earth's surface are sufficient to create waves that can penetrate to the required depth. When conducting research on the ocean floor, large air guns are fired, which use compressed air to generate shock waves.

Why are geoscientists interested in atom bombs?

Since 1945, the world's nuclear powers have tested hundreds of these weapons. The primary aim of tests was to improve the destructive power of the bombs, but they also provided an insight into the Earth's structure. Large explosions shake the Earth in the same way as an earthquake. Seismic waves pass through the interior and are scattered and reflected in a way that reveals the composition of deep zones. Unlike earthquakes, geoscientists knew exactly when an atomic bomb explosion was to take place so were able to prepare their measuring instruments to receive the waves.

Since the Nuclear Test Ban Treaty of 1996, some nuclear powers have halted tests. However, India, North Korea and Pakistan have not yet signed the agreement, while eight other nations – including the USA and China – have not yet ratified it. For this reason, the network of seismic wave sensors is now used to detect nuclear tests, like the one conducted by North Korea in October 2006.

Will the continents eventually get back together one day?

When the German scientist Alfred Wegener published his theory of continental drift some 90 years ago, it caused controversy. We now know that he was right, and that the Earth's great landmasses really do move. About 200 million years ago, they came together to form Pangaea, the supercontinent. This later broke up, and gradually the continents arranged themselves into the pattern we see now.

The processes that cause continental drift are still active. In 100 million years' time, Antarctica will have moved into warmer areas and become ice-free. In 250 million years, all the continents will have drifted back together again to create a new supercontinent, Pangaea Ultima. Scientists believe that there will be a vast and very hot desert at its centre.

Why do the continents move?

Most wood is not as dense as water and so can float on its surface. It is the same with the continents. Although they are largely composed of granite, they actually float – not on water but on the molten rock of the Earth's mantle. The density of the molten matter is $3.5\text{g}/\text{cm}^3$, considerably more than that of granite at $2.7\text{g}/\text{cm}^3$. The ocean floors, which are also part of the Earth's crust, float on the molten mantle as well. Since they have a density of about $3\text{g}/\text{cm}^3$, somewhat greater than that of the continents, they sink deeper into the molten rock.

The tectonic plates don't just float, they also move. The reason for this lies in the Earth's mantle, which is almost 3000km deep. Temperatures at its base are several thousand degrees, causing currents to rise slowly towards the surface. As they do so, they gradually cool and sink back into the depths. This creates a giant conveyor belt of circulating molten rock, which provides the force that moves the continents.

How fast do the continents move?

The speed at which continental plates move changes over the course of time. It also varies from plate to plate. Under normal circumstances, a plate can move at a rate of anywhere between 2cm and 10cm a year.

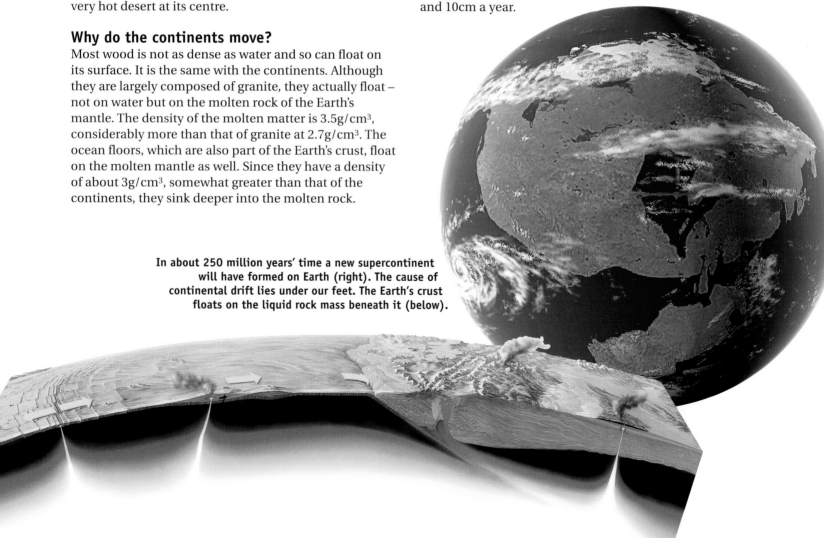

In about 250 million years' time a new supercontinent will have formed on Earth (right). The cause of continental drift lies under our feet. The Earth's crust floats on the liquid rock mass beneath it (below).

The weight of water stored in the great ice sheets that cover Greenland and Antarctica has pushed the surface of the land beneath below sea level. Despite being the second largest glacier system in Europe, the Garner Glacier in Switzerland (above) is not large enough to make the continent sink by any appreciable amount.

Can glaciers cause the continents to sink?

Continents of solid rock float on a sea of molten magma, and as they do so they sink a small way into this liquid rock. If the weight of a continent increases – as a result of glacier formation, for example – it sinks deeper into the molten layer beneath.

Glaciers weigh nowhere near enough to cause the continents to sink significantly. Because glaciers in most of the world's high mountains and at the South Pole are melting, the continents are slowly rising rather than sinking. The process is happening slowly over a period of thousands of years, at a rate of only a few millimetres each century. A similar effect has been observed in parts of Scandinavia. For the thousands of years that have elapsed since the end of the last Ice Age – when the entire region was covered with glaciers – this part of Europe has been rising by about 1cm every 100 years.

How does a geyser work?

Geysers are a stunning sight as they shoot huge jets of hot water high into the air. They are only seen at a few of the world's active hot springs, with the main ones in: Yellowstone National Park, USA; Dolina Geizerov, Kamchatka, Russia; and, El Taito, Chile. The eruptions can occur at regular intervals or be unscheduled, depending on the way in which water flows into the underground caverns where it is heated by volcanic activity. The pressure gradually builds up as water is heated to temperatures above boiling point. Only when the temperature is well above 100°C do steam bubbles start to rise through narrow channels leading to the surface, blasting fountains of mineral-charged water and steam high into the air.

Are there cold-water geysers?

There are geysers that produce fountains of cold water, but their eruptions cannot be compared with 'normal' geysers. The eruption is not caused by hot water under pressure but by a chemical reaction. This occurs when carbon dioxide released in caves dissolves in water, creating aerated carbonic acid.

Natural carbon dioxide-driven, cold-water geysers are rare. The only two known are the Cold Water Geyser in Yellowstone National Park and the Salton Sea geyser in California. Others, such as Andernach, Germany, have been caused by boreholes puncturing pressurised gas- and water-filled sedimentary rocks (artesian aquifers).

A geyser develops as volcanic activity heats underground water, creating steam under pressure. When the pressure is sufficient, the water is forced up through a constricted pipe to shoot skywards in a spectacular jet.

Cold-water geysers look like steam-driven geysers, although the carbon dioxide-laden water often appears more white and frothy.

How do we know...

...that the Earth's heat keeps the ice off Iceland's pavements?

The rising price of crude oil is not such a concern for the people of Iceland. Almost all of the country's electricity and heating requirements are met by hydroelectric power, using heat from the depths of the Earth. Because the energy gained this way is easily accessible and available in large quantities, it is possible to keep the streets of the capital Reykjavik free of ice in winter with a hot-water circulation system built into the ground.

Iceland's Svartsengi power station produces geothermal energy, with its waste water supplying a thermal spa.

Where is the Earth's magnetic field located?

We can detect the Earth's magnetic field all around the planet. The Earth has a north and a south pole, just like a small bar magnet. The needle of a compass will point in a north–south direction, with the arrow pointing northwards. This this will happen anywhere on Earth, apart from areas close to the poles. But because opposite poles attract, the geographic North Pole is actually the magnetic south pole and vice versa. On closer examination, we find that the magnetic field deviates slightly from the geographic north–south axis – at present it tilts away from the Earth's rotational axis by about 11.5°. To make navigation more accurate, maps usually show the small correction that needs to be made to take into account the local difference between magnetic and geographic north.

by the movement of molten iron in the Earth's outer core. It protects our planet against the charged particles of the solar winds.

What does the magnetic field tell us about the Earth's history?

The Earth's magnetic field changes constantly, which provides valuable information about our planet's formation. For millions of years, volcanic activity in its interior has been forcing molten rock to the surface, where it cools and solidifies. As it starts to cool, any minerals in the still liquid rock that can be magnetised – such as magnetite – align themselves with the magnetic field as it is at that time. When the rock hardens, these tiny magnets are locked in position. After spending years studying changes in the magnetic field, scientists can now tell when a sample of volcanic rock was formed.

Can the magnetic poles wander?

The Earth's magnetic poles move all the time. During our planet's multimillion-year history, the magnetic poles have swapped places many times, with the result that the magnetic south pole is presently located in northern Canada. Because of this movement, a compass needle shows only the approximate direction of the geographic poles. The latest measurements of the Earth's magnetic field, made with the help of satellites, show that the magnetic south pole – to which the north arrow of a compass points – is moving about 90m westward every day, or around 30km/year.

How do we detect the Earth's magnetic field?

Magnetic fields are invisible, however certain metals respond to magnetic fields. As every schoolchild knows, iron filings in the vicinity of a magnet form themselves into tight rows, linking the magnet's north pole to its south pole. These rows indicate how and where the magnetic field runs.

A compass needle works on the same principle. The magnetic north pole of the compass needle always points exactly to the magnetic south pole of the Earth's magnetic field. With the help of sophisticated equipment it is possible to determine the strength as well as the direction of a magnetic field.

Thanks to the magnetic field captured in solidifying lava on Mount Teide, a volcano on Tenerife, scientists can uncover information about the island's history.

True or False?

There is a huge dynamo in the Earth's core

At the centre of our planet there is a geodynamo, which is the source of the Earth's magnetic field. The dynamo is created by the constant movement of vast quantities of molten iron in the Earth's outer core. The high temperature in the hard, metallic inner core (about 5000°C) heats the mass of molten iron and drives it up towards the Earth's mantle. There it cools and sinks back towards the core. This cycle is called a convection current. Because the liquid iron is conductive, the movement produces (induces) an electric current and it is the induction of the geodynamo that produces the Earth's magnetic field.

Do birds and animals make use of the Earth's magnetic field to find their way around?

Migratory birds, homing pigeons, turtles, sharks and probably whales are all able to use the Earth's magnetic field to find their way. These creatures have a special sense organ containing magnetic particles, such as tiny fragments of iron or nickel, that work the same way as a magnetic compass.

This discovery does not completely solve the mystery of their remarkable 'sixth sense'. The precise location of the sense organ in individual species has yet to be determined, and the nature of the biochemical processes that take place there is not fully understood. Recent studies of the magnetic sense in migratory birds have led scientists to conclude that the organ is situated in the right eye. The information gained by the right eye is processed in the left half of the brain, and this is thought to provide feedback to the light receptors. The birds may be able to virtually 'see' magnetic fields.

The magnetic organ in sharks is located in their nose. It works like a sensitive aerial that reacts to electromagnetic fields. Sharks are able to detect weak magnetic fields and an electrical potential of just a few millionths of a volt.

What do the northern and southern lights have to do with Earth's magnetic field?

Without the Earth's magnetic field, there would be no northern or southern lights – the aurora borealis and aurora australis. It is only in regions close to the poles that the magnetic field is aligned in such a way that it allows cosmic radiation – containing charged particles from the Sun – to penetrate deep into the Earth's atmosphere. The particle showers – known as solar wind – contain electrons, protons and heavy ions. These can be travelling at speeds of up to 3 million km/h when they collide with molecules of water, oxygen and nitrogen in the atmosphere, causing them to light up in a brilliant display of patterns and colours.

At times of increased solar activity, the solar wind may become so strong that the spectacular light show can even be observed in regions far from the poles. Sightings have been reported from Greece, Australia and mainland USA.

A shimmering curtain of auroral light hangs in the night sky over Norway.

The perils of living on a restless Earth

Earthquakes shake buildings and people regularly, although the massive shocks that result in thousands of deaths are rare. On average, there are about three earthquakes every day, proving that the Earth is still an extraordinarily dynamic planet. Despite decades of research, no one can predict earthquakes with any degree of reliability. What we do know is that the risk of earthquakes is highest where tectonic plates meet and interact with one another.

What happens when the Earth quakes?

Over 90 per cent of the world's earthquakes are caused by movements between tectonic plates. These plates move like ice floes on water and travel several centimetres each year (see p. 256). As they move, the compacted rocks grind against each other and can become locked together. This causes pressure to build up gradually until the forces involved exceed the strength of the rocks. The rocks rupture and the Earth's crust shifts suddenly – causing the ground to quake. Seismic waves spread from the area of the break at a speed of about 13 000km/h, and can cause damage hundreds of kilometres from the area directly above the centre of the quake.

Earthquakes can be caused by factors other than just shifting tectonic plates. Volcanic eruptions and collapsing mine shafts can both produce earthquakes, and some geophysicists suspect that earthquakes can be caused by the collapse of underground caverns that have been emptied of the vast oil and gas deposits they once contained.

How are earthquakes measured?

The magnitude of an earthquake is expressed by a number on the Richter scale. This system for measuring earthquakes has no upper limit, and was devised by the American seismologist Charles Francis Richter (1900–1985). It is based on the measurement of earth tremors with a seismograph. The scale begins at 1, which indicates a weak but noticeable earth

Some of the after-effects of an earthquake at Roermond in the Netherlands on 13 April 1992. This was the strongest earthquake in Central Europe since 1756.

Devastation caused by the San Francisco earthquake of 1906, which resulted in 3000 deaths.

tremor. Every subsequent point indicates a tenfold increase in the strength of an earthquake. A quake of magnitude 8 is very powerful. Richter made no provisions for an upper limit to his scale, even though the physical characteristics of the Earth's crust make an earthquake stronger than 9.5 virtually impossible.

The more recent 'moment magnitude scale', which was introduced in 1979, is more accurate than the Richter scale. Instead of measuring the energy released, it takes into account other data, including the length of cracks in the Earth's crust. The 'moment' is a mechanical measurement of the movement of bodies following the release of a force. Like the Richter scale, the moment magnitude system uses seismographic data. Minor earth tremors create cracks in the Earth's crust that are no more than a few hundred metres long, while those produced by major shocks can extend for several hundred kilometres. Scientists calculated that an earthquake in Chile in May 1960 had the highest moment magnitude ever recorded of 9.5.

The Mercalli intensity scale can be used without any measuring equipment, but is relatively inaccurate. It judges the strength of an earthquake by its perceptible and visible effects. The 12-degree scale is named after the Italian vulcanologist Giuseppe Mercalli (1850–1914), who devised it at the beginning of the 20th century. At that time accurate measuring instruments did not exist. At level 6, cracks appear in walls; at level 7, chimneys fall off roofs; and, at level 8, gable fronts collapse.

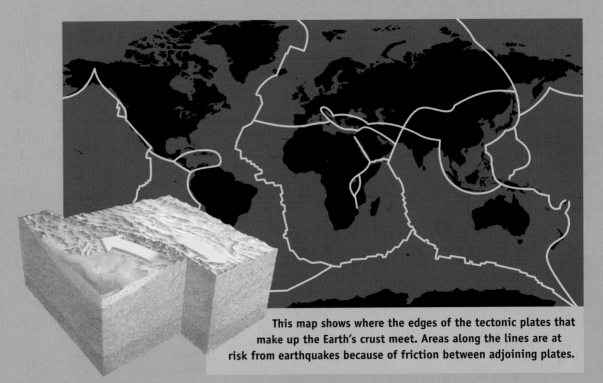

This map shows where the edges of the tectonic plates that make up the Earth's crust meet. Areas along the lines are at risk from earthquakes because of friction between adjoining plates.

Which cities are at greatest risk from earthquakes?

Cities built on joint lines between tectonic plates are at the greatest risk. Builders must follow strict codes in such US (Los Angeles and San Francisco) and Japanese (Tokyo and Kobe) cities. At greatest risk are older, unregulated cities, such as Istanbul in Turkey and Jerusalem. Within the next decade or so, a big quake in Istanbul is predicted to kill tens of thousands in building collapse and fire.

Will it ever be possible to forecast earthquakes accurately?

For decades, geoscientists have been trying to understand the factors that cause earthquakes. With the help of networks of sensors they are able to determine how far beneath the Earth's crust an earthquake is formed. They know where there is friction between areas of rock and the direction in which the Earth is likely to move in an earthquake. However, they have not achieved one of their main aims, which is to forecast earthquakes reliably. Some experts doubt that this will ever be possible. Although there have often been warnings, it has been rare for earthquakes to turn out as predicted by the scientists. New early warning models are mostly based on a close analysis of powerful earthquakes in the past, but because of wide variations in the Earth's crust, this kind of analysis is rarely applicable to other regions.

Scientists are trying to get the best out of modern sensor and data transmission technologies. An early-warning system now being developed for Istanbul will allow warning signals to be sent by radio to the city. If a shock occurs 100km away, a broadcast warning can overtake the destructive seismic waves, giving the city at least a few seconds to prepare. This is long enough to cut off gas mains and shut down power stations.

Can animals sense earthquakes?

There are often reports of animals becoming restless shortly before the start of an earthquake, although many doubt the existence of this supposed ability. Chinese scientists have proposed the use of snakes to predict earthquakes because they are highly sensitive to the smallest movement in the ground. The animals are to be monitored by cameras at a snake farm in Nanning, southern China. If the creatures start to behave oddly, a warning will be issued to earthquake stations in the province of Guangxi. Researchers will have to wait for the next major earthquake to find out whether this approach will work or not.

Why is there more to water than meets the eye?

Water in its purest form almost never occurs in nature. Water molecules, which are composed of two hydrogen atoms and one oxygen atom, are mixed with a range of substances such as salts, micro-organisms, gases and other foreign matter. In order to produce drinking water, all the substances that are harmful to humans need to be filtered out.

Why does sea water vary from place to place?

Although most of the seas are interconnected, their salt content and water quality differs greatly. What makes the difference is the rivers that flow into each sea, and the shape and geological structure of the seashore. Currents tend to carry water in one direction only, so there is no thorough and uniform mixing. The types of water found in isolated inland seas, such as the extremely salty Dead Sea or the Caspian Sea (both of which, strictly speaking, are lakes), are distinctive.

Why is sea water salty?

Most types of rock contain salts – and many salts are water soluble. The salts are washed out of the rocks by rain and meltwater, and streams and rivers carry them to the sea. On average, salts – mostly sodium chloride – make up 3.5 per cent of seawater mass, but there are huge variations: the Baltic contains between 0.2 and two per cent salt; the Dead Sea 28 per cent. Solar radiation causes the water to evaporate, further concentrating the solution. In some regions, evaporation over millions of years has resulted in the formation of vast salt flats. The world's largest, the Salar de Uyuni in Bolivia, is over 10 500sq km in area and is estimated to contain ten billion tonnes of salt. Fresh water is also salty, although the salt content is too low to be noticeable.

On average, icebergs last for three years, although some can last for as long as 30 years.

Solar evaporation ponds for salt production beside the Great Salt Lake in Utah, USA.

Why are the seas blue?

Water is transparent, but the sea looks blue to us. This is not because it reflects the colour of the sky, but because of the composition of sunlight, which is made up of the colours of the rainbow. When it strikes the surface of the water, the water absorbs all of the colours except for blue – the blue is reflected, and this is what our eyes see.

Seas are not always blue. Other colours occur as a result of the various substances that can be found in sea water. Microscopic algae give the water a green hue, and if a large quantity of algae are present – as they are in the Baltic Sea in summer – the water turns a greenish-brown. The chlorophyll contained in some blue-green algae colours the water red, and yellow is seen when there are large concentrations of sand and clay particles in the water, such as in the Yellow Sea between China and the Korean peninsula.

Is the Dead Sea really dead?

The extremely high salt content in the Dead Sea creates a hostile environment, one in which almost no animal or plant life can survive. But the Dead Sea is not entirely lifeless. It is populated by micro-organisms, especially bacteria, which break down saltpetre (potassium nitrate) and sulfur.

The Dead Sea could 'die' one day. It is already the lowest-lying sea in the world. The surface of the water lies 396m below sea level and is still sinking. People who live along its only source of water – the River Jordan – are diverting increasing amounts of water for agriculture and human consumption. If this continues at the present rate, the Dead Sea could dry up within the next 50 years or so.

True or False?

Melting icebergs cause sea levels to rise

A melting iceberg floating in the open ocean poses no flood risk. A body that floats in water, and is only partly immersed – like an iceberg – only displaces a certain quantity of fluid, which depends on its weight. It corresponds exactly to the amount of water that the melted iceberg would produce. This natural law was discovered by the Greek mathematician and physicist Archimedes. According to the Archimedes Principle, the weight of a body that is submerged in fluid is reduced by the weight of the volume of the fluid it has displaced. Therefore, the body is subject to hydrostatic buoyancy – just like an iceberg floating in the ocean.

Because of buoyancy caused by the Dead Sea's high salt content, it is practically impossible for swimmers to dive beneath its surface.

How important are the tides?

If there were no tides, our coasts would look quite different. There would be no mangrove forests, salt marshes or mudflats. It would be good news for shipping, as sea captains would find navigation easier and wouldn't have to spend hours waiting for the tide to be right before entering or leaving harbours.

What causes the ebb and flow of the tides?
The tides are caused by the gravitational pull of the Moon and the Sun on the Earth. The Moon's gravitational attraction is twice that of the larger but more distant Sun. The Earth's centrifugal force, which is created by the planet's rotation, causes movement in the Earth's water masses, too. This phenomenon is comparable to the spin cycle in a washing machine during which clothes are flung outwards.

The speed of the incoming tide once endangered pilgrims crossing the mudflats to the Benedictine abbey on Mont Saint Michel, off the north coast of France.

Half moon (waxing)

The Moon's gravitational pull

Earth

Sun

The Sun's gravitational pull

The Moon's gravitational pull

Half moon (waning)

Depending on the position of the Sun, Moon and Earth in relation to one another, there is either a low tide or a high tide. On the side of the Earth closest to the Moon, the Moon's gravitational pull is more powerful than the Earth's centrifugal force and hence the water is pulled towards the Moon, creating a high tide. On the other side of the Earth, which is facing away from the Moon, the centrifugal force of the Earth is stronger than the gravitational pull of the Moon and again there is a high tide. The high tides take water away from the regions in between, where there are low tides.

How great can the tidal range be?
The tidal range describes the difference in water levels between low and high tide, and the terms turning tide and rising tide are used to describe the period during which the water falls or rises. In nautical terminology, the maximum and minimum levels are known as high and low water.

The tidal range can vary widely from place to place. While it is only around 200mm in the Baltic Sea, it is as high as a house in the Bay of St Malo on the English Channel, and regularly reaches 14m. This is because the Channel is relatively narrow, causing the tidal wave to slow down and the water dams up. The greatest tidal range in the world has been measured on the southern shore of the Bay of Fundy on Canada's Atlantic coast. During a spring tide (see below) the tidal range reaches 21m. The smallest tidal range in the world is in the Baltic, Mediterranean and Caribbean seas.

What are spring tides and neap tides?
We experience particularly high tides when the Sun, Moon and Earth are all aligned during the period when there is either a full Moon or a new Moon. At these times the gravitational forces of the Sun and the Moon are combined, and the result is a spring tide. The term 'spring' has nothing to do with the season, but possibly originated from the idea that the tide 'springs' up at certain times of the year. There is a full Moon and a new Moon in a regular cycle of approximately every 14 days because it takes the Moon 27.32 days to orbit the Earth.

When there is a half Moon, and the Sun, Moon and Earth are at right angles to each other, the tides are lower than average. The gravity of the Sun and the Moon is pulling the seas and oceans on Earth in different directions. The consequence is a neap tide, when the tidal range is around 50 to 60 per cent lower than it is at the time of a spring tide. The term 'neap' comes from an Old English word meaning pinched, narrow or scanty.

When the Earth's oceans are pulled in different directions, the range between high and low tide is at a minimum. These are called neap tides. When the Moon and Sun pull together there is a 'spring' tide.

Are tides the answer to our energy problems?

Energy produced by the tides has been successfully harnessed at La Rance in the north of France, where a tidal power plant was built in the 1960s. It works in a similar way to a pumped-storage hydroelectric plant. A 750m-wide dam was built at the mouth of the River Rance, where a vast reservoir holds back the incoming tide – which can be as high as 12m – and later releases it. The water passes through 24 two-way turbines, producing electricity both as it enters and leaves the reservoir.

Other plants that use the tides to generate energy work like wind turbines, using water currents instead of air currents, but they are still at the trial stage. A rotor with a generator is installed on a stable mast set on the seabed. When water flows past the rotor blades it turns and the generator produces electricity.

What's so special about mudflats?

What makes mudflats special is the fact that they are flat, with a gradient of only one metre per kilometre. The tidal range in these areas is relatively large, and water regularly retreats for several hours creating a unique environment for plants and animals.

The world's longest stretch of mudflats is the North Sea tidal shallows. They extend from Den Helder in the Netherlands to Esbjerg in Denmark and cover an area of about 8000sq km. Most life carries on just below ground level, where innumerable shellfish and worms find refuge from attack by hungry birds.

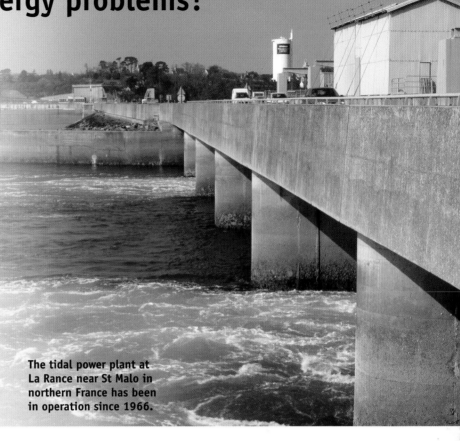

The tidal power plant at La Rance near St Malo in northern France has been in operation since 1966.

A variety of animals can be easily seen in areas of tidal shallows.

Why are there currents in the oceans?

Constant winds – such as the trade winds – blow incessantly across vast stretches of open ocean creating currents. Other factors are the shape of the seabed, water temperatures and differences in density between warm and cold water masses, which increase and decrease at different depths.

The global system of maritime currents moves vast masses of water around the surface of the planet. In one second the Gulf Stream – which makes its way from West Africa to South America and from there through the Gulf of Mexico into the North Atlantic – transports 100 times more water than all the rivers on Earth put together.

Why are tsunamis so dangerous?

Tsunamis are usually caused by distant earthquakes on the seabed, so they can appear suddenly, even on the calmest, sunniest days. With mountainous waves as high as 10m to 50m, they can devastate coastal areas in moments. Even observant locals have no more than a few minutes to escape after they notice the typical sign of an approaching tsunami – water receding from the seashore, sometimes by several hundred metres. It is difficult to set up an effective early-warning system because in the open sea the waves are only a few centimetres high and don't reach their full, fatal size until they enter shallow waters close to the shore.

The last major tsunami occurred on 17 July 2006 following an underwater earthquake off the coast of the Indonesian island of Java, claiming 700 lives. The tsunami on 26 December 2004 was also triggered by an underwater earthquake and killed at least 231 000 people in eight Asian countries. This makes it one of the worst tsunami catastrophes ever recorded.

Are tsunamis only caused by earthquakes?

The majority of tsunamis are caused by earthquakes on the ocean floor, where they are known as seaquakes. For a tsunami to develop, the quake needs to be at least a magnitude 7 and must cause the ocean floor suddenly to shift vertically. Only about one in every 100 earthquakes produces a tsunami.

Volcanic eruptions and sudden landslides in coastal regions can also unleash tsunamis. Meteorite strikes and even atomic bomb explosions can be responsible for triggering this phenomenon.

The largest tsunami ever observed occurred in 1958 in an Alaskan fjord. There the waves were funnelled into great walls of water between 150m and 520m high.

How fast do tsunamis move?

The enormous speed of a tsunami is what makes it so dangerous. There is barely any time for detection or to warn imperilled coastal populations so that they have time to leave the area for higher ground. The speed of a wave is determined by the depth of the sea. The deeper the water, the faster the tsunami. The Pacific Ocean is about 5000m deep, and waves travelling across it can reach speeds of about 800km/h. A tsunami may take only a few hours to travel thousands of kilometres and cross entire oceans. Waves produced by even the severest storms rarely travel faster than 100km/h. As water close to a coast becomes shallower, the tsunami wave slows down. At the same time its wavelength – the distance between successive crests – gets shorter and the wave begins to grow rapidly in height. The water that makes up a tsunami – and any other ocean wave – does not actually move through the ocean. The water molecules pass their momentum on to adjacent molecules.

A warning sign in Sri Lanka points the way to the closest tsunami shelter or high ground.

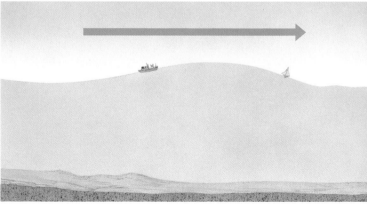

At the Tsunami Alert Centre of the meteorological institute in Kuala Lumpur (Malaysia), attempts are being made to detect seaquakes and the tsunami they cause as early as possible.

How a tsunami is formed: a seaquake produces movement of the water (left) and, almost unnoticed, the wave this creates begins to move forward (centre), building up to an enormous height as it approaches the coast (right).

Where do tsunamis occur most frequently?

Earthquake-prone areas to tend to be at greatest risk. Most tsunamis occur in the Pacific Ocean, which is encircled by a belt of volcanoes, known as the Ring of Fire. Here, on the edges of the continental plates that surround the Pacific, earthquakes and volcanic eruptions are common.

Are any other coastlines under threat?

Most coastlines are at risk. Tsunamis do not occur in the Atlantic or Indian Ocean as frequently, but this makes them even more dangerous as the people are caught unawares. In the Atlantic, a likely starting point for a tsunami would be the Canary Islands off the west coast of Africa. These volcanic islands are subject to earthquakes that have triggered landslides in the past that set the sea in motion. About 300 000 years ago, part of the Canary island of El Hierro fell in to the sea and sent a tsunami racing across the Atlantic. When it reached the east coast of what is now the USA, it hurled rocks as big as houses several hundreds of metres inland. Scientists now warn that an eruption on the island of La Palma in the Canaries could cause a landslide resulting in a tsunami. The last major Atlantic tsunami, following a magnitude 8.7 earthquake, devastated the city of Lisbon, Portugal, in 1755, killing more than 100 000 people. Active volcanic islands in the Caribbean arc have similar potential to generate tsunamis.

Coastlines of smaller seas, such as the Mediterranean, are not completely safe. The Messina-Reggio magnitude 7.1 earthquake of 1908 generated a tsunami that devastated the towns of Italy's Messina Straits.

How reliable are tsunami early-warning systems?

Since the disaster of 26 December 2004 in the Indian Ocean, scientists have been seeking ways to warn local populations in time. The Indian Ocean nations have agreed to install an early-warning system – a network of special microphones known as geophones that can detect seaquakes acoustically. The readings can then be carried electronically to any endangered coastline at a faster rate than the tsunami can travel. This should provide people with valuable time in which to prepare.

UN scientists are building a warning system off the coast of Sumatra in Indonesia. The network of sensors with its special buoys and satellite technology should help to avert tragedy. One of the greatest challenges is to avoid false alarms, since not every earthquake causes a tsunami.

Why is the deep sea still such a mystery?

We probably know less about the depths of the ocean than we do about the surface of the Moon. It is difficult to penetrate the cold, dark depths of the sea. In the past 30 years, increasing numbers of diving expeditions have ventured into these last, unexplored regions of our planet. Scientists are discovering amazing geological formations and previously unimagined forms of life.

How does the world look from the seabed?

The deep sea is an extreme habitat – it is cold and dark and the pressure is exceptionally high. Just surviving in such conditions is difficult, which is why human beings have been slow to discover more about this world kilometres below the surface. For now, the region still belongs to the fantastic creatures that have adapted to this hostile environment over millions of years. Some deep-sea fish have neither eyes nor gas bladders (also known as swim bladders), as they would not be able to see anything in the dark and a gas bladder would be damaged by the high pressure.

How deep does the light of the Sun penetrate?

Sunlight can only penetrate to a maximum of 300m below the surface of the water. It becomes darker and darker as you descend until total darkness reigns. The average depth of the world's oceans is a little over 4000m, so most of these habitats are permanently pitch black. Rays of light in water spread out, refract and scatter differently to the way they do in the air. Colours under water are hard to distinguish, with red disappearing at 5m, orange at 10m and yellow at 15m. Some deep-sea dwellers have developed ways of producing their own light, using a phenomenon known as bioluminescence. Angler fish belong to this group and have an appendage like a fishing rod, tipped by a luminous lure, in front of their mouths. This attracts small fish, which are snapped up as soon as they come within reach.

What causes marine luminescence?

When strange patterns of blue-green light appear on the surface of the water, the cause is generally marine luminescence. The creature usually responsible is the large, single-cell plankton *Dinoflagellatum noctiluca miliaris*. Marine luminescence is impressive on dark nights when there is a new Moon. It is created by a chemical reaction between two enzymes. If the plankton is disturbed by the movement of a wave, a passing ship or a swimmer, the two enzymes react with oxygen from the air to produce a bluish-green light.

Colonies of dinoflagellates produce the mysterious marine luminescence that sometimes bathes sea water in a blue-green light.

The small luminous tip on the head of the angler fish lures prey directly into its mouth.

Creature from another world: this deep-sea squid is only 40mm long and almost transparent.

What kinds of life forms can survive in the depths of the ocean?

Photosynthesis is impossible without light, so there are no plants in the deep sea. Inhabitants of the region that are vegetarians must make do with the remains of plants that have died and sunk to the bottom. The menu for carnivores is somewhat more varied, since they can feast on their deep-sea companions.

Most deep-sea fish grow to lengths of no more than 30cm – probably because of the limited availability of food. Many deep-water fish have virtually transparent bodies or are black to enable them to hide more effectively from their enemies. Most have eyes that are extremely large, or alternatively they may have none at all. Some deep-sea fish have developed unusual sensory organs to track down their prey. They may be sensitive to smells or vibrations. Scientists estimate that, because most deep-sea creatures are microscopic, there are at least a million as yet undiscovered species lurking in the depths.

What are black smokers?

In 1977, the deep-sea submersible *Alvin* discovered the underwater world off the Galapagos Islands and a series of undersea hydrothermal springs at a depth of 2000m. Scientists observed that mysterious vents – which later came to be known as black smokers – were emitting a continuous stream of black 'smoke'. The cause was water heated by hot magma from the Earth's mantle, containing sulphur compounds and other minerals.

The temperature around the vents was about 300°C. Living in their vicinity the team discovered many different life forms, including bacteria, worms and crabs. Perhaps the strangest inhabitants were the tube worms. They have neither mouths, intestines nor stomachs and should not really be able to exist since they cannot feed themselves. In a unique symbiotic relationship, they survive because their insides are lined with bacteria that feed on the sulphur-rich water and transform it into nutrients that provide nourishment for the worms.

Minerals such as sulphates released by black smokers provide the starting point for a strange food chain that flourishes deep beneath the sea.

:273

... that human beings have reached depths of nearly 11,000m?

In 1960, Jacques Piccard and Don Walsh took their bathyscaph *Trieste* to almost the deepest known point of the sea – the Mariana Trench, east of the Pacific island of Guam. At a depth of 10 960m, the two explorers still came across fish and other life forms. Only *Kaiko*, an unmanned Japanese submersible vessel, has gone deeper, reaching a record depth of 11 034m in the Mariana Trench. Ordinary submarines like those used by the world's navies can only dive to depths of between 300m and 400m, as they are not designed to penetrate deep water but to move as silently as possible. One exception is the former Soviet Union's nuclear submarine, known to NATO as the Alpha Class, which can reach depths of 1000m. Their hulls are made from titanium to withstand the immense pressure.

In the 1950s, the Swiss physicist August Picccard built the deep-sea submarine *Trieste* (seen here as a model), which his son Jacques (top right) used in 1960 to dive into the Mariana Trench.

Is it true that more ships sink in the Bermuda Triangle than anywhere else?

This is debatable, and many of the numerous attempts to explain the supposed phenomenon belong in the realms of fantasy. Scientists have come up with a plausible theory as to why ships might be lost without trace in this area. On the seabed in the Bermuda area, there are deposits of methane hydrate, a gas that forms into ice. If deposits are damaged by an earthquake, large bubbles of gas could make their way to the surface where they might cause a sudden and dramatic reduction in the density of the water, causing ships to lose buoyancy and sink.

According to another theory, electromagnetic fields might be disturbing the navigational instruments on ships, causing them to lose their bearings in bad weather. However, no electromagnetic irregularities have so far been registered in the area.

The Bermuda Triangle lies in the western Atlantic between the Bermuda Islands, southern Florida and Puerto Rico.

Why are rivers the Earth's lifelines?

A network of thousands of rivers stretches across the Earth's continents. Along with the planet's lakes and man-made wells, they meet the global need for drinking water. Their importance to the history of human civilisation, from the first settlements to their use today as transport routes, cannot be overestimated.

How de we get clean drinking water?

Tap-water has to meet stringent requirements in most developed countries – sometimes even more stringent than those for bottled water. Tap-water may contain various minerals, but should be cool, appetising, odourless and colourless. In Central Europe, most of the fresh drinking water is drawn from ground water, either via natural springs or man-made wells. While most spring water is usually so clean that it needs no further purification or special processing, water treatment plants have to go to much greater lengths with water drawn from lakes, reservoirs and rivers. Purification plants have to deal with pollution caused by nitrates from the liquid manures and chemical fertilisers used in agriculture, as well as a range of poisonous chemicals used by industry. Chemical, physical and biological purification technologies are all used.

In dry coastal regions, such as the Arabian Peninsula, most drinking water has to be recovered from the sea. Desalination plants process the salt water to make it fit for human consumption.

According to United Nations statistics, about one-sixth of the Earth's population does not have access to safe drinking water. Every year, nearly two million people die from diarrhoeal diseases caused by contaminated water. The UN aims to ensure that the number of people living without clean water by 2015 is half what it was in the year 2000.

Where does spring water come from?

Water bubbles out of springs naturally. Spring water is rain that accumulates in the earth and gradually percolates down through the soil and crevices in rock. The water penetrates further and further down until it reaches a layer of impermeable matter, such as clay. The water then flows along above the impermeable layer until it reaches the surface of the ground and emerges as a spring. Impurities are filtered out of the water as it makes its way gradually through the ground, and minerals are added as they dissolve.

Why are some lakes and rivers salty?

Rivers contain fresh water, with a salt content of less than 0.1 per cent. The salt content can vary, depending on the minerals to be found in the rocks through which the river runs, and in some areas can even make the water unfit for human consumption.

Although some rivers originate from springs, their water does not stay as pure as spring water. They collect a lot of surface water and pollutants on their way to the sea. Fresh river water and the salty sea water mix in the river estuary. This mixture is known as brackish water and contains between 0.1 and one per cent salt. If a river flows into the sea where there happens to be a large tidal range, the brackish water can advance up to 50km inland. Even further inland it is possible for brackish water to form as a result of water leached from natural salt deposits in the ground. Salt, and a number of other minerals that can dissolve in lesser quantities in flowing water, can get carried into inland lakes. When there is a lot of evaporation, the mineral salt left behind in increasing quantities can build up to make some lakes very salty.

At the purification plant, pollutants are removed from the water. The resulting sludge can be used as fertiliser.

The Ostriconi estuary on the island of Corsica provides a diverse habitat for plants and animals.

There are many salt lakes around the globe. The 10 500sq km Salar de Uyuni in Bolivia is the largest and, in the dry season, it is transformed into a salt desert. The lake with the highest salt content – 35 per cent – is Lake Assal in the Republic of Djibouti. The Dead Sea, which, despite the name, is a lake, has an average salt content of 28 per cent, while the Great Salt Lake in Utah, USA, contains 25 per cent salt. The Aral Sea in Central Asia is gradually drying up and its salt content is now about eight per cent.

What are the Earth's mightiest rivers?

Earth's mightiest rivers are those that traverse vast continents and carry thousands of tonnes of gravel, sand and clay into the ocean daily. They build enormous deltas of sediment at their mouths, so heavy that they actually push down into the sea floor. The world's ten most powerful rivers, in terms of the volume of water that they discharge, are: the Amazon; Congo; Yangtze; Orinoco; Brahmaputra; Parana; Mississippi; Mekong; Ganges; and, Danube. Although the Nile takes 11th place on this list, it is the world's longest river at 6700km, compared to the Amazon's 6500km.

These giant rivers occur in response to the tectonic activity found along the edges of the Earth's crustal plates. Mighty rivers flow away from massive collision mountain ranges, such as the Brahmaputra and Ganges draining the Himalayas, or the Danube flowing from the Alps. Other rivers, like the Amazon, Orinoco and Parana, flow from tall and active volcanic mountain ranges, such as the South American Andes.

Do all rivers flow into the sea?

As part of the global process that keeps water constantly in circulation, all rivers ultimately flow into the sea, although most minor rivers flow into larger ones first. Some develop into mighty waterways, like the Danube, Mekong, Chang Jiang or Amazon.

In desert regions, river levels can sometimes fall so low that the water dries up before it can reach the sea or a larger river. In karst regions, where the rocks (such as limestone) are extremely porous, a whole river can simply disappear underground, although this water will bubble up to the surface again through a spring somewhere along the way, and continue on its journey to the sea.

Exploring the
everyday

Human beings soon become accustomed to the world they grow up in. We take it for granted that aircraft fly, some cheese has holes in it, dust is grey, glue sticks, coffee keeps you awake, fire burns – and so do your eyes when onions are being chopped – mobile telephones drop out and washing machines gobble up socks. But why? The everyday world offers many fascinating puzzles – and the explanations often cause us to look at things in a different way.

Why can't we do without physics and chemistry in the kitchen?

Everything we do in the kitchen has something to do with physics or chemistry, from cooking our food to freezing it. Rising and falling temperatures, even just mixing and stirring, can cause changes to occur. The molecules that make up foodstuffs, the water or oil they are cooked in and the air in the environment all undergo chemical reactions. Even top chefs make use of our better understanding of the physics and chemistry of cookery to improve their creations.

What is freezer burn?

When frozen foods develop dry, white or brownish spots around the edges – known as freezer or frost burn – it is usually because the food's packaging was not airtight or was too large. The food has not actually gone off, but it is inedible. This kind of damage occurs mostly with frozen foods with a high water content – such as meat and vegetables – when the surface of the food comes into contact with air. The frozen fluid turns into its gaseous state and escapes. The oxygen in the air oxidises these porous areas, destroying flavourings and nutrients. As they thaw, the dried-up spots absorb little water, which leaves them tough.

Why is it better to snap-freeze food?

Snap-freezing reduces the temperature of foods to at least -18°C in just a few minutes. Only tiny ice crystals form in the individual cells, which means that the cell structure is not harmed and vitamins, nutrients and colourant molecules are left undamaged. When food is frozen slowly in a conventional freezer, large ice crystals can form and damage the cell walls. This makes foods mushy and discoloured after thawing. Depending on which foods are to be snap-frozen, the industry uses either liquefied gas, plate freezers, freezing chambers or freezing tunnels.

The company DuPont developed non-stick coatings and coined the name Teflon.

What happens to a soufflé when the oven door is opened?

A soufflé doubles or trebles in volume in the oven. The batter contains tiny air bubbles that inflate when heated. The water in the batter also forms bubbles as it evaporates, and pushes the batter upwards. Each bubble is encased in protein molecules that thicken at temperatures of 42°C or more, binding the whole structure together. However, it takes a little while for all the protein in the soufflé to coagulate, and the process progresses slowly, from the outside in. If the oven door is opened too soon, the temperature in the air bubbles drops suddenly and the unstable protein mixture collapses. The damage cannot be undone and the coagulation process will continue before the air in the bubbles has had a chance to expand again.

True or False?

Ice cream can take the heat

Fried ice cream is delicious. In Asia, it is served wrapped in pastry, while elsewhere it is rolled in coconut flakes or cornflake crumbs. The balls of ice cream must be kept extremely cold before being wrapped in puff pastry or coated in egg and crumbs, after which they must go straight back into the freezer. This ensures the ice cream survives the 15 seconds it has to spend in the deep-fryer or frying pan without melting significantly.

Why doesn't food stick to non-stick pans?

The molecules of an uncoated metal frying pan and the molecules of frying food are strongly attracted to each other. This creates bonds between the metal surface and the carbohydrates or proteins in the food, and this chemical attraction becomes visible when the food sticks to the pan and burns. Frying with butter or oil reduces sticking because these fats provide a temporary intermediate layer. The long-term solution is to use a pan with a permanent layer on its surface, such as Teflon or a similar synthetic material. The compounds that make up fried eggs and other culprits have nowhere to stick on these surfaces. The non-stick coating is held firmly to the pan because the coating's molecules cling to the countless imperfections on the rough metal surface. However, the molecules don't stick to each other all that well, which is why non-stick pans scratch so easily.

Why does egg white stiffen when whisked?

Egg white is 90 per cent water. It also contains long protein molecules that keep a proportion of the water molecules loosely connected. In their normal state, the proteins are in the form of dense spheres, comparable to balls of wool. When egg white is whisked until it stiffens, two things happen: air bubbles form in the water and protein mixture; and, the protein molecules unravel. Because they consist of hydrophilic (water-loving) and hydrophobic (water-repelling) parts, they attach themselves to the boundaries between the water and the air bubbles and so encapsulate the air bubbles – with the result that the egg white becomes stiff. If the egg white is beaten too much, the water can detach itself from the proteins again and the process will fail. Traces of egg yolk can also cause problems. Its molecules weaken the coating around the air bubbles making it harder for the proteins to attach themselves. Egg yolk and other fats only cease to be a problem once the egg whites are stiff.

Why does tomato sauce get stuck in the bottle?

Tomato sauce is a highly viscous paste made from tomatoes, spices and other additives. It can only move if it is shaken or if sufficient force is applied. When liquid leaves a bottle, the space at the bottom has to be replaced by air bubbling through the water from the open end of the bottle. Tomato sauce is too viscous to allow air bubbles to pass through it easily, so a vacuum forms in the bottom of the bottle when the sauce moves away from it. This creates a pressure difference that works against the force of gravity, which is pulling the sauce downwards.

There are several ways of getting the tomato sauce out. Plastic bottles are flexible and can be squeezed to eliminate the vacuum. Sauce in rigid bottles can be helped along by hitting the base of the bottle, sometimes with disastrous results. The third method helps with both types of container – simply insert a drinking straw deep into the bottle so that air can pass through the tomato sauce to the bottom.

Why must brandy be warmed before it can be set alight?

Brandy and other strong spirits contain at least 15 per cent by volume of ethyl alcohol (ethanol). Ethanol is a short-chain alcohol, which is highly flammable. But it is not the liquid alcohol itself that burns but the mixture of air and alcohol floating above it. Even at room temperature some of the pure ethanol evaporates. In alcoholic drinks, the alcohol is diluted with water and other substances.

The higher a drink's alcohol content, the easier it is for combustible vapour to form; if the brandy is warmed, there will be even more ethanol molecules available to feed a flame. The opposite is true for chilled alcoholic drinks.

Like alcohol, motor fuel also consists of hydrocarbon molecules. Petrol evaporates readily, which is why the vapour in a half-empty tank is highly flammable. Diesel and heating oil are made up of long-chain molecules that evaporate less easily, although the higher the temperature the more these oils evaporate and mix with air.

Why do jelly babies grow larger in water?

Jelly babies made with gelatine swell in cold water, while those made with gum arabic, agar, pectin or starch fall apart. The fact that gelatine jelly babies swell up has nothing to do with osmosis. In osmosis, water passes through a membrane to balance out different concentrations of dissolved particles, such as salt or sugar. However, jelly babies are not wrapped in a membrane. Gelatine is able to absorb large quantities of water because the water molecules attach themselves to particular structures of the protein molecules in the gelatine, increasing their volume. Chemists call this gelatification.

After three days in water, a jelly baby is nearly twice its original length – but its flavour leaves much to be desired.

What gives sparkling water its sparkle?

The addition of carbon dioxide gives mineral water a refreshing sparkle. This requires high pressure and relatively cool conditions, for only then will the gaseous carbon dioxide dissolve – forming carbonic acid – and be evenly distributed in the water. In an unopened bottle of mineral water, the pressure is four to five times higher than it is in the ambient air. When a bottle is opened, the pressure is reduced and some of the carbon dioxide escapes. This is visible in the form of bubbles that come fizzing up suddenly. Carbonated mineral water fizzes when pressure in the bottle is reduced, but not all the carbon dioxide bubbles escape at the same time. This is because, although carbon dioxide is lighter than water – which is why the fizz travels upwards – it is heavier than air and slows down when it reaches the air.

In the case of naturally sparkling mineral water, the water absorbs carbon dioxide as it seeps through volcanic rock. As the water travels to the surface, the fizz escapes but the gas is collected and added to the water just before bottling.

Why does a slice of bread always fall buttered side down?

This is simply due to the height of the table from which the buttered bread has fallen. Most dining tables are about 1m high. A slice of bread will usually slip off the edge of the plate sideways and this – in combination with gravity – will cause it to turn on a horizontal axis. If the bread were to fall from a low stool it will not turn far enough and so would land with the unbuttered side down. The same would happen if the bread fell from a table that was 2m high, because it would complete a full revolution. But most dining tables are just the right height for the drop to be sufficient for half a rotation – which means bread often lands on its buttered side.

How do food moths find their way into closed containers?

Moths that attack food are either flour moths, which are about 1cm long, or the slightly smaller Indian meal moths. It is not the moths themselves but their larvae that cause trouble in the pantry. The hungry caterpillars are attracted to food with a high carbohydrate content such as grains, cereals, nuts, sugar and dried fruits, although tea and coffee are also at risk.

The grubs follow their sense of smell and can cover distances of several hundred metres. They then munch their way through paper and thin plastic packaging. When newly hatched, they are small enough to work their way along the screw threads of jars. Not even airtight containers made of glass, metal or heavy-duty plastic can keep them out. If the food is contaminated with moth eggs prior to purchase, the larvae will hatch in their own private food supply.

How can we eradicate these household pests?

Once food is infested with moths it is inedible. Dispose of it and clean the cupboard thoroughly to prevent future infestations. Traps baited with pheromones to lure the male moths will stop them reproducing. Ichneumon wasps are used in some warehouses to track down and destroy moth eggs.

What exactly is 'soya meat'?

Soya meat is what the food industry calls 'textured soya', and it is used as a meat substitute. It has a similar protein content but no cholesterol. Herbs, spices and flavourings can be added to give soya a meaty taste, since it has barely any flavour of its own. Soya meat can also be made into different shapes, with variations in texture and composition designed to make it resemble a variety of meat products. Textured soya is produced from ground-up soya beans that have had up to 95 per cent of their fat content removed by being pressed repeatedly. The ground soya is then mixed with water and passed through an extruder – a kind of mincer. Depending on the machine's setting, the soya emerges in different forms: mince, strings of sausages or thick ropes which can be sliced to make 'steaks'.

Soya is a versatile plant that can be made into all sorts of foods, from oil and sausages to tofu.

True or False?

It will soon be possible to produce food without plants or animals

All foods ultimately consist of fat, protein and carbohydrate molecules, but their basic building blocks are individual atoms. We should be able to assemble these atoms to create molecules artificially. However, this would be complicated and expensive. Scientists can already take single muscle cells and propagate them in large numbers under laboratory conditions. The same process could be used to grow meat in a laboratory so that there would no longer be any need to slaughter animals. Before these cell cultures can be made into anything resembling a traditional steak, they must grow into a three-dimensional form, undergo strength training – just the same as real muscle tissue – and be marbled with irregular areas of fatty tissue. We should not expect to see an affordable product on our supermarket shelves soon.

How is instant coffee manufactured?

When we make a cup of instant coffee we are actually brewing it for the second time. This is because the manufacture of instant coffee begins with making coffee. Roasted and ground coffee beans are boiled in pots, just as they would be for mocha or Turkish coffee, only much more intensely. When most of the water has evaporated and the coffee grounds have been removed, the thick coffee solution that remains is dried. There are two ways of doing this: spray-drying and freeze-drying. With spray-drying, the wet coffee solution is sprayed through a fine nozzle into a drying tower, from the bottom of which hot, dry air rises. The remaining water evaporates and the instant coffee powder collects at the bottom of the tower. With freeze-drying, the coffee extract is chilled to -40°C in minutes. The slabs of iced coffee then pass into a warm vacuum, where sublimation takes place – the solid ice turns straight into water vapour, bypassing the liquid stage. All that is left behind is the familiar dry, water-soluble powder.

How do we find out how many kilojoules are contained in a food?

A kilojoule is a unit of energy. It also refers to the energy value of food which, when digested, can be used by the body. The older term calorie is still used in some parts of the world.

Kilojoule content is established by physical means. The food is pressed into a solid pill into which is inserted a fuse. This sample is then set alight inside a 'bomb calorimeter', into which oxygen is pumped until the food sample is burnt. The heat this generates passes to the calorimeter's steel container and from there into a tank of water. The rise in temperature in the precisely measured quantity of water shows how much heat was generated during combustion.

This value is, however, only theoretical, since human digestion cannot be equated with purely physical combustion. Firstly, this method ignores the fact that the dietary fibre that burns in the calorimeter is excreted unused by the human body. Secondly, it takes no account of the fact that the metabolism of a human being (or any other animal) processes food with the same physical energy content at different speeds and with different degrees of efficiency. And not everyone's metabolism uses food in the same way.

What's the difference between long-life and fresh cow's milk?

Long-life milk will keep for a long period of time. Fresh cow's milk is pasteurised – heated at ultra-high temperatures to kill bacteria – and then homogenised to ensure that the fat globules are evenly distributed throughout the milk. This stops them forming cream on top of the milk too quickly. If the treated milk is quickly packaged under sterile conditions, it will keep for several months, even at room temperature. Ultra-high temperature processing (or UHT) is a form of pasteurisation. Milk is heated for a few seconds to a temperature of more than 140°C and then rapidly cooled to about 4°C. Ordinary pasteurised milk, on the other hand, is only heated to 75°C and then cooled. Ultra-high temperature processing destroys some of the milk proteins and vitamins and turns the lactose to caramel. This alters the flavour, but it also makes long-life milk easier to digest than regular milk.

How do we know...

... how the hole gets into macaroni?

Macaroni dough is pushed through the nozzle of a pasta-making machine. The nozzle attachment for macaroni has a disc at its centre that leaves only a narrow ring for the dough to pass through. The disc is secured to the nozzle by two or three thin brackets, which obstruct the dough and split the macaroni lengthways. After the dough has passed the brackets, it continues to move through the nozzle where it is pressed back together, with the result that a continuous tube of pasta emerges from the machine. The trick is to dry the soft macaroni without allowing the dough tubes to collapse. This is done by suspending the strands of pasta vertically, so that they are only squashed together at the point from which they are hung, and this part can be trimmed off later, after they have dried off.

Macaroni is dried for several hours at a temperature of 100°C before it is packaged.

How do the holes get into cheese?

Cheese is made by adding rennet – a combination of various enzymes – to milk so that it coagulates. In the process, some of the protein and fat turn to 'curds', which separate from the rest called 'whey'. The curds are then pressed into moulds and stored for a time until they mature.

In some cheeses, such as Tilsit, irregular holes appear because the curds are poured loosely into a mould and not pressed. More whey drains away during storage, leaving cavities in the cheese. Large, spherical holes like those in Emmental – known as 'Swiss' cheese in some countries – develop in cheeses that are packed into moulds and develop a firm rind. The holes are produced by the lactic acid bacteria present in raw, unpasteurised milk. To make cheeses with particularly large holes requires the addition of propionic acid bacteria, which consume the lactose and produce carbon dioxide as the cheese matures. The small bubbles join with larger ones, and as the rind develops and the curds harden, they are unable to escape. The holes vary in size according to how the cheese is matured and the number of bacteria present.

Why are mouldy cheeses mouldy?

Cheeses like Stilton or Camembert are injected with harmless, edible, mould cultures under the most stringent hygienic conditions. Once the mould has spread, there is no room left in the cheese for any harmful alien moulds. If cut cheese is left uneaten for a while the 'good' mould can grow over the cut surface or spread to other types of cheese stored nearby.

The large holes in cheeses such as Emmental – or 'Swiss' cheese – are created by lactic acid bacteria, which is present in raw, unpasteurised milk.

How do baking powder, sourdough and yeast work?

These all serve to 'relax' dough and make it rise by producing bubbles of carbon dioxide. Baking powder creates carbon dioxide by means of a chemical reaction, while with sourdough it is produced by bacteria, and with yeast the job is done by the tiny, individual yeast fungi.

Baking powder contains two salts that react with each other. These are usually sodium bicarbonate and an acid salt. The chemical reaction occurs as soon as water and heat are introduced – which is why dough only rises once it is in the oven. However, the volume of yeast dough and sourdough increases prior to baking. Lactic and acetic acid bacteria in sourdough produce carbon dioxide from sugar at temperatures between 20°C and 35°C. Single-celled yeast fungi create carbon dioxide when fed with sugar and water and exposed to temperatures of about 32°C. This is why sourdough and yeast dough have to rest in a warm place before baking. The heat of the oven at any temperature above 45°C will kill off the bacteria and yeast fungi.

How are 'best-before' dates determined?

Best-before dates are often seen as equivalent to 'use-by' dates, but the two terms differ in one important detail. The best-before date indicates that the product can be kept at least until the specified date, provided it has been correctly packed, transported and stored. Food manufacturers are obliged to find out the average period after which their products will spoil in sealed packaging, and the best-before is set well before that time. Once passed its best-before date, food does not have to be thrown away immediately. It may continue to be sold and used, provided it has been examined and found not to have spoilt. In contrast, products that have a use-by date have to be removed from circulation immediately that date has passed. This is why perishable foods like meat are given a use-by date. The same applies to the 'expiry' date, which is mainly used on pharmaceutical products.

The best-before date is only valid for food in sealed packaging. Once opened, bacteria can quickly spoil food.

What kinds of temperatures are found in the domestic kitchen?

From icy, below-zero temperatures in the freezer to the heat of a gas flame – a broad range of temperatures can be found in any kitchen!

1500°C	The hottest point inside a gas flame
700–1000°C	The temperature of a grill – the elements of which are red hot
180–250°C	Baking or roasting in a hot oven
170–200°C	Deep-frying in hot fat
120–180°C	Shallow frying in the frying pan
100°C	Boiling point of water at sea level – the boiling point is lower at higher altitudes
80°C	Casein curdles in hot milk and forms a skin
61°C	Egg white begins to set – the yolk starts to set at 65°C, and at 85°C egg white is hard
0°C	Freezing point of water, and of most foods
–6 to –24°C	Refrigerator temperature (fridge compartment -6°C; freezer compartment -24°C)

Why shouldn't olive oil be stored in the refrigerator?

The mono- and polyunsaturated fat content of an edible oil is what determines the temperature at which it will solidify, what its burn point is in the frying pan, and how quickly contact with oxygen will turn it rancid. Olive oil consists of up to four-fifths monounsaturated fatty acids, which makes it ideal for frying. The oil keeps best at about 15°C, and in the refrigerator it will turn cloudy with flaky lumps until its temperature rises again. Whether or not this spoils the flavour is a matter of opinion. Other oils with higher proportions of polyunsaturated fatty acids are more sensitive to heat and oxygen, and unrefined salad oils keep longer under cool conditions. All edible oils are best stored in the dark or in dark-coloured bottles.

Why doesn't honey have to be stored in the fridge?

Honey is widely thought to be the only food that does not spoil. Its high sugar and low water content makes it an environment in which bacteria and other micro-organisms cannot multiply. Honey also contains certain enzymes and aromatic acids that are effective against microbes. Doctors may even use honey to help serious wounds heal more quickly – in the form of honey bandages, for example. On average, honey consists of 70–80 per cent sugar and only 15–20 per cent water. Added to this are small amounts of pollen, flavourings, minerals, vitamins, enzymes and other proteins. Unlike sugar – which is pure sucrose – honey contains various kinds of sugar, especially glucose, fructose and small amounts of sucrose and dextrin. The proportions of the sugars determine how long it will take for a particular honey to crystallise and harden. And even the fridge cannot prevent that from happening.

What gives honey its honey colour?

The colour comes from the colour of the nectar or honeydew collected by the bees. The term 'honey coloured' usually refers to a warm, slightly ochre shade of golden yellow, although different varieties of honey have different colours – from pale yellow rapeseed honey, through reddish-brown chestnut honey and dark brown forest honey. Lime honey has a greenish tinge. The standard RAL colour chart, used throughout Europe, lists the typical hue of acacia honey under the designation honey-yellow RAL 1005, between golden yellow and maize yellow. The transparent and luminous colour of all honey fades as – after weeks or months, depending on the variety – it crystallises and sets.

Why do human beings cook their food?

Although humans could survive on a diet of raw food, cooking and smoking food extends the menu available to us as these processes destroy inedible, bitter and toxic substances. Some plants are indigestible until their constituents are broken down by cooking. Heat also kills off any germs and smoked foods keep for longer than fresh food.

Air-drying delicately flavoured Parma ham for the luxury end of the food industry.

Why do we eat – and enjoy it?

Food guarantees our survival. It provides the body with vital nutrients and supplies us with energy. To enable the body to use food, during the process of digestion various organic compounds – protein, fat and carbohydrates – are released. These provide the energy that the body uses to move and think, to fuel such automatic physical processes as breathing, blood circulation and maintaining a constant body temperature. Minerals, vitamins and trace elements are essential for fine-tuning the metabolism, and the water in food provides the body with a basic supply of liquid.

Eating satisfies us and makes us feel good. From being breastfed as babies, taking in food is linked to the release of neurotransmitters in the brain that are responsible for pleasure and happiness. Thanks to their unusually high number of nerve endings, the lips function as an erogenous zone, which is stimulated to some extent by eating.

A full stomach triggers a feeling of wellbeing in the brain. Various ingredients in food go one step further – the best-known being chocolate, which directly releases the 'happy hormones' endorphin and serotonin. Wheat beer contains vitamin B, which is considered beneficial to the nervous system, while hot and spicy foods work indirectly. As soon as the taste buds feel the heat they emit pain-relieving neurotransmitters, which have a mood-enhancing effect.

Does chocolate always give you spots?

Chocolate on its own does not produce spots – except as a result of an allergic reaction to cocoa. Ordinary spots arise because the skin produces too much sebum. Sebum builds up in the hair follicles and forms unsightly blackheads or becomes inflamed and creates pimples. A shift in the body's hormonal balance is mainly responsible for excess sebum, which is why pimples and acne are most likely to develop during puberty.

Recent studies show that diet does have some influence, although fatty foods, which have always been thought to be a culprit, don't actually cause acne.

This is a problem that rarely occurs among native populations with diets rich in fat, such as the Arctic Inuit. In the western world, acne is a widespread skin complaint and members of native populations don't usually develop it until they move into a modern city.

Western eating habits can encourage skin impurities. Scientists suspect that a diet rich in simple carbohydrates is the indirect trigger for acne. The consumption of large amounts of sugar leads to a continuous release of insulin, which creates a shift in the hormonal balance and thereby stimulates the production of sebum.

This would suggest that chocolate does at least encourage spots, particularly the high-sugar, low-cocoa varieties.

However, chocolate won't do any harm to those whose diet is based on complex carbohydrates and small amounts of sugar and white flour.

Provided you eat a balanced diet, chocolate won't give you spots.

Why is every little bean heard as well as seen?

Pulses such as beans, peas and lentils are notorious for producing gas in the intestinal tract. Like the fibre in cereals, they contain a large proportion of indigestible carbohydrates, or oligosaccharides. Because the enzymes in the gastric juices and the small intestine are unable to split them into smaller sugar molecules that the body can use, they pass unaltered into the large intestine where specialised bacteria are waiting to break them down into methane and carbon dioxide.

Small amounts of gas in the intestine usually escape through the mucous membranes and leave the body silently via the lungs. Larger amounts, however, expand and bubble in the bowel before they find their way out either through the mouth or through the anus – usually accompanied by an odour.

Artificial sweeteners can also cause flatulence if they are taken in large quantities. After all, the reason they have the advantage over fattening sugar is that they are not digested. This can mean that gas builds up in the intestine if we eat too many sugar-free sweets.

Current research shows that beans and other legumes lose most of their indigestible fibre if they are fermented for long enough – at least two days – before eating. It is claimed that fermentation even improves the flavour.

Reach for a fruit yogurt rather than chocolate or sweets if you fancy something sweet.

What makes us eat – hunger or appetite?

Hunger does not care what the body gets to eat so long as it gets something. Hunger is triggered by physical signals, including an empty stomach, falling blood sugar levels and metabolic changes. Serotonin and other hormones regulate these signals in the brain's control centre for hunger, situated in the hypothalamus, or interbrain.

Appetite allows people to eat even when they feel 'full'. This desire is dependent on the food on offer, its colour, the way it is prepared and how it is perceived by an individual. Appetite tends to be psychologically driven, specifically by the limbic system that also controls emotional behaviour and learning processes. Thinking about certain foods or their aroma can stimulate the appetite.

The greater the choice and the bigger the portions, the more people will eat – regardless of how hungry they are. People are more likely to tuck into food they know will taste good. If a favourite dish – such as fried chicken – has been dyed blue it is likely to be left untouched. The same applies to crunchy snacks of insects for the average westerner, although in some cultures they are considered to be a great delicacy.

Will a pickled herring cure a hangover?

A pickled herring – a flavoured fish popular in some parts of Europe – is said to cure a hangover. It is most effective against headaches and nausea if eaten prior to the consumption of alcohol, but will not cure a hangover. A hangover is caused by the breakdown of alcohol in the body. During this process water, minerals and vitamins are eliminated, which causes dehydration. In addition, alcohols produced in fermentation – which give drinks their particular flavour – degrade into toxic intermediate products that cause the heart to pump less oxygen to the brain. Different types of alcohol produce different levels of cytokines. These are the neurotransmitters that cause weakness, nausea and headaches. Studies have found that brandy can cause worse hangovers than red wine, rum, whisky, white wine, vodka or even pure alcohol.

Eat high-protein, fatty foods like fish, nuts or cheese prior to consuming alcohol. These top up mineral levels and hold the alcohol in the stomach for longer, where it begins to break down. Consuming plenty of water or fruit juice during a night out, and also before going to bed, will prevent dehydration and dilute the alcohol, thereby reducing overall consumption. The morning after is best dealt with by eating a breakfast high in minerals and vitamins and taking outdoor exercise. Painkillers tend to do more harm than good when mixed with residual alcohol in the body.

Can cola and pretzels help cure diarrhoea?

Cola and pretzels help to restore water, salts and other nutrients to the body. The salt in the savoury snack and the sugar in the drink replace the large numbers of electrolytes lost through diarrhoea, which the body needs in order to function properly. Cola restores the body's fluid balance, while the sugar and white flour in the pretzels provide easily digestible calories to help the weakened metabolism recover.

The combination of sugar and salt is also good for restarting the intestinal process that supplies essential sodium to the cells. Cola and pretzels together have a good psychological effect on children, who view them as treats. Receiving a treat cheers them up and helps speed recovery. Those who don't like cola and pretzels can try sweet tea and plain biscuits.

Cola and pretzels can help the body recover from a bout of diarrhoea.

True or False?

Mushrooms & spinach should not be reheated

There is no reason not to reheat these foods, but they should never be kept warm or left to stand at room temperature for long periods. The ancient caution predates the advent of modern refrigeration. Mushrooms have a high protein and water content and, in lukewarm conditions, provide an excellent breeding-ground for bacteria and other microbes. This can be prevented by quickly cooling leftover mushroom dishes and reheating them to at least 70°C before serving.

The problem with spinach is the nitrate it contains. This vegetable absorbs large quantities of nitrate as a result of modern farming methods. It is only hazardous when it is turned into nitrite by bacteria – most likely at room temperature – since nitrite turns into nitrosamines which are carcinogenic.

Can it really be healthy to have worms in your stomach?

Tapeworms in the intestine could be used as a slimming aid. Beef and pork tapeworms can live inside human beings for up to 25 years, and most infected people will not experience any problems. During the infestation, people lose weight because the worm lives on the remains of food in the intestine, preventing it from being absorbed by the body. Only occasionally does its presence cause abdominal pain, flatulence or a rumbling stomach. Sufferers can be unaware they are harbouring a tapeworm, unless they happen to see parts of it following excretion. Medication will in most cases eradicate tapeworms.

Worm infestation should be taken seriously because there are other kinds of tapeworm, as well as roundworm and threadworm, which can trigger highly unpleasant symptoms. Their larvae can penetrate the lungs and other organs where they can cause serious damage. The healthiest way for worms – highly prized in some African and Asian cultures – to find their way into the intestines is for them to be eaten boiled or fried.

Why does asparagus make urine smell?

Asparagus contains the amino acid asparagine, which contains sulphur compounds. This has a mild smell of asparagus but enzymes in the stomach break down the sulphur compounds. British biochemists have identified six different sulphur compounds in human urine that are responsible for the rancid smell.

Not everybody 'sulphurises' asparagus. Only 50 per cent of Europeans have the necessary enzyme in their stomachs. The predisposition to this characteristic is a dominantly inherited trait, which means that one parent passes it on to all of their children – even if the other parent is able to eat asparagus without unpleasant consequences. The gene responsible for this has yet to be identified.

Another British study has discovered that this distinctive scent is only perceived by some people. So some people produce the sulphurous smell and don't know anything about it. It seems that only ten per cent of the population has a sense of smell with the necessary degree of sensitivity.

Why does chopping onions bring tears to your eyes?

Onions contain sulphur compounds that are released as soon as the knife cuts into the vegetable. The substance that causes the greatest trouble is a compound known as propanthial S-oxide, which rises into the air and combines with water in our tear ducts to form stinging sulphuric acid, which is what makes us cry.

The only way to stop onion chopping being such a tearful business is to prevent the oxide from getting into the eyes. This could be done by cutting the onions under water, turning your face away or wearing protective goggles – even wearing contact lenses can help. The more coarsely you chop the onion and the sharper the knife the less oxide is released. It is not clear whether it is helpful to chill the onion prior to chopping it.

On average, smokers are slightly lighter than nonsmokers – but they are certainly not healthier.

True or False?

Salt can be deadly

Too much sodium chloride – or common table salt – can paralyse the body. An excess of salt is fatal at concentrations above 3g/kg of body weight. At this point there is so much salt in the bloodstream outside the cells that it draws water out of the cells. This process of osmosis is triggered when the concentration of salt on both sides of the cell membrane is not the same, in an attempt to rebalance the salt concentration. Because the cell membrane does not let salt in, water has to be released to dilute the over-salted blood; because the cells cannot work without fluid, the body ceases to function and the victim dies.

The healthy body does require extra salt, taking what it needs from food. Between 50g and 300g of salt can prove fatal to an adult. A large tablespoonful (25g) is enough to kill an infant.

Does smoking help to make you slim?

Nicotine from cigarettes travels straight to the brain where it links up with nicotinic acetylcholine receptors and sets in motion a series of hormonal and physical reactions. These include the dulling of hunger pangs and a reduction in the appetite, so it is true that smokers eat less. The fact that smokers' hands and mouths are engaged in the business of smoking also means they are less likely to snack. At the same time, nicotine makes the heart beat faster, raises blood pressure and increases intestinal functioning, so that the body uses more energy. This is why smokers weigh on average 2kg to 3kg less than nonsmokers.

The reason people put on weight when they give up smoking is because smoking releases hormones in the brain's reward centre, known scientifically as the *nucleus accumbens*. People who try and quit smoking therefore have to look elsewhere for the accustomed feeling of reward or comfort, and many turn to food instead. At the same time, cutting out cigarettes improves a person's sense of taste and smell, which makes many new nonsmokers enjoy their food more so that they eat larger quantities than before. In most cases, they only put on a few kilos which they can lose by finding other distractions and exercising more.

Why do we burp?

Belching is a way of removing unwanted air from the digestive tract. Babies must be 'burped' because they are not able to coordinate feeding and breathing properly. A baby may inadvertently swallow air that can cause unpleasant pressure in its oesophagus and stomach, which results in discomfort and crying.

Adults also take too much air into their stomachs if they eat or drink too quickly, and some people have been known to swallow huge quantities of air.

When we burp after drinking a carbonated drink, such as sparkling mineral water, beer or champagne, the cause may be carbon dioxide, which escapes from the carbonated drink into the stomach. In rare cases, faults in the mechanism that closes the link between the oesophagus and the stomach can lead to belching, which may also bring up some stomach contents as well as air.

Apart from ruminants, such as sheep or cattle, most animals cannot belch. The ruminants mainly bring up methane, a gas which is released by bacteria in their digestive tracts.

Why do we feel thirsty?

Thirst develops in the brain, when the body tells it that there is too little water or too much salt in the blood. In both cases, the cells are deprived of water, which can disrupt both the intricate processes taking place in individual cells and the overall function of the larger vital organs. When the proportion of water falls by half a per cent – this is the thirst threshold – the body sends a message to the brain's thirst centre, where the hypothalamus produces the anti-diuretic hormone – or ADH – which slows down the excretion of water from the kidneys. The kidneys control the body's fluid and electrolyte balance and can be badly affected by long periods of dehydration, which can leave them unable to function properly. The production of saliva in the mouth is also reduced, which is why a dry throat is a sign of thirst.

Heavy sweating and fever, diarrhoea and high levels of alcohol consumption lead to dehydration. If a human being loses more than 15 to 20 per cent of his or her body weight in the form of water they will die of thirst.

Why shouldn't quiz contestants drink cold water?

Although thirst makes it more difficult to think, taking a drink at the wrong time can also slow down thought processes – especially if the drink is cold water. Scientists in Britain conducted an experiment that showed that thirsty people were better able to perform an intellectually demanding task if they took a sip of water immediately beforehand. They performed ten per cent better than the group that was left to go thirsty. However, subjects who weren't thirsty and drank a small amount of cold water during the course of the task came off 15 per cent worse. The researchers suspect that the temperature of the water was linked to the loss of concentration. Cold water has to be heated in the stomach to bring it up to normal body temperature, thereby reducing the energy available to the brain for intellectual activity.

. . . the civet gut can improve coffee's flavour?

Indonesian 'kopi luwak' is thought to be the world's most expensive variety of coffee, with 1kg costing as much as 1000 euros ($US1400)! The reason for this is the exotic production method, which keeps annual output to less than 230kg. Prior to being dried and roasted, the coffee beans pass through the digestive system of the Asian palm civet. The civet's gastric juices attack the beans and enzymes in them break down various proteins – processes that transform the aroma of the coffee beans. Canadian food scientists have tried to uncover the secret of this special coffee. They examined the coffee beans under an electron microscope and noticed tiny dents in their surfaces. These probably contain traces of the digestive secretions that lend the coffee its particular flavour. A similar result may be achievable on a larger scale in the laboratory by using lactic acid bacteria to create a controllable fermentation process. Mass production would reduce the price of this gourmet variety, but also its exoticism.

Why does coffee help to keep us awake?

The caffeine in coffee and black tea passes into the bloodstream and affects the central nervous system. It competes with the neurotransmitter adenosine, which would ordinarily attach itself to certain receptors in the nerve cells to calm them and prevent them becoming overloaded. Caffeine has a similar structure and fits into the same receptors. It therefore blocks them and prevents the calming adenosine effect, which leads to increased alertness and an ability to concentrate.

Constant and heavy coffee consumption causes the nerve cells to react by creating more receptors. Therefore, the same amount of coffee will, in the long term, have a less pronounced effect. Suddenly giving up caffeine can cause temporary physical withdrawal symptoms. Coffee and tea have the greatest stimulating effect if they are drunk in small quantities throughout the day.

Why do older people sometimes drink too little?

The elderly often lose their natural sense of thirst, or they suppress it to avoid having to go to the toilet so frequently. The human body is composed of 50 to 60 per cent water but, from the age of about 60, that percentage decreases by about five per cent, which is why more wrinkles appear with age.

The kidneys control the body's water and electrolyte balance, and in older people they become less adaptable. They continue to excrete large amounts of water and sodium, even if insufficient water has been taken in.

They also react less to ADH, the most important neurotransmitter in the control circuit between the brain and the body's water content. As a result, many older people simply forget to drink enough.

This knowledge is no help if people are deliberately drinking less because they suffer from incontinence or a weak bladder. This lack of fluids can have serious health consequences, which is why experts advise older people to do pelvic floor exercises and wear incontinence pads.

Functional foods

Increasing numbers of advertisements c▮
that, as well as their usual ingredients,
products contain health-promoting addi▮

The term 'functional foods' is applied to a range of products that are nourishing and also claimed to contain additives that promote health or prevent disease. Vitamins, minerals, fatty acids, bacteria, fibre and other ingredients are added to ordinary foods and drinks to promote good digestion or to bolster the immune system. Some foods are said to reduce blood pressure, reduce the risk of diabetes or even fight cancer. Medical researchers are also working on incorporating vaccines into food.

How long have there been functional foods?

The crews that sailed with Christopher Columbus were plagued by scurvy – a lack of vitamin C – but it wasn't until the 18th century that sailors discovered that citrus fruits such as lemon could help prevent this deficiency disease. Later still it was discovered that the cause of the condition was the unbalanced diet on board ship. From that time on, people sought to remedy the situation with preserved fruit and pickled vegetables. The 19th century saw the introduction of the first foods designed to fortify invalids, such as wholesome malt beer, malt lozenges and Ovaltine. In the 1920s, fluorine and iodine were added to salt in an attempt to protect the public from tooth decay and goitre.

Since 1993, the FOSHU (Food for Specific Health Use) label has been used in Japan to market health-promoting foods that are claimed to be effective against osteoporosis, cancer, intestinal obstruction or heart attacks. In the USA and Europe, food labelling is not permitted to refer to its therapeutic properties, and claims made in advertising slogans remain a problem area.

Are functional foods effective?

Most functional food products promise more than they deliver. The amounts of beneficial substances in these foods are usually negligible, and medical experts dispute whether artificially added vitamins and minerals are as effective as those found naturally in fruit and vegetables. Dietary fibre will always pass undigested through the intestines and speed up digestion, whereas omega-3 fatty acids work differently when taken as a supplement than when consumed directly in the form of fish or linseeds. The health-promoting effects of some functional additives have been demonstrated in the laboratory – but not in humans.

What supplements are available in the shops?

Manufacturers distinguish between several product groups. Minerals and vitamins are designed to meet daily requirements, prevent deficiencies and strengthen the immune system. Dietary fibre is aimed at improving digestion, and probiotic bacteria cultures are claimed to support the 'good' bacteria in the gut. To reduce cholesterol levels and guard against cardiovascular diseases, it is said that plant sterols and stanols should be taken regularly in large quantities, as well as omega-3 fatty acids in fish oil. In their natural form, antioxidants – vitamins C and E, for example – are thought to protect the body's cells and so prevent cancer – as are isoflavones in soya products, lignans in cereals and isothiocyanate in cabbage. Indigestible sweeteners, which have long been established in the market, replace sugar in

chewing gum and other sweets, thereby indirectly preventing tooth decay.

Does probiotic yogurt improve digestion?

We know that these microbes survive the journey through the gastric juices and reach the intestine, but the intestines contain so many bacteria that the incoming microbes won't have an easy time finding a corner to settle in. If large amounts of the probiotic bacteria are consumed on a regular basis, they will be able to populate a niche. The same is true when the intestines have to be repopulated after a course of antibiotics. However, yogurt is packed with protein and calcium, so is a healthy food to eat whatever the arguments for and against probiotics.

Stand by for the marketing onslaught – sweets full of vitamins that are actually good for you!

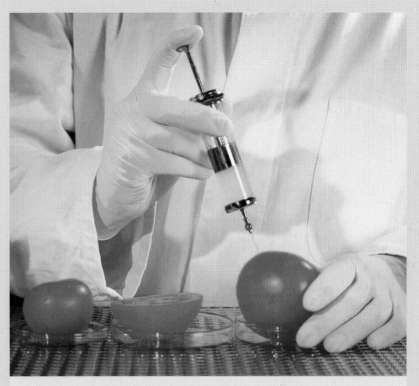

It would be convenient – but less interesting – if vitamins could be extracted from fruit and injected into tasty, non-perishable fruit gums or drinks. In reality, it isn't so simple. There are serious doubts about the efficacy of vitamin additives in comparison to those consumed naturally.

Is there such a thing as a healthy sweet?

The healthiest sweet food is fresh fruit. It contains a natural combination of vitamins, antioxidants and long-chain sugars only, which the body does not break down immediately. Sweet foods that have not grown naturally can all be categorised in terms of their degree of unhealthiness. A sweet that contains almost no sugar, little fat and added vitamins and minerals is least unhealthy. However, the effect of these added ingredients is unknown, given the quantities in which they are added and that they are not natural food products.

How are functional foods manufactured?

Major food manufacturers have a number of laboratories where their products are tested and they develop new versions. They also conduct a number of tests to establish how to include additives. These are extracted from natural products and delivered as raw materials. Sometimes they are simply added to products but in other cases it is a more complicated matter of food design. Examples include virtually fat-free ice cream, which tastes as creamy as the real thing as a result of the addition of insulin and chicory fructose.

Why do flags flutter?

The behaviour of everyday objects can usually be explained by physics. But apparently simple phenomenon – such as a length of fabric fluttering in the breeze – can be complex and hard to describe mathematically. What physicists do know is that air swirls down the length of the fabric, turning into chaotic turbulence at its end – where the flag flutters.

Why do kettles sing?

Kettles emit numerous subtle sounds, ranging from hisses to shrill whistles, while water is being heated. The actual sounds depend on water temperature. At the beginning of the heating process, tiny air bubbles burst at a rapid rate. These are dissolved in the water, and then travel upwards. When the water reaches about 65°C, steam bubbles form on the bottom of the hot kettle and collapse in the cooler upper layers of water with a loud, deep, plopping sound. These bubbles don't reach the surface until the water reaches a temperature of almost 100°C, and then the muffled plop is replaced by a gentle bubbling sound. The escaping steam passes through a whistle – if the kettle is fitted with one – which creates the shrill sound that signals that the water is boiling.

How does whistling with two fingers work?

Whistling with the aid of fingers makes a sound that is much louder than it would be if created using lips alone. The fingers ensure that the expelled air vibrates at the required frequency. The physical principle that underlies both types of whistling is the same: a cavity and a rapid flow of air that creates turbulence as it strikes an opening. The cavity is provided by the whistler's mouth, and the air travelling through it produces an acoustic vibration that, with practice, results in a relatively pure – and penetrating – note. Experienced whistlers can produce a powerful sound with just one finger, or by shaping their tongue in just the right way.

Why is there a hole in the front of a guitar?

The hole in the body of a guitar – and any other similar stringed instrument – provides an escape route for sound waves. A guitar has a wooden body to serve as an amplifier and to give the instrument its characteristic timbre. If the sound waves were unable to escape, a guitar would be barely audible because the strings when they are plucked or strummed produce only small vibrations. It is only when the vibrations are transferred to the wooden body and the air inside the body – and then released through the sound hole – that they are amplified enough to be clearly audible to an audience.

Electric guitars don't need either a hole or a sound box. Instead, an electric 'pick-up' catches the vibrations of the strings and feeds them into an amplifier, which forms part of the instrument.

In large crowds, at demonstrations or sporting events, people can make themselves heard by whistling loudly through their fingers.

Why do door hinges creak?

The creaking sound made by a rusty hinge is produced by the same process that gives rise to the screech of chalk on a blackboard or a rich note from a violin. It occurs when two surfaces with microscopic roughness repeatedly catch for fractions of a second as they rub against one another – in what is known as the 'stick-slip effect' – causing rapid vibrations. The sound becomes audible when a resonating surface such as a door, a blackboard or a violin is linked to the source of the friction – a hinge, some chalk or a violin bow and strings – so that it vibrates in response and amplifies the sound. Creaking hinges can be fixed by lubricating them with oil, which separates the rough surfaces that are causing the friction.

Why do Velcro fasteners make a ripping sound when they are pulled apart?

Velcro fasteners are made up of plastic loops of varying thickness. On one side of the fastener they lie in a tangle of thin threads, while on the other they are arranged in a regular pattern of stiff, rounded loops that have been cut away on one side to form a hook. The ripping sound is produced when the two enmeshed strips are pulled apart. The tips of the hooks – which are stuck in the tangle of thin loops – are stretched until they pull clear and spring back to their original shape. The ripping sound is due mostly to the smack of plastic on plastic as the hooks snap back into place. Attempts to develop a silent Velcro fastener have not been successful.

How loud is...?

How loud a sound seems depends on the size of the vibration in the air, but also the ear's ability to perceive it, which depends on the frequency of the vibrations

0dB	Only just perceptible to the human ear
20dB	Whispering
50–60dB	Normal conversation
60–75dB	Radio at low volume
70dB	Passing car
85dB	Ear protection advisable
95dB	Passing lorry, trumpet played loudly
105dB	Disco
115dB	Pneumatic drill, woodwind instruments played loudly
130dB	Jet aircraft, Formula 1 car race
150dB	Gunfire; deafens within half a second
200dB	Major explosion; fatal, destroys pulmonary alveoli

Is house dust always grey?

When viewed under a microscope, dust is made up of all sorts of different particles, which can be almost any colour. The overall colour of dust depends on how many and what type of particles enter a house through open windows or on the clothes or shoes of inhabitants. In desert regions, or in the pollen season, dusty surfaces can acquire a red or yellow sheen. However, dust does usually appear to be grey because it is a mixture of many particles. Each particle is tiny – down to a hundredth of a millimetre in size – and they do not generally have smooth, monochrome surfaces. Some particles absorb light, some allow light to pass through, while others scatter it in all directions. The overall result is a dull, uniform grey.

House dust is made up of microscopic particles of all shapes and colours.

Why can't we see the end of a ray of light?

A 'ray' is an idealised narrow beam of light – a line showing the path of the light. You see a beam of light directed straight at you because virtually all the light reaches your eyes. You can see a beam of light from the side if the light stays on for a while and some of the light is scattered off dust or fog in the air.

Light can be described as electromagnetic waves that, unlike sound waves, can travel through a vacuum. The waves travel in straight lines at 300 000km/sec (in a vacuum) until they are absorbed or deflected. When you switch a torch on at night you can't see the front of the beam advancing out into the darkness – or the retreating end of a beam when you switch the torch off – because the radiation is moving so rapidly.

The chimney of this chemical factory belches reddish smoke into the atmosphere.

Why can smoke be different colours?

Smoke is made up of small particles and droplets of water, all suspended in warm air. Its colour depends on the material that is burning and the temperature of the fire. Thick, black smoke contains large quantities of soot particles, which are usually the product of mineral oil or other forms of carbon burning at low temperatures. The soot absorbs light, which makes it appear black.

A cooler fire, damp fuel or even tobacco will produce carbon particles, ash and vapour, which reflect the light and appear grey or white. Fire-fighters are trained to recognise the nature of a burning substance by the colour and quantity of the smoke emitted. Furniture tends to burn with a grey smoke, but will be darker if it contains a lot of plastic. Yellow or brown smoke can mean that the fire is short of oxygen and that hydrocarbons are not burning completely. Opening a door in those circumstances would introduce oxygen and cause an explosion. Greenish or reddish smoke usually indicates burning chemicals. Breathing it in could be dangerous.

When is glass transparent?

Whether a material looks transparent to the human eye depends on whether it allows visible light (electromagnetic radiation with a wavelength of between about 400 and 700 nanometres) to pass through it or not. If these waves are not reflected from the surface (or scattered by it, as they are with frosted glass) but are able to penetrate the material, it is their interaction with the atoms that makes the difference in transparency.

For a material to appear transparent depends on its structure and the wavelength of the prevailing light. If the glass contains atoms of other elements – compounds of Antimony, for example – it may block the progress of light waves and appear opaque.

For visible light, some glass is transparent but this is not true for parts of the infrared or ultraviolet spectrum. Other materials may be transparent to other types of electromagnetic radiation. Infrared light can shine through silicon, while X-rays can look deep inside the human body and radio waves can penetrate solid brick walls.

Is it true that glass is not solid but extremely viscous?

Glass is a molten mass that has hardened. Its molecules are untidily arranged in an amorphous order. This is because the hot, fluid glass is cooled rapidly. But it is a myth that glass is a viscous fluid at room temperature, and that its molecules gradually move downwards, propelled by the force of gravity. It has been claimed that the proof of this can be seen in large, medieval glass window panes, which are often thicker at the bottom than at the top. Research has found that the panes have not changed over the centuries, but that the thickening was the result of production methods in the Middle Ages. Medieval glaziers knew that their panes were uneven and usually set them in their frames with the thicker end at the bottom, although some church windows have been found with panes that are thicker at the top. Recent measurements show that the glass lenses in very old telescopes have not changed by even a fraction of a millimetre over time. Only when glass is heated does it become fluid once more.

Why do mirrors show left and right in reverse, but not top and bottom?

A flat mirror doesn't reverse anything – it is only the brain of the viewer that misinterprets the image. It places the viewer in the position of the person in the mirror, and suddenly the wedding ring worn on his or her left hand appears on the right. A concave mirror will show a reversed mirror image if the object is correctly located relative to the mirror – at a point beyond the mirror's focal distance.

They may look as different, but glass is made from quartz sand.

Why don't clothes look the same in the changing room as they do outside?

Fitting rooms are lit by incandescent light bulbs or fluorescent tubes, while exterior spaces are lit by the Sun. Although we think of all three as white light, they are made up of slightly different combinations of wavelengths (colours), although the differences are not obvious unless seen side by side.

Physicists assign each light source a 'colour temperature', while common usage sometimes assigns a quite different 'temperature'. We talk of incandescent bulbs creating 'warm' light, while fluorescent light is 'cold' and the light of the midday Sun is 'neutral'. As each of the different combinations of light waves strike clothing, some is reflected and reaches the retina of a viewer's eye. The light spectrum that reaches the eye will be different depending on the light source. This is especially noticeable on shopping trips, when colour might be important. The brain usually balances out slight colour variations. If it 'knows' that a dress is turquoise, it will appear just as turquoise in the midday Sun as it does at sunset.

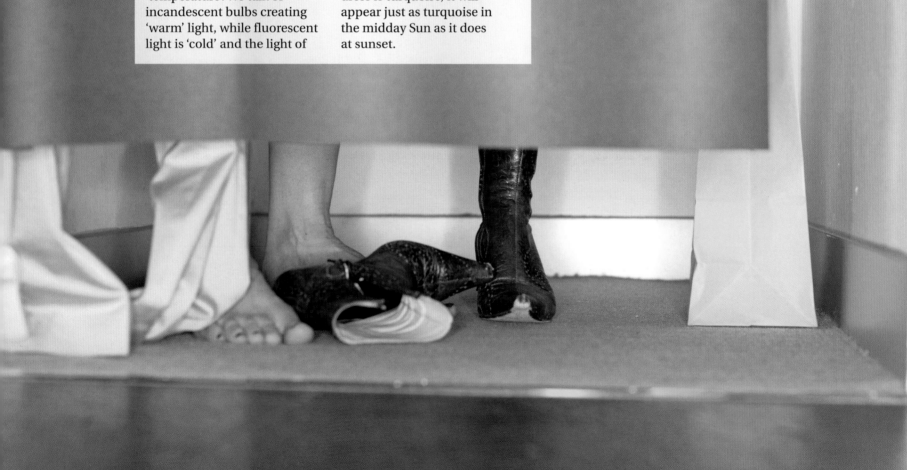

How bright is...?

This table shows the illuminance – the amount of light falling on a surface, taking into account the way the human eye works – provided by various sources. It is measured in lumens/m² (lux)

0.001 lux	Starlight
0.01–0.1 lux	Full Moon
0.1 lux	Poor street lighting
1–100 lux	Sunset
50–200 lux	Domestic living room
200–300 lux	Commercial office
1000–5000 lux	Shop display window
2000–5000 lux	Dull, cloudy day
5000–10 000 lux	Operating theatre
25 000–50 000 lux	Hazy day
50 000–100 000 lux	Bright sunlight

How do smokers blow smoke rings?

This requires smoke in the respiratory tract, practice and lips set in a circular shape. The trick is to breathe out small amounts of air to create circular swirls. The air flows from the throat through the mouth and is slowed down slightly at the sides when it reaches the lips, although it continues to travel unhindered through the centre of the mouth. This creates a ring of turbulence. The air in the ring rotates relatively quickly and so traps smoke particles, thereby making its shape visible. If this happens in draught-free surroundings, the smoke rings may last several minutes.

The Earth can also produce smoke rings – which are occasionally seen when a volcano erupts – and so can jet aircraft and guns when the conditions are right.

Why do grease spots make paper translucent?

Paper consists of chaotic layers of long-chain cellulose molecules. The surface appears matt white because it reflects light and scatters it in every direction. It only becomes shiny when fat – or water – smoothes out the uneven patches and sends all the beams of light in the same direction.

Paper is made more opaque by adding more paper fibres to increase the scattering of light.

To make the paper less opaque requires a reduction in its scattering properties. This can be achieved by making the paper thinner. Alternatively, grease on the paper lays long fat molecules between the cellulose of the paper and smoothes out the sudden change in optical properties between the two. As a result, the light is scattered less and the paper becomes more translucent.

How do you drink through a straw?

First, you must reduce the air pressure inside your mouth – you must suck. The fluid in a glass or bottle is usually subject to atmospheric pressure – both on the surface of the fluid inside the bottle and on that inside the straw. By sucking on a straw, you remove air molecules from the top of the straw, thereby reducing the internal pressure. The air pressure on the rest of the fluid in the bottle has remained the same so is now greater than that inside the top of the straw, with the result that the liquid is pushed up the straw.

True or False?

Long balloons are harder to inflate than round ones

It takes a little more lung power to blow up long balloons than round ones – provided they hold the same amount of air and the rubber is the same thickness. As a balloon is inflated, the person blowing it up is working against the rubber's 'reset force' – caused by its molecules trying to return to their original position. The strength of this force is proportional to the surface area of the balloon. Since a long balloon with the same capacity as a round one has a larger surface area, it produces more resistance when being inflated.

How does an iron work?

Textile fabrics can be smoothed with heat, moisture and pressure, even if they are badly wrinkled. Textile fibres – unlike those in paper – are not arranged haphazardly on top of one another. Instead they are arranged in an orderly structure by spinning, weaving or knitting. Plant fibres, such as those used to make cotton fabrics, have to be broken down and softened prior to processing, while animal fibres, such as wool and hair, are naturally flexible.

Pressure and heat on a flat surface smoothes the fabric and produces creases where needed. The key weapon in banishing creases in some materials is moisture. It penetrates the fibres and makes them swell, loosening the grip between them. While modern steam irons deliver moist air directly to garments, it used to be necessary to sprinkle water on before the hot iron was applied. After swelling, the fabric fibres are dried by the hot iron as it smoothes it, fixing the fibres in the desired position.

Modern steam irons provide all that is necessary for achieving smooth fabrics and perfect creases: heat, moisture and pressure.

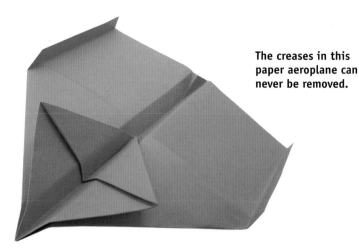

The creases in this paper aeroplane can never be removed.

Why can't creases be taken out of paper?

Paper is tolerant of rough treatment. As long as it isn't creased or folded, it can be twisted and rolled up and will still return to its original flat shape. Paper is made up mainly of cellulose fibres, most of them obtained from pulped timber. The slender fibres are flexible enough to bend, and when paper is rolled up they are pushed slightly closer together. However, the resulting curl can be removed by bending the paper back and pressing it out flat. When a sheet of paper is creased or folded, many of the tiny wood fibres are broken.

Soft paper contains textile fibres obtained from rags and is less prone to creasing. But even so-called wood-free types of paper contain up to five per cent wood fibres, which will break if enough pressure is applied.

How does electric current 'flow'?

Just as water flows through a pipe, electricity can be thought of as electrically charged particles – usually electrons – flowing though a wire or other conductor. Most electrons – whether they are in air, plastic or metal – are constrained so that they are only ever able to move around the nucleus of the atom of which they are a part.

In metals, some of the electrons are able to jump easily between the atoms. A battery tries to drive electrons in one direction (from the negative terminal to the positive terminal, since electrons carry a negative charge). If the battery isn't connected to anything, the electrons end up unevenly distributed, with an excess at one of the battery terminals (the negative one), and a deficiency at the other terminal (the positive one). If the two terminals are then connected to each other by a metal wire, a current will flow.

The surplus electrons don't just whiz from one end of the wire to the other. The whole pack of evenly distributed but loosely held electrons already in the wire are propelled along its length. They move forward through the wire, so that some of them are pushed out at the other end and back to the battery (doing some work on the way if the battery is part of a device such as a radio). The individual electrons move slowly during this process, but the signal that the motion has started travels at almost the speed of light.

What is fire made of?

Fire is a combination of chemical and physical changes in which substances interact to release heat, light, smoke and ash. Whether it is a fire in electrical insulation, an oil fire or a domestic fire, they all require three ingredients: combustible material; oxygen; and, a thermal trigger. If one of these is absent, the fire goes out. This principle is the basis of fire-fighting, which aims to cool the flames by smothering them, removing oxygen or depriving them of fuel.

Combustible material consists mainly of carbon and hydrogen, and can be solid, liquid or gaseous, from hydrogen to wax, oil, wood and plastic. Even metal will burn provided the temperature is high enough. Sufficient quantities of oxygen are usually supplied by the air, and the energy to light a fire is produced by friction or sparks.

Why will a piece of flint produce sparks?

Flint produces sparks because it is so hard. Pieces of flint are only striking tools, and the sparks themselves come from a sparking material such as pyrite – a compound of iron and sulphur – steel, or a composite metal with iron content. In the Stone Age, a firelighter was often a lump of pyrite, a firestone to strike it on and some dried moss or the tinder fungus sometimes known as touchwood. In medieval pistols, the flint created the spark by striking iron, while in today's disposable cigarette lighters a friction wheel creates tiny sparks from a composite metal that ignites the flow of gas. Expensive lighters use piezoelectric quartz crystals to produce sparks.

Why are candle flames yellow?

The colour of a flame is the result of several factors – the temperature of the flame, the chemical composition of the fuel that sustains it and the eye's ability to perceive it.

The predominant yellow colour comes from burning carbon molecules. After the heat has stripped the wax from the wick, the carbon initially clumps together into soot particles, which then burn further up with a bright light fed by oxygen in the air. The presence of some metal atoms in a flame can give rise to a rich range of colours. However, what an observer sees depends partly on the structure of the human eye, which is less sensitive to some colours than others. Reds appear less intense than yellows or greens, and so the flame may appear to be mostly yellow.

Can sugar be turned into plastic?

Primary school experiments demonstrate that chemistry is indispensable in our everyday lives. Chemistry allows us to develop new materials and control and alter a broad range of important properties of materials – everything from smells to colours, tastes, hardness and durability. Resins and plastics are easily produced from sugar, although it is cheaper to use oil.

How does glue stick things together?

Glues use two forces to connect the items to be joined together: adhesion, which is the relationship between the glue and the item to be glued; and, cohesion, the structural relationship between the glue molecules themselves. Good adhesion requires glue, at the point of application, to be as liquid as possible and to develop chemical bonds with the material to be glued. Good cohesion requires the glue to become as solid as possible after curing and to develop into a few, very long-chain molecules.

The most popular glues are solvent adhesives, in which the molecules of the adhesive (the actual glue) are dissolved in organic solvents or water. The name applies to general-purpose glue, superglue, paste and glue sticks. While the solvent evaporates in the air, the adhesive molecules can develop chemical bonds with each other. Hot-melt adhesives work best at temperatures of about 120–180°C, when it is at its most adhesive. Cohesion improves as the glue cools down.

Pressure-sensitive adhesives stay soft and sticky at almost all temperatures – they are used on labels and sticky tape and only need gentle pressure to adhere. Two-component adhesives can cure because their two components – a binder and a hardener – react chemically with each other once they are mixed together. The so-called superglues – usually cyanoacrylates – need moisture in order to cure. The glue gets this from the surrounding air and then hardens in seconds as a result of a chemical chain reaction.

Why doesn't glue stick to the inside of the tube?

Solvent glues consist of about 30 per cent adhesive and 70 per cent solvent, the latter component being organic molecules or water. Only when the solvent is able to escape can the remaining molecules stick together. This can happen in the tube itself, if there is a large enough hole. Tubes are usually made of aluminium or a synthetic material with a smooth surface that provides the adhesive molecules with few sticking opportunities.

Superglue in a tube consists of the adhesive only but this, too, requires contact with air from which it captures the water that starts a chain reaction among its molecules, causing it to harden. This can happen inside the tube if moisture gets in.

Why does sticky tape stick more firmly to paper than Post-it Notes do?

Post-it Notes and sticky tape both make use of pressure-sensitive adhesives that stay soft all the time and work through the force of adhesion only. While Post-it Notes can be peeled off and reused, sticky tape can only be peeled from very smooth surfaces. The difference lies in the form of the adhesive molecules.

In sticky tape, the molecules lie randomly on a carrier layer and their attraction to the molecules of the other surface is effective throughout its length. The attraction is so strong that when the tape is pulled off it can take paper fibres with it. On uneven, microscopically fissured surfaces there is an additional effect which reinforces the adhesive strength. Tiny air bubbles are trapped during adhesion and expand when the tape is pulled off. Low pressure is created within these bubbles, and this needs to be overcome.

On Post-it Notes, the adhesive forms tiny balls with a diameter of 30–50 micrometers. A bed of another adhesive holds them onto the note, while still allowing them to move. Thanks to the spherical shape of the balls, the adhesive forces work only at the point of contact. When pressed more firmly they spring back, and only a small area of adhesive touches the paper. The many small adhesive points provide barely any adhesive force to prevent the notes from being peeled off. Since they are able to rotate in their bed of adhesive, the bonding area remains 'fresh' for longer.

Natural adhesives

Humans have been using adhesives for thousands of years, although most were copied from nature

Hot-melt adhesive	Bees process wax at body temperature to make their honeycombs. The wax subsequently hardens in the air.
Pressure-sensitive adhesive	Spiders' webs contain threads with an adhesive that never hardens. Some carnivorous plants make use of adhesive in droplets to catch their prey.
Solvent adhesive	Termites have glands on their heads that produce an adhesive that they use to bond soil and bits of plants in order to construct extremely hard structures. Swallows, nuthatches and some other birds also use their own secretions as nest-building adhesives.
Wallpaper paste	Paper wasps chew up wood shavings to produce a cellulose adhesive with which they build their papery nests.
Superglue	Insects, like the lacewing, lay their eggs on leaves and ensure they stay in place with a drop of protein adhesive that cures on contact with air. Mussels are also firmly attached to rocks or pieces of wood with an adhesive that hardens even underwater.
Wood glue	The sticky, white milk in some types of rubber plants has a similar composition to the wood glue used by carpenters.

Swallows use natural adhesive – clay or soil mixed with their unusual saliva – to build extremely strong nests.

How was ash able to help on washdays?

Fat and ash make clothes dirty, but those two substances also provide a way of getting them clean again. The alkalis in ash can be combined with fatty acids to make soap. The manufacture and widespread use of soap did not take off until the 19th century. Previously – and as early as ancient Roman times – clothes would be soaked overnight in a tub of water with the addition of ash, soda from rocks or urine to loosen the dirt.

It was later discovered that fat could be turned to soap by the addition of ash. Vegetable and fish oils, as well as tallow obtained by boiling cattle and pig bones, were suitable fats. During the soap-making process, 'surfactant' molecules are formed that reduce the surface tension of the water, thereby helping to mix fat and water. These molecules are water insoluble at one end, and water soluble at the other. In large numbers they are good at breaking down dirt stains and wrapping them up in such a way that the water soluble ends point outwards, after which the whole thing is simply rinsed away.

Soap boilers were in charge of soap production until the process was industrialised in the 19th century.

How do laundry whiteners work?

Whitening agents or optical brighteners must not be confused with bleaches or stain removers. An optical brightener's function is preventative, and its aim is to make whites appear as bright as possible. The problem inherent in white textile fibres is that they start off with a slight tinge of grey. Over time they tend to yellow slightly because the fibres change and absorb most of the blue light waves from sunlight, which makes reflected light appear more yellowish.

To counteract this, textile and detergents manufacturers add fluorescent dyes, which transform invisible ultraviolet light into visible blue light. This fills the gap in the spectrum and further strengthens the blue content, which means our eyes perceive a 'bright white'.

These whitening agents come out in the wash, which is why they have to be added to every load of washing. These substances are not easily bio-degradable and so their use has become controversial.

Why do vinegar and lemon juice make good cleaners?

The acetic acid found in vinegar or citric acid from lemon juice helps to remove the stubborn 'lime' stains that build up around taps in some areas. Lime stains are formed by calcium and magnesium ions contained in tap water in the form of salts. The 'harder' the water, the more salts it contains. These salts stay behind when the water molecules evaporate and are best removed with acid. A weak acid, such as vinegar or lemon juice, is sufficient to break the salts down so that they can be washed away.

True or False?

Tonic water glows under disco lights

This interesting effect is even noticeable when tonic water is exposed to sunlight from the side, causing the optical brighteners within it to give off a faint blue glow.

The substance responsible is quinine, which also provides the distinctive bitter flavour, even after a 50 000-fold dilution. Quinine – an alkaloid that originally came from the bark of the cinchona tree – reduces fevers, has antiseptic properties and is the classic remedy for malaria. Like detergent brighteners, quinine molecules in tonic water fluoresce in response to ultraviolet light, providing the blue glow in a disco.

Why is it so difficult to remove blood and grass stains from clothes?

Grass, cherry or red wine stains are more difficult to remove than shoe polish or lubricating oil because, unlike fats, they don't provide anything for detergent surfactants to enfold and eliminate. If pigment molecules are not immediately diluted and washed out, they 'get stuck' between the fibres once they have dried. Some well-known household remedies, such as aspirin, salt, starch and ethyl alcohol, may help remove the stains, and special detergents promise similar results. Bloodstains are an exception – they contain protein molecules that coagulate at 42°C. Consequently, if hot water is used in an attempt to get rid of a bloodstain, the heat will make it adhere more firmly to the fabric.

The following advice applies to any stains that are still visible after rinsing – if the pigments won't shift, they have to be made invisible with a stain remover. Stain

removers are based on the principle of oxidation. They release highly reactive oxygen molecules that form colourless compounds with the pigment molecules – although this can also take some of the colour out of the fabric itself. Centuries ago, the same effect was achieved by laying the laundry out on a lawn. The short wavelength radiation in sunlight would shine on the damp linen and release oxygen molecules that oxidised away some of the stains.

Laying the laundry out on the lawn in an attempt to shift stubborn stains.

Why is red cabbage sometimes blue cabbage?

Red cabbage should properly be known as 'purple cabbage', because its colour is a mixture of red and blue. The soils in which it grows and the way it is cooked also helps to determine which of the two colours dominates. In some parts of the world the vegetable is known as red cabbage, whereas in others they speak of blue cabbage. The colour is caused by violet anthocyanin molecules, which react like a pH-value indicator for the environment, turning red in acid conditions and blue in neutral to alkaline conditions. They may even turn a greeny yellow in very alkaline environments. This cabbage is therefore redder when it grows in acid soil, and its leaves are strongly tinged with blue in alkaline environments. The colour is also affected by cooking – especially if vinegar and lemon juice are used, as they are in some parts of Germany.

Why do some spectacles turn dark in sunlight?

The first so-called photochromic lenses appeared on the market in the 1960s. These were made from glass that contained molecules of silver halides, which undergo a chemical change when exposed to the ultraviolet light in sunlight. The silver halide is split into silver and halogen atoms and the silver particles cause the lenses to darken, absorbing some parts of the visible light spectrum. When the lenses are no longer exposed to ultraviolet light, the silver and halogen recombine to form transparent silver halide molecules and the lenses lighten accordingly.

One of the problems with glass photochromic lenses was that the thicker parts of prescription lenses would darken more than the thinner parts, since the chemical responsible for the change was distributed evenly through the glass. This was overcome following the introduction of plastic photochromic lenses in the 1980s.

When paper leaves the factory it is snowy white, but this changes over time.

Why does paper turn yellow with age?

The degree to which paper turns yellow depends on its lignin content. Lignin comes from the wood used to make the paper. This giant molecule ensures that the cell walls of plants are hard and stable but it decays in the Sun's ultraviolet light. The more lignin paper contains, and the longer it is exposed to sunlight, the more yellow it will become as a photochemical reaction with oxygen reduces lignin to its smaller component parts. These absorb some of the blue light, which is why the complementary colour yellow dominates in reflected light, making the paper appear yellow. Lignin is not responsible for paper becoming brittle and crumbly – the paper's acid content is to blame for that.

Paper manufacturers have various methods for maintaining the whiteness of their product. High-quality paper contains barely any lignin these days – it is wood-free. The addition of blue pigment aims to deceive the eye, because blue and yellow combine to make white. Some papers contain brighteners, similar to those found in soap powder.

Why don't euro coins go rusty?

An electric current should begin to flow when two different metals come into contact, especially if they are immersed in salt water. There is a potential difference (a voltage) and one metal passes electrons to the other metal while it corrodes.

The one and two euro coins are made up of two different metals, so it would be logical to expect a flow of electricity and consequent corrosion. However, the gold- and silver-coloured portions of the coins are made up of similar alloys and so don't really have any significant potential difference. The gold-coloured part contains 75 per cent copper, 20 per cent zinc and five per cent nickel, while the silver part is made up of 75 per cent copper and 25 per cent nickel only.

Why do apples and avocados discolour once they have been cut?

Many cell walls are destroyed when a knife cuts through fruit, and this sets off a chain reaction. Colourless organic compounds called phenols in the cell react with enzymes in the fruit as well as oxygen in the air. After several interim stages, this causes brown pigments to form. The amount of these pigments increase with the number of destroyed cells, rises in temperature and time. Rapid cooking halts the process because it destroys the enzymes. Sprinkling lemon juice on the fruit will also slow down the process, because the vitamin C molecule in the juice is an antioxidant and keeps oxygen away.

This reaction is also the reason why raisins are the colour they are and why fruits and vegetables 'bruise' when their cell walls are damaged. All that is required is for the skin to be punctured so that sufficient oxygen can pass through to the bruised area.

How do 'click-on' hand warmers work?

A small plastic bag containing a transparent liquid in which a small metal disc floats acts as a transportable heating pad. Clicking the disc sets off a chain reaction, and the liquid turns into solid crystals. Within a minute the material becomes hot and solid and continues to give off heat for a long time afterwards. The material in the pad can be returned to its liquid form by immersing the pad in water that has been heated to 70°C. The secret of these latent heat stores lies in the material they contain – a mixture of water and sodium acetate-trihydrate. At temperatures above 58°C it is a liquid and below that it turns into a supersaturated solution that should crystallise out. In airtight conditions and if carefully handled it can remain liquid when cool. In this state the material is highly susceptible to any disturbance. Just the click of the metal disc provides a point of crystallisation, and the salt's ions abruptly begin to build a crystal mesh and to embed water molecules. This process releases all the energy that was added during liquefaction, and as a result the pad heats up.

Hand warmers are most welcome, especially at outdoor events in winter.

Why don't plastic kitchen utensils melt?

Plastic bowls can melt when placed close to a heat source such as a toaster, but some plastic cooking utensils remain solid and intact even if placed on a hotplate. This is because not all plastics are the same. Whether a synthetic material is derived from oil or starch molecules, it is always made up of polymers – long chains of organic molecules, with repeating structural units. The characteristics of synthetic materials depend on the structure of the polymers. The three principal groups are thermoplastics, duroplastics and elastomers.

Plastic bowls are made of the most common of these materials, the thermoplastics. Their string-like molecules are arranged next to one another and can slide past each other when heated. Consequently, a heated object deforms and the new shape sets on cooling. This characteristic is useful for the manufacture of cheap, mass-produced goods, and for producing sophisticated workpieces. Kitchen utensils are made of a duroplastic, with chain molecules that mesh tightly and irreversibly. Additional heat will not melt the material, and cooling won't make it malleable again either. Duroplastics can only be worked mechanically once they have hardened, which makes them suitable materials for heat-critical components as well as for hard coatings and varnishes. The molecular chains in elastomers engage in wide meshes, which are barely affected by heat. This makes them ideal as a basic material for rubber bands and car tyres.

Realistic artificial snow for theatre productions poses a challenge for technicians.

How is theatre snow made?

On the ski slopes, artificial snow comes out of a snow gun that produces real snowflakes made of frozen water. On stage, in a department store window or on a film set, this solution would make things soggy, so plastic snowflakes are used. However, Hollywood snow is expensive and not biodegradable. For this reason, scientists developed artificial snow made of potato or maize. The authentic-looking flakes made of foamed starch fall as lightly as the real thing and, with a little moisture, can even be used to make snowmen and icicles. The next rainfall washes away the environmentally friendly raw materials.

In the theatre, fire safety is particularly important so a new kind of artificial snow made of foamed polyethylene is used. This is fire-resistant, less expensive and more realistic than previous types of indoor snow. Although the individual flakes don't look like snow crystals, the way they fall as well as the size and distribution of the flakes is close enough to the natural stuff to convince an audience.

How do sunscreens work?

Sunscreens can to used in two ways to protect against sunburn. Physical filters or pigment filters keep unwanted ultraviolet radiation away from the skin by providing an intermediate layer that reflects the rays. The pigments used – such as titanium oxide or zinc oxide – are not perceptible to the naked eye. Their numbers increase with the light protection factor, and at high levels they lie on the skin like a thin layer of pigment. Protection is also provided by chemical filters, which are completely unobtrusive. Once applied, their molecules are absorbed by the outer layer of the skin where they intercept the ultraviolet light by means of a chemical reaction before it can cause any damage to cells. In order to be effective, chemical filters must be applied about half an hour before exposure to the Sun.

Where do fireworks get their colours from?

Rockets that emit showers of sparks for entertainment have existed since the days of ancient China. In the 9th century BC, charcoal was mixed with saltpetre and sulphur to make black powder that produced magical exploding lights in the sky. Today, magnesium compounds provide the basic raw materials. Magnesium burns with an intense white light, as does aluminium. If strontium compounds are added, the flame turns scarlet. Borax – a compound of sodium, boron and oxygen – provides the greens; barium a yellowy green and lithium the purple. Orange hues are derived from calcium and sodium compounds, copper results in a light blue, while iron powder ensures there are sparks and glitter effects. The pyrotechnician's art consists of mixing raw materials – some of which are poisonous and others that are explosive – in such a way that the display occurs safely high overhead and not close to the ground.

Getting the mixture right – a firework's colours depend on its contents.

True or False?

Sugar in the petrol tank of a car will stop it working

Most foreign matter has no effect on modern cars, as the fuel has to pass through several filters on its way from the tank to the engine. Sugar would not reach the filters, let alone pass through them, because it is not soluble in petrol. It would sink to the bottom of the tank. The only way of stopping an engine with sugar would be by adding so much that it covers the tank outlet, preventing any petrol from reaching the engine. Sand would do that job just as well.

The old movie trick of sticking a banana into the exhaust pipe to stop a car is also a myth. The pressure created by the exhaust fumes would forcably expel the obstruction.

Do car tyres always have to be black?

The black colour of tyres was originally the result of the production methods employed. Rubber – which is a pale yellow, milky substance obtained from rubber trees – was the raw material first used for the manufacture of tyres, although synthetically produced versions tend to be used these days. In its natural state, latex rubber is sticky and only acquires the required degree of hardness, elasticity and wear-resistance after it has been vulcanised (a process that ties the rubber molecules together) and soot has been added.

For brief periods during the 20th century there were passing fads for white tyres, and tyres with a blue tread also appeared on the market for a short time. In both cases, the colour was only a layer that had been applied to the black rubber, and it tended to come off.

Only recently have manufacturers been able to replace the soot with a light-coloured silicate and to make tyres in any colour by the addition of a range of pigments. However, tyres in novel colours are still rare because the silicate process is expensive. Market research indicates that most consumers prefer black tyres because they consider them to be more reliable.

When does fish smell fishy?

Fresh fish doesn't smell, a smell only develops when fish has been in transit for a long time or after long periods in the freezer. The smell is caused by micro-organisms that settle on the fish – specialised bacteria that transform some amino acids into amines by releasing carbon dioxide. This causes the simplest and most prevalent amino acid in fish – glycin – to turn into the simplest organic amine, methylamine – which has a strong smell of fish.

The longer a fish is exposed to air the more active the bacteria are, and the stronger the odour. However, there is a simple remedy – a slice of lemon. The fruit acid in lemon provides a chemical counterpart to the methyl-amine alkali, reacting with it to become amino salt. This captures the smell, because only free amine is volatile and able to give rise to the objectionable smell.

Why do all rubbish bins smell the same?

Not every rubbish bin stinks. This is because the smell depends on a bin's contents and how often it changes. When a bin does stink, it smells of 'rubbish bin'. Although every mixture of garbage emits hundreds of different smell molecules, depending on its composition, it only takes small amounts of gases from decomposing animal matter to superimpose themselves over everything else. The brain's smell centre works economically. To recognise 'typical' fragrances in perfume and other complex aromas, it is sufficient to perceive the major single fragrances – known as the lead substances.

In rubbish, these lead substances are primarily amines, ammonia compounds and hydrogen sulphite molecules as well as occasional quantities of methane. If the rubbish contains meat, fish or egg, the putrefactive bacteria will in time turn the protein into the foul-smelling amines putrescine and cadaverine. High levels of nitrogen – from fruit, vegetables and animal products – release foul-smelling ammonia compounds, also a preliminary stage in the creation of amines. Methane-producing bacteria will feel particularly at home if rubbish has little contact with air.

Overall, humidity and warmth provide particularly good conditions for the microbes that produce strong smells. Preventative measures include placing the bin in the shade, mixing the refuse well and adding dry matter.

How does perfume work?

Perfume and other smells travel directly to the brain. They are made up of hundreds of odour molecules that are so small and weigh so little that they disperse well in air and remain active for a long time. This is how they arrive at the roof of the nasal cavity, which is where the olfactory receptor neurons of humans and animals are located. Most odour molecules are carbon compounds that attach themselves to the receptors that stick out of the surface of the nasal cavity. Human beings have almost 350 different receptor types, and they are capable of distinguishing between the thousands of different smells produced by combinations of odour molecules.

Perfume and other fragrances trigger instant, often unconscious reactions. Experiencing a smell as pleasant or unpleasant is culturally dependent and learnt, with a 'bad' odour automatically triggering disgust or fear in an individual, while 'good' smells elicit elation, stimulation or relaxation.

Why am I able to recognise granny's larder by its smell?

Granny's larder – or any location you know well from childhood – often has a specific blend of odours. This combination of odour molecules fixed itself in your brain during childhood. The more frequently the receptors sent your brain the same analysis and evaluation, the more intensive the effect in the neighbouring hippocampus, where memory is located. As a result of this proximity, your brain automatically reacts to the smell of granny's larder – or any similar combination – with associations and memories that were stored in childhood.

Why does catmint have such a powerful effect on some cats?

Catmint is the common name for a group of small flowering plants that have a narcotic effect on over 50 per cent of cats. The animals meow, tremble, run and roll around as if drunk for a few minutes. This sensitivity is inherited, and only occurs in adult cats, including tigers and lions. The trigger for this behaviour is the essential oils and terpene molecules in the mint, particularly nepetalactone, actinidine, iridomyrmecine and matatabilactone – which are inhaled, although it is not clear what precise effect they have on the brain of a cat. In the case of human beings, catmint tends to have a calming and pain-relieving effect.

The so-called 'scaredy-cat' plant has the opposite effect on cats, dogs and other mammals. A member of the colourful *Coleus* family, it also releases essential oils on contact, producing a smell that animals find repulsive. The common wormwood plant also repels dogs.

Sensitivity to the smell of catmint triggers apparently crazy behaviour in the majority of cats.

Is it true that geraniums will keep gnats out of a room?

There are about 250 kinds of geraniums – the popular name for members of the genus *Pelargonium*. On contact, the fragrant *Pelargoniums* release essential oils from tiny hairs on their leaves, which break off when touched. Lemon-scented geraniums are considered to be particularly effective for keeping gnats away, as well as wasps and other insects.

Laboratory experiments have shown that catnip also has a repellent effect on gnats, while various *Plectranthus* hybrids – relatives of the 'scaredy-cat' plant – also have a repellent effect on a range of insects.

Do you really understand everyday technology?

Good technology is easy to use – one twist or push and the safety lock opens, the car starts, the oven heats up, the light comes on and the camera is focused. But how does it all work? What are the mechanisms and phenomena behind modern devices? We need to understand the problem that has been solved by a gadget before we can understand how useful it is.

What does a Chinese typewriter look like?

Printing a newspaper in China, Japan and Korea requires several thousand characters. No keyboard could contain them all, which is why mechanical typewriters in those countries are fitted with rolls that sit above large type cases with carefully arranged characters. Typists take about two months to learn how to control the arm in order to pick out the required character and press it onto the paper. Once expert, they achieve considerable speed because each keystroke produces a full word.

A Chinese typewriter requires considerably more machinery than the typewriters westerners are used to.

Why are the numbers 1, 2 and 3 at the top of the dial on a telephone, but at the bottom on a pocket calculator?

The reason for the different keypads is historical. Electronic pocket calculators are the direct descendants of mechanical calculators and cash registers. They used lever gears for which it was best to have the lower numbers, including zero, at the bottom. This resulted in a nine- or ten-key panel that continued to be used for mechanical office calculators. Column one had 0 at the bottom and 9 at the top; the next column 10 at the bottom to 90 at the top; then from 100 to 900, and so on. It wasn't until electronic calculators appeared that the number of buttons was reduced to ten.

The people who developed the telephone conducted a user study. In the early 1960s, when the first push-button telephones were seen in the USA, various keypad arrangements were put to the test. Two provided good results, and so the decision was made not to have two rows of five numbers, but instead to go for the arrangement that is common today. Here, too, the 0 is at the bottom – after 9, which was familiar from the old dials. This variation also turned out to be more natural for use with the alphabet.

Why shouldn't you put metal objects in a microwave oven?

The interior of a microwave oven is a closed metal space, which reflects the electromagnetic microwaves without heating up itself. The size of the interior of an oven is designed so that the microwaves from the source – and those reflected by the walls – add to and intensify the total microwave field. Metal items in the cooking chamber would screen off the rays and prevent the waves reaching the food that is to be cooked. Moreover, the waves would also cause the electrons in the metal to move, which could cause electrical discharges in the form of sparks. Thin foil and gold-rimmed plates especially should not be placed in a microwave oven, because their thin metal layer heats up well and may even evaporate.

Why can't a microwave oven be used to cool things?
Microwaves heat up food and other items that contain water. This is possible because water molecules are weak dipoles – they have a positively charged and a negatively charged end. In the microwaves' alternating field – which changes direction rapidly – the water molecules are constantly being rearranged. This means that they begin to move quickly – just as they would if they had been placed on a hotplate. In order to cool water, the movement would have to be slowed down – energy would have to be withdrawn from the molecules. That can only be achieved through sudden expansion or contact with a colder object. While heating with the help of microwaves occurs inside the food being cooked, cooling by contact with a colder object can only occur slowly, starting from the outside and gradually working in.

Why do egg cookers need so little water?

The trick is pure physics, and has to do with heat conductivity. Water in a saucepan transports heat directly to the egg, so that the hot, fast-moving water molecules pass their energy to the molecules of the eggshell and then on to the egg's interior. As a result, the egg white and the yolk set. In an egg cooker, the heat transfer is accomplished using steam instead of water. Again, the fast-moving molecules pass heat to the egg, colliding with each other and with the shell to pass on some of their energy. The more eggs there are to be cooked, the smaller the amount of water that is required because less space has to be filled with steam.

Why doesn't the ink in a ballpoint pen leak out?

At the heart of a ballpoint pen is a tiny sphere – the ball point – which is trapped in a socket at the tip of the ink chamber. The viscous ink in the chamber is pulled in a downward direction by the force of gravity, but cannot flow unaided out of the tiny gap between the socket and the ball.

The ink at the bottom of the ink chamber sticks to the ball, which is made of brass or steel and is often slightly scored. When the ball turns in its socket, it takes some of the ink out of the chamber and moves it onto the paper, where it adheres even better than it does to the ball. The socket is slightly narrowed towards the front to keep the ball from falling out, although it is loose enough to allow space for the ink. Vertical grooves on the inner surface of the socket also help the ink to flow around the ball. This arrangement also ensures that the ink does not dry out in the chamber. Since the ball is always covered by a layer of ink, this means that air cannot penetrate the ink chamber.

True or False?

Eggs explode in microwave ovens
Eggshells cannot withstand the rapid heating caused by microwaves. Conventional cooking causes the heat to travel slowly from the outside to the inside, and the steam that gradually forms can escape through the pores. Some people stick a pin in eggs to provide an additional escape route. In a microwave oven, steam is generated rapidly throughout the egg's interior, creating sufficient pressure on the shell to make it explode. The same principle is used when making microwave popcorn.

How do security labels help to prevent theft from shops?

There are two main categories of security labels: mechanical and electronic. Mechanical systems involve a container of pigment that escapes when opened with force. The colour makes the goods – especially clothes – unusable. Most security labels, however, work on electromagnetic principles, reacting to electromagnetic fields emitted by transmitting gates in the doorways of shops. Among these, most commonly used is the acousto-magnetic adhesive paper label, in which embedded metal strips begin to vibrate in response to a transmitted radio pulse and emit their own alternating signal, which is detected and causes an alarm to sound. Electromagnetic labels are a second type of paper label that produce detectable changes in the alternating magnetic field generated by a transmitting gate, again leading to an alarm sounding. Both types can be activated and deactivated when the goods have been paid for. A third type are radio frequency labels that have to be destroyed to stop them triggering an alarm signal. These square stickers contain a silvery spiral and an electric oscillating circuit that extracts energy from the aerial transmitted field and so draws attention to itself.

How do breathalysers work?

When drivers are asked to 'blow into the tube', an electro-chemical sensor measures the amount of alcohol (ethanol) present in the driver's breath. Ethanol reacts chemically with an electrode, which is generally made of platinum, and oxidises to form acetic acid. The acetic acid in turn releases a few electrons and ensures that a current flows. The higher the amount of ethanol in the breath, the higher the flow of current. To be sure that the reading actually corresponds with the blood alcohol content, the last drink must have been consumed at least 20 minutes prior to the test, to give the oesophagus time to 'ventilate'. Tobacco smoke in the breath, or acetone in the breath of a diabetic, can also affect the reading.

How does a gun silencer work?

Silencers work best on weapons that eject their ammunition at speeds that are slower than that of sound, as a supersonic bang would make an additional noise. The remainder of the bang comes from the hot gases created by the exploded gunpowder, which is what drives the bullet from the barrel. The more these gases can be slowed down, the less noise there is.

Silencers are usually made up of two areas: an expansion chamber and a number of baffles. In the expansion chamber, the hot gas suddenly has more room to move and is able to expand and become cooler – and the molecules collide less frequently. This principle is also used for the silencers on motor vehicle exhaust pipes. The baffles in the silencer cause further deceleration by causing the hot stream of gas to swirl and – because they are made from metal – they absorb part of its thermal energy.

Silencers are a necessary part of every secret agent's stock-in-trade – at least as far as the movie-going public is concerned.

What is a machine?

Whether it's a block and tackle, a pocket watch or a gear system, all mechanical machines are combinations of six basic elements – or simple machines. These are the lever, the wheel and axle, the screw, the inclined plane, the wedge and the pulley. They change the size and direction of a mechanical force and transform one type of energy into another. Simple machines are known to have existed as far back as Ancient Greece.

Can washing machines 'eat' socks?

Socks appear to get lost in the wash because their absence is noticeable by the fact that one odd sock is left behind. Manufacturers argue that missing socks do not disappear in the machine but after the wash. This is because small items often get tangled up inside larger ones, like pillow cases or jumpers. This is caused by the drum's movement, which changes direction regularly. As a result, larger items of laundry that have contact with the wall of the drum swing round more quickly than smaller ones that swim freely in the water. Smaller pieces of washing can also be left behind in the drum or the washing basket.

On occasion, an item of washing is 'eaten' by a washing machine. In top-loaders, items can slip into the gap between the drum and the machine's housing. In front-loaders, this gap is covered by a rubber seal, but during the spin cycle a small gap can appear as a result of the speed at which it is revolving. Because low pressure is created, it is sometimes possible for small items to slip under the drum.

How can clay houses be earthquake-proofed?

Scientists and engineers looking for ways to construct earthquake-proof buildings tend to concentrate on modern concrete and high-rise buildings. However, most of the housing that is destroyed by earthquakes consists of small buildings made from air-dried clay bricks. Construction engineers have now discovered a simple, low-cost way of making clay brick houses earthquake-proof as well. This is done by drilling holes into the walls and bracing them with a framework of wired bamboo struts. This kind of structure survived simulated earthquakes that exceeded magnitude 8 on the Richter scale in tests conducted on shake tables.

How do cranes make themselves taller?

When constructing a high-rise building, large loads have to be added to the growing structure. At the beginning of a construction project, when the building is still quite low, a tall crane would be unstable. This is why cranes that grow with the building they are helping to construct were developed. The mast consists of square, strutted metal elements and a climbing module – a frame with a telescopic hydraulic system – that is located just below the head of the crane. If the crane needs to grow taller, a new section of mast is brought onto the site while an electric motor powers the hydraulic system to push the head of the crane upwards so that the new mast section can be inserted into the gap.

Why would we want to buy 'intelligent' clothing?

The term is controversial, but clothing fitted with microchips can, with some justification, be termed 'intelligent'. Tiny processors can be inserted into hip seams, sensors and washable cables can be built into fabrics, and loudspeakers can be inserted into the collars of garments. Power for electronic components can come from solar panels fitted onto the shoulders of wearers or batteries that are charged by movement alone.

While it is already possible to integrate MP3 players, GPS systems and microcomputers into garments, developers forsee a much wider range of applications for so-called intelligent clothing. It might be useful for certain medical applications. Underwear could be fitted with electrodes to check vital functions, or knee bandages could warn wearers against making potentially damaging movements. Paramedics could wear jackets that store data from an examination or provide patient records at the site of an accident. Bicycle couriers could make telephone calls without needing to use their hands, and radio-controlled locks that would open automatically as they approach their bikes. It might be possible to develop garments for horse or bike riders that would blow up like an airbag immediately after an accident, providing spinal protection. While intelligent clothing may have novelty appeal at first, in a few years' time fashion designers will ensure that it doesn't stand out. There are bound to be buyers for chic evening gowns with built-in back-massagers.

A computer worn like a piece of jewellery may not be available just yet, but people are working on the concept.

How do swimsuits allow the Sun's rays to pass through?

So-called 'Tan Thru' swimsuits are designed to allow sunbathers to obtain an all-over tan, while preserving their modesty. This is achieved thanks to a synthetic fabric that reveals countless small holes when stretched, allowing about two-thirds of the Sun's rays to penetrate. Sunbathers appear to be decently clad thanks to the fabric's pattern, which has an optical structure that diverts the eye, tricking the brain into overlooking the holes.

An inventor has also developed a special synthetic suit to enable skiers to tan more than just their faces. Transparent foil allows sunlight and the ultraviolet rays responsible for tanning to penetrate. The suit also traps heat and prevents it from escaping, keeping skiers warm even when the air temperature drops to -30°C.

These modern-day seven-league boots are claimed to activate 90 per cent of a wearer's muscles.

Are these the modern version of 'seven-league boots'?

Human beings have always dreamt of overcoming their anatomical limitations and covering great distances quickly, which is probably why 'seven-league boots' appear in fairy tales. These days, we rely on technological developments rather than magic to help us be faster and get further and higher.

The idea for the modern seven-league boots came from observations of kangaroos. Scientists realised that the tendons on their hind legs act like springs, enabling the animals to take long, energy-efficient jumps. The initial prototypes looked like steel kangaroo legs and had to be strapped to the lower part of the body. This system was of interest to the military because it enabled a single individual to carry much heavier loads than normal.

Smaller and lighter versions have now become available as a sports and leisure item. The latest of these metal structures are strapped to the calves and, with practice, people can use them to jump distances of up to 4m and heights of up to 2m.

Can a diving suit deter a shark attack?

A special diving suit has been developed that should cause sharks to lose their appetite for a potential human victim from some distance away. Prey build up electric fields as they swim, and inside a shark's nose there are finely tuned sensors that can pick up the most subtle of electric fields, which is why sharks can detect a potential meal with ease. The revolutionary new suits don't disguise the electric field, but make use of it. Thin ceramic fibres embedded into the diving suits' conventional neoprene produce a small electric charge with every movement. The charge is directed via metal electrodes to both ends of the suit where it is discharged into the water, causing a large electric field to build up around the diver. This produces muscle spasms in any nearby shark, with the result that they keep their distance.

Could humans 'breathe' water like a fish?

Unless you are using a snorkel close to the surface, you will have to take a bottle of compressed air in order to breathe underwater – or make use of recently developed artificial gills. These gills extract air from the water and deliver it to the diver. At the heart of the system is a centrifuge on a small tank that can process about 200 litres of water per minute. By reducing the pressure in the tank, the air escapes from the water, just as gas bubbles appear when a bottle of sparkling mineral water is opened. One litre of water contains between 1.5 per cent and 2.5 per cent air, of which about 34 per cent is oxygen. A smaller version of the artificial gill is due to be integrated into a diver's vest. Instead of being dependent on compressed air bottles, divers will rely on battery power to keep them alive.

True or False?

We have clothing that makes us invisible

Harry Potter's invisibility cloak exists only in fiction. Physicists have demonstrated how a block of copper could be made invisible, but it was only invisible when viewed using waves in the infrared range, which spread around the copper as if it were not present. To mimic an invisibility cloak would require a technical trick. By recording images of a scene behind a garment and then projecting the film directly onto its front, it is possible to give the impression of being able to see through the wearer. This trick only works when viewed from particular angles, and when the ambient light is not too great.

How are landmines detected?

Finding landmines requires either well-protected heavy machinery that detonates the mines as it passes over them – or someone coming close to the hidden mines on foot. With both methods it is possible for mines to be missed, especially those that are buried deep in the ground. It is also the case that metal detectors are becoming obsolete because synthetic materials are increasingly widely used in the manufacture of bombs. Attempts are now being made to 'see' into the ground and to develop equipment that can detect molecules of explosives in the air. There is already a prototype mine-detection system that sends powerful sound waves into the ground from a distance of a few metres. This causes the mines to vibrate gently, and this sound response can be picked up using sensors. This system can reveal the type of mine as well as its location. Nuclear magnetic resonance – widely used to produce images of internal human organs – is also useful for detecting concealed explosive devices. This kind of detector uses electromagnetic waves to find the nitrogen molecules present in most explosive devices.

In Afghanistan, landmines are still killing about 100 people per month – almost two decades after the Afghan War.

Can animals help in detecting mines?

Researchers have turned to animals in the quest for a more efficient method for detecting mines. The animals in question are not dogs, but insects – specifically honey bees. Their particularly acute sense of smell is able to detect the presence of explosive, even when mines are buried underground, and the bees are capable of distinguishing between different types of explosive. Scientists are hoping that it may be possible to train these creatures to home in on the source of specific odours when they are in flight. Tests using cockroaches and wasps are also being planned.

How do night vision devices work?

If humans want to be able to see in the dark they will have to emulate those animals that are able to do it well naturally, such as cats and snakes. Cats can't see in total darkness, but the merest hint of light is all they need. Their eyes have a mirror on their rear surface that refects light rays back onto the retina, getting the most out of what little light is available. Image intensifiers also make the best use of available light, but they do it by using highly sensitive photocells to intensify the available light electronically.

Night vision devices for use in conditions of complete darkness generally make use of infrared radiation – just as snakes can detect the infrared (heat) radiation given off by their warm prey. Passive devices register any warmth that is present in a scene. They can recognise living beings and warm machines, so long as their temperature is different from the ambient temperature. Active devices illuminate the environment with infrared radiation – which is invisible to the human eye – and then analyse the reflected signal.

What use is satellite technology to farmers?

In the not too distant past, the only way a satellite could help a farmer was by providing a weather forecast. Innovations have meant that space technology can offer farmers a lot more help – especially if they are looking after large expanses of land. A method has been developed in Australia, where the quality of grazing land for cattle can be checked with the help of satellite data. Satellite sensors report on the state of the biomass in predetermined areas, which means that dairy farmers are able to react promptly by providing additional fertiliser when the need arises. Other researchers are working on 'guiding' farm animals from space. Thanks to special collars, cattle can be located and prevented – by a gentle electric shock – from crossing virtual boundaries. At the same time, the collars can document contact with other cattle or visits to the milking machine. Farmers who are prepared to spend a lot of time in front of a computer can even control a tractor from there, using a GPS system to keep the vehicle on course as it ploughs the soil automatically – at least until the fuel runs out.

How can robots help organic farmers?

Without chemical herbicides, the battle against weeds can take huge amounts of time and patience – or at least it did until weeding robots came along. Tests in Sweden demonstrated that with the help of a camera and a computer these machines were able to distinguish between crop plants and weeds. However, the system only works when plants grow in rows, as is the case with crops such as sugar beet, lettuce, carrots and cabbage. When a weed is detected, the robot extends its weeding tools and removes it mechanically. In the fight against pests, farmers can now resort to a British robot fitted with GPS and an infrared system. It roams around fields looking for snails and slugs. Once located and identified, the robot's grasping arm collects the creatures and places them in a container with a fermentation chamber. Here bacteria turn the snails and slugs into biogas, which feeds a fuel cell that produces electricity that helps to keep the robot working – until the slugs run out.

Do robots make good scarecrows?

Crows and pigeon are clever, and soon work out that a scarecrow never moves from its position, or that noises meant as deterrents are repeated at regular intervals. Farmers and others are resorting to robots as scarecrows, such as the 'tightrope' robot from Ireland, which helps to keep birds away from power lines. The gadget scares the birds with noises and flashes of light, receiving its power directly from the power lines, which means that it can work at night. Sensors protect it from overheating, frost damage and power spikes in the network.

In the southern states of the USA, birds are a particular pest to fish farmers. Pelicans like to help themselves to fish stocks and can cause heavy losses. Warning shots make little impact, so a robot alligator was developed instead. This mimics a real alligator, ramming itself into hunting waterfowl and bombarding them with water. Two rotating paddles move the robot across the water, while sensors and a camera detect birds. A solar cell on the robot's back charges batteries during the day so the machine is able to keep watch at night. Whether the birds are bright enough to learn that they can ignore the fake alligator has yet to be seen.

Security

Turn off

House

Health

Life in an 'intelligent' home

The contemporary home needs to provide its inhabitants with more than just privacy and protection from the natural elements. In our modern, technological world, we place ever-increasing demands on the family home. Research is being conducted to develop the 'intelligent' home, with the aim of relieving residents of a variety of mundane tasks as well as saving energy and providing entertainment. The house will also have to be able to learn and adapt to individual preferences. This is achieved by numerous sensors and microchips built into equipment and switches. Decisions that can't be made automatically by the computer can be supplied by a human operator – and there is always the option of controlling the home remotely by telephone.

Isn't this all just wishful thinking?

Around the year 2000, the first families started moving into 'e-homes', 'intelligent houses' or 'networked test homes' around the world and began living a high-tech lifestyle. Every year, several thousand more completely networked buildings are constructed worldwide. Although they are still expensive, prices will drop when demand increases sufficiently for mass production to become worthwhile.

A wide range of modules that can be integrated into existing homes are gaining acceptance. These innovations cannot be too complicated to use, because any technology that turns out not to be user-friendly will fail in the wider market. Any technology designed to interact with a large number of people of different ages and from different backgrounds has to be simple and self-explanatory.

Are intelligent homes more secure?

Alarm systems have been standard in some homes for a long time. The latest systems do more than react to suspicious movements. Where windows and doors form part of a network, the system is able to transmit the building's ground plan, details of a break-in and initial photographs of burglars to the police, and close off the affected room from the rest of

The intelligent home's user-friendly control centre can be used to monitor a wide range of individual functions.

Choose the week's recipes from the Internet and leave it to the 'intelligent fridge' to ensure that the ingredients are there when needed.

the house. A modern locking system means that bunches of keys that can be lost or stolen are a thing of the past. Instead, a fingerprint or eye scan is all that's needed to obtain entry. Access could be extended to include guests or restricted to exclude unwanted persons without having to have new keys cut or locks changed.

Burglars are not the only threat that intelligent homes can protect us from. A networked building has an overview of the number of people present in the building and which pieces of equipment are in use. If a hotplate or tap has been left on, the system triggers an alarm when the last person in the building prepares to leave. Damage from fire or burst water pipes can also be reduced because fire alarms and water sensors can alert occupants, the fire service or owners by telephone if they are absent.

New technology can help the elderly with medical care. There is clothing that monitors pulse, blood pressure and other vital functions, even calling the doctor if required and sending out an alarm if the wearer has had a fall and cannot move. A building can perform the same functions as special clothing but in a more discreet manner. Video systems can recognise falls and a lack of consciousness and notify the emergency services. In Japan, there are even toilets that, as well as checking body temperature and blood pressure, discreetly determine whether the urine passed contains the correct levels of minerals, and thus whether the

'Living Tomorrow Amsterdam' – an architectural project and museum – showcases futuristic design and technology to demonstrate what living and working may be like in the future.

kidneys are working properly and sufficient quantities of liquids have been consumed.

Will technology help the planet?

The main reason for introducing these new technologies is to make our lives easier and to save energy. Heating systems can be powered by solar power plants and other environmentally friendly sources of energy. Light sensors can monitor light levels and switch the lights on just as it is getting dark – and only if there is someone in the room. Windows can cloud over in the heat of summer to keep rooms cool, and in winter blinds can be lowered automatically at dusk to help conserve heat. Heating systems can be set to run on economy when buildings are

empty, but can be turned up by telephone when someone is returning home unexpectedly. They can have a bath waiting for them or a previously prepared macaroni cheese in the oven ready for dinner.

Shopping can be made easier by allowing the fridge and pantry to monitor food supplies. Radio Frequency Identification labels on packaging indicate when stocks of milk or coffee are running low, allowing the system to order fresh supplies from the supermarket automatically. These can be delivered to a cooling area accessible from the street.

Music and TV systems provide entertainment that adapts individual preferences. Flat screens integrated into walls serve as TVs, computers or

digital wallpaper, while loudspeakers in each room provide background music.

Won't we feel constantly under surveillance?

This new technology could monitor the residents of a house so closely that it may seem to leave them with few choices. It will be vital to have off-buttons and alternative settings so that everything can be controllable by hand if necessary. It is also important that links to the Internet or the fire service are voluntary, as is the extent to which technology is installed in a home. Completely networked homes will remain the exception for decades to come, although individual applications may grow in popularity if people see them as being useful.

Which is the most economical form of transport?

The most economical form of transport is one that goes furthest using the least energy. It may be a toboggan, which slides down a slope unaided – not counting the climb back to the top – a glider that can stay aloft on favourable updraughts for hours once airborne, or a solar-powered car that can cross a continent using only energy obtained from sunlight. The lighter the better seems to be the key.

Why is the output of an engine measured in horsepower?

Towards the end of the 18th century – when steam engines replaced horses for extracting water from coalmines in Scotland – James Watt compared the two systems. He determined that a strong horse was able to carry 150lb of water over a distance of 220ft in one minute – 67.5kg across 66.7m. Therefore one British horsepower (hp) was as much as 33 000 foot pounds of work per minute, which is 745.2 joules per second or 745.7 Watts. With the spread of motorised vehicles, which were taxed on the basis of their horsepower, the units developed slightly differently in different countries. In Germany – where the horses were apparently not quite as strong – one horsepower was defined as 735.5W. Incidentally, a real horse is able to sustain an output of between 10hp and 20hp during a race.

Can I build my own car?

Car manufacturers spend years developing models so that they are ready for the road before going into mass production. Private individuals can do this as well, but it will take much longer working from scratch. A private car maker will need a generous source of funding and the technical ability to make a car that will be passed by local authorities as roadworthy. Skilled mechanical engineers may be able to build themselves a functioning engine, and could then resort to prefabricated parts or buy a kit car for the body. The same applies to building your own boat or aircraft. Aeroplane construction is probably the most demanding, but there are kits available that allow amateur enthusiasts to build simple, ultralight aircraft.

One of the entries in a local youth and science competition was this ride-on lawnmower powered by a wood gasification unit.

Can a car be made to run on wood?

Built for research purposes, cars that run on wood – using the same principle as the wood pellet heating systems employed in a few residential buildings – are back on the roads in some countries. However, other more familiar fuels, including petrol, diesel, gas or – more recently – hydrogen, are more efficient. Wood has only been used to run vehicles when there is a shortage of other fuels . This was the case during the two world wars, when petrol and coal were reserved for use by the armed forces in some European countries. During this period, private tractors, heavy-goods vehicles, buses and cars were all adapted to make use of local timber, which had to be converted into gas and methyl alcohol so that it could be used to run an engine.

Which are the safest colours for cars?

The lighter the colour of a car, the more visible it is and the lower the accident rate. But the conclusions of different studies diverge when it comes to detail. In the USA, blue and yellow are considered safest: blue in daylight and fog, yellow at night. Grey proved to be the worst. In Germany, white is considered safest, except in snow or white sand, where bright yellow or light orange are preferable. Dark green was seen as the worst. If we include silver as a colour, the picture changes. According to a study conducted in New Zealand, the rate of serious accidents for drivers of silver cars is only half that of white car drivers. The results for red, blue, yellow and grey cars were comparable to those for white cars. The cars most prone to having serious accidents were brown, black or green.

How does a car run on petrol and gas at the same time?

It is not usual for two different types of engine to function simultaneously in a vehicle, but they operate at different times. Non-nuclear submarines use diesel engines when running on the surface, but when underwater they are electrically powered; motorised bicycles switch between electric and pedal power. In cars, the usual combinations are petrol and electricity or petrol and gas. In the first case, the vehicle always has two engines – one a petrol engine and the other an electric motor. In the second case, a single engine is sufficient to deal with the switch-over between petrol and gas, although the engine has to be connected to two tanks. The standard four-stroke engine runs as well on petrol or alcohol mixed with air as it does on gas, including biogas, natural gas or hydrogen. An ordinary vehicle can therefore be adapted for gas by the addition of a tank in the boot. Fine adjustment to the engine optimises its performance with different fuels.

Why does putting petrol in a diesel tank cause damage?

Using the wrong fuel nozzle can cost you thousands of pounds. The engine of a diesel vehicle that has been filled with petrol by accident should not be started. The only exception is when the amount is very small, since petrol will mix with diesel.

Petrol damages diesel engines because a diesel engine works under extremely high pressure, with the viscous diesel also serving to lubricate the sensitive fuel-injection system. Petrol is not an effective lubricant, and so the high-pressure pump and valves will wear out rapidly, with the added risk that fine metal fragments will travel through the system and cause additional damage. The result will be a loud 'hammering' from the engine, which is likely to be ruined. If there is a filling station error, it is cheaper to have the vehicle towed away and professionally cleaned. Pumping out the wrong fuel and cleaning the tank and pipes will cost a few hundred pounds.

In 2003, the Dutch solar vehicle *Nuna II* set a record in the World Solar Challenge by travelling 830km in a single day.

Can we look forward to a zero-litre car?

A zero-litre car is one that works without using any combustible fuel at all. In the foreseeable future, this will only be possible through the use of solar power. Every two years, the Australian World Solar Challenge demonstrates that modern, lightweight, solar-powered vehicles can cover considerable distances on the flat at speeds of up to 130km/h (they must not exceed the speed limit). Distances of more than 800km have been achieved in a day.

Vehicles that use a minimal amount of combustible fuel are also taking part in competitions. In the Shell Eco Marathon, the aim is to travel as far as possible on only 1 litre of petrol – or the equivalent amount of hydrogen, solar power or diesel. In 2006, a Swiss fuel cell-powered vehicle set the world record for energy efficiency with a distance of 5134km per litre. At this rate, about 8 litres of fuel would be sufficient for it to travel once around the globe. However, the test vehicle weighed just 30kg, so these minimal consumption studies are somewhat divorced from reality. Nevertheless, some of the ideas developed for these competitions may turn up in more efficient, environmentally friendly road vehicles.

How can driver microsleeps be prevented?

Without a companion to keep the driver awake, he or she may need an 'electronic passenger'. This consists of sensors that register when the driver's grip on the steering wheel or pressure on the accelerator pedal slackens. In addition, cameras are fitted with software that recognises when eyelids are closing more frequently and for longer periods, or when the driver's head drops forwards. In this way, technology is able to recognise the threat of a microsleep far earlier than the driver, and is able to do something about it, either by emitting loud warning sounds or, less commonly, by squirting water into the driver's face. There are even systems that will try to keep a driver alert and awake. These include one that chats and tells jokes, changes the radio channels unexpectedly and asks questions. If the answers are slow in coming or incorrect, or if the driver's voice lacks intonation, the system interprets this as signs of tiredness and triggers an alarm.

The alertness control system is designed to recognise when a driver is at risk of falling asleep.

Is it possible to run a car on air?

Rising petrol prices are accelerating the search for alternative fuels. Scientists are working on cars that run on compressed air or other gases. When compressed air expands, the force can be used to drive an engine piston. This kind of compressed-air car does not emit exhaust fumes and is almost silent. However, this kind of fuel is only truly environmentally friendly if the power required to compress the air in the first place is obtained from green energy.

The Air Car contains heavily compressed air in gas bottles that is used to drive a gas expansion engine. It is designed to achieve top speeds of about 100km/h. Depending on the system being used, filling the tanks can take a few minutes to several hours. Another project uses liquid nitrogen as its fuel. This gas, which makes up 78 per cent of the air, is liquid at -196°C and expands 700 times when heated. Although there are no fuel stations for liquid nitrogen, it is widely used in industry.

Is there technology that can help us find parking spaces more quickly?

Two requirements must be met to achieve this: a network of sensors that registers vacant parking spaces; and, a receiver system for drivers. A system developed in the USA combines both. The parking areas of a city centre can be fitted with tiny cameras or small radar probes that constantly check whether a car parking space is vacant or occupied. This information is then wirelessly transmitted to a local, decentralised network that can be accessed by the driver's navigation device. The screen then shows the closest parking space and directions for reaching it.

How does a car park itself automatically?

Many people find parallel parking a nightmare, but modern technology may soon be able to make the process easier or even take it over completely. The first commercial vehicles featuring parking assistance systems are already available. A combination of sensors and an on-board computer will park a car faster and with more precision than the average driver. Ultrasound sensors are fitted to the front and rear bumpers, and these sensors measure the distance to the kerb and to obstacles or other vehicles. They begin by looking for a parking space using the side sensors to monitor the edges of the road for large enough gaps. The on-board computer uses GPS to check that a gap is not a side street or crossroads, but it's up to the driver to notice any driveways, which must be kept clear. If the gap is large enough, the parking system takes over the steering completely. It is guided by the kerb or, if this is missing, by other parked vehicles. Then it asks the driver – either audibly or on the display unit – to drive forwards or backwards and to operate the accelerator and brake pedals. If the car gets too close to another vehicle, or if a pedestrian walks into the empty parking space, the car stops automatically.

We may not have to wait long before technology can help us find the closest parking space.

Are there aircraft that never need to land?

Several projects have come close to producing a craft that remains permanently airborne. *Helios*, NASA's unmanned lightweight aircraft, caused a stir in 2001 when it established a new altitude record for a non-rocket-powered craft. With a wingspan of some 75m and a weight of almost 600kg, it was powered by a solar- and fuel-cell system. During the day, the solar cells powered the propellers' electric motors and split water into hydrogen and oxygen, which were stored individually and used as fuel to keep the aircraft flying at night. Then the two gases were recombined in the fuel cells to become water, producing electricity in the process.

It is possible for a laser-powered aircraft to spend days aloft without fuel, fuel cells or batteries. However, this can only be achieved at low altitudes, since visual contact with Earth is needed. The aircraft draws its energy from a laser on the ground. The aircraft's underside is fitted with special solar cells that transform the laser light into electricity. An initial test flight with a model plane weighing only 300g was successful, and larger applications are being considered.

How does a glider manage to stay in the air for hours at a time?

Flying is all about the air underneath an aircraft's wings. In motorised flight, propulsion is provided by propellers or jets. The forward movement causes air to slip over the specially shaped wings at a faster rate above than below. This creates low pressure above the wing and an upward force on the wing. The heavier an aircraft, the greater the need for propulsion to ensure that it has enough lift to stay in the air. Lightweight aircraft can glide slowly through the sky with no propulsion at all. If they come across favourable thermals – bodies of warm air that rise from the ground having been heated by the Sun – they can use them to gain altitude. Practised gliders travel from thermal to thermal and don't come down to land until evening, when the Earth starts to cool down.

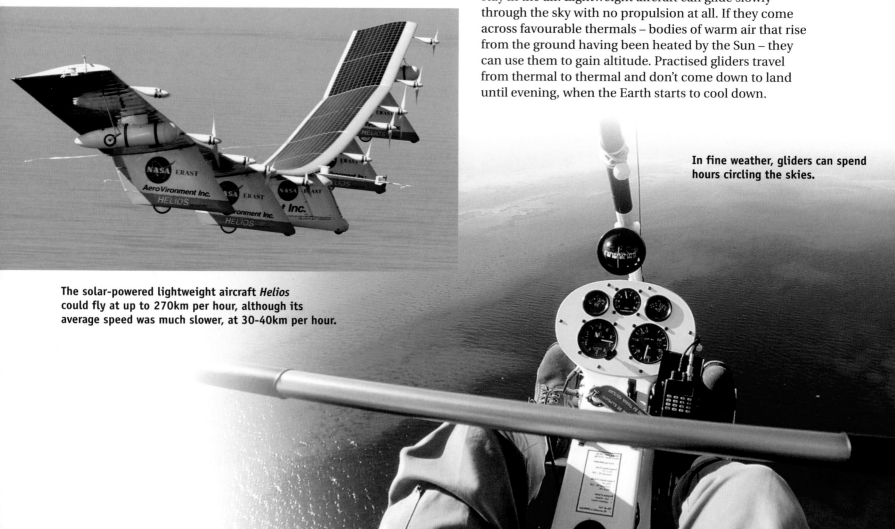

The solar-powered lightweight aircraft *Helios* could fly at up to 270km per hour, although its average speed was much slower, at 30-40km per hour.

In fine weather, gliders can spend hours circling the skies.

What's the quickest way to get passengers seated?

The faster everyone is seated, the greater the savings made by the airlines. Studies have demonstrated that it's quicker to achieve this by not having a system – that is by allowing passengers to choose their own seats, as they do on low-cost airlines. The standard method allows people to board the aircraft freely, but makes them find their way to allocated seats. This produces too many blockages, because the people who board first often obstruct the aisles. Filling the plane from the back to the front is not much more effective, even if people are asked to board separately according to their row number. The fastest method for narrow fuselaged planes is for people with window seats to board first, followed by centre seat holders and finally people with aisle seats.

Why do aeroplanes drop into air pockets?

In calm weather, aeroplanes glide because – thanks to the curved shape – air flows faster over the tops of the wings, creating low pressure and producing an upward force. The aircraft receives additional lift when wind streams in from below against the wings. However, in a storm there is turbulence, which causes sudden pressure differences – the air pressure can drop suddenly, and fewer air molecules provide less lift. The airstream can also suddenly change direction, and the wings may encounter a powerful flow of air from above which pushes them downwards. Depending on the size of a storm area, a plane may drop a few metres in a short time, which causes passengers who do not have their seatbelts fastened to feel their seats fall away beneath them. In an instant the plane can have traversed the air pocket, or may come across a rising wind, and so the aircraft rises and the passengers feel the seats pushing them up. Regulations stipulate that aeroplanes must be able to withstand vertical squalls of 17m/sec without difficulty.

Why don't aeroplane tyres burst on landing?

The tyres on a large jet aircraft have to be able to withstand huge forces during take-off and landing – even when the load is distributed among many sets of wheels. Even with smaller aircraft, the wheels and tyres are subjected to considerable stress, especially during landing as the wheels are not rotating prior to touchdown. If they were, they would be able to discharge some of the energy when rolling. Instead, they are static on touchdown, which causes part of the rubber and the reinforcing material that makes up the tyre to be worn away immediately. The tubeless tyres – like the tyres on a racing car – are filled with nitrogen at high pressure because this gas is a fire-retardant. Some types of tyre are fitted with a valve that opens if the pressure is too high to allow nitrogen to escape slowly. However, jumbo jet tyres are under the greatest pressure at take-off, when a fully loaded aircraft with full fuel tanks exerts a weight of up to 25 tonnes on each of the main undercarriage tyres. Aircraft tyres are replaced after about 120 take-offs and landings and remoulded between eight and 10 times.

True or False?

Aircraft can crash if they get too close together

If two aircraft come too close to each other, the airstream produced by the plane in front could push the following plane sideways. This is due to 'wake turbulence' in the air behind an aircraft, especially that created by the tips of the wings. Turbulence can be strong, stay in the air for several minutes and follow an aircraft for many kilometres. If a second plane comes too close, the turbulence can disturb the airstream around its wings causing it to plummet.

What is the fastest way for ships to travel across water?

The more streamlined the hull of a ship, the faster it will cut through the water. Over half the resistance ships have to overcome arises from friction with the water. The less hull surface there is in contact with the water, the less resistance there is to overcome. This is why catamarans and flat-bottomed speedboats,

which more or less slide over the water, were developed. Attempts to further minimise water resistance led to the development of hydrofoils, where only the foils under the hull float in or on the water, and to air-cushioned vehicles or hovercraft, where ventilators blow air under a 'skirt' so that the craft can float on top of it. All these vehicles can achieve top speeds of about 100km/h. Modern experimental Wing-in-Surface Effect ships – also known as surface skimmers – sit midway between an aircraft and a conventional ship. They exploit the fact that if they can travel fast enough, a layer of air is formed between the craft and the surface of the water. This makes high speeds possible, but only if the waves aren't too high.

Is swimming in treacle slower than swimming in water?
This question has become a favourite of physicists, and it is claimed that even Newton was asked for his opinion about it. A pool filled with a viscous liquid and 16 swimming volunteers provided the answer. Although syrup slows down swimming, this effect is almost cancelled out by another, which comes into play when swimmers are the size of human beings. Each swimming motion in the syrup provides greater propulsion. Swimmers were no more than four per cent slower in the viscous liquid, regardless of whether they did the breaststroke, freestyle or butterfly. As with water, the more streamlined the swimmers' body form, the faster he or she was able to move through the treacle.

Water resistance has been minimised as much as possible in hydrofoils, which accounts for their speed.

Why do ice-skates glide on ice?

The less friction there is, the faster it is possible to glide and a layer of water minimises friction. Skates glide on ice because ice always has an extremely thin layer of water on its surface, even at the lowest temperatures. The special, chamfered form of the skate runners ensures that water collects beneath them for the skates to glide on, while the sharp outer edges of the runners help the skater to stay on track. There is also the often mentioned – although less important – 'pressure-melting effect' to take into account. This is the result of the weight of the skater's body being

concentrated on the narrow area of the skate runners, causing the ice beneath to melt temporarily. The melting is minor, and depends on body weight and the ambient temperature.

Skis also produce a thin, smooth film of liquid on top of the snow – an effect that is more noticeable on powder snow than on icy snow. If the underside of a ski is black rather than white, it will move faster down a slope. The black underside is better at storing heat and therefore helps to melt the snow. Scientists discovered this when they attached heat sensors to the undersides of skis.

Why do skateboards fly with the skater?

When a skaterboarder performs a spectacular jump, the skateboard appears to stick to his or her feet, although one would expect gravity to keep it on the ground. In skateboarding lingo, this is called doing an 'ollie' – a name derived from the trick's inventor – and requires a great deal of practice. During an 'ollie', the board is not connected to the feet but is heading upwards following a skilful catapulting manoeuvre. The important part of the trick is the timing of the moves during take-off. The back foot is used to give the skateboard's bent end or 'tail' a quick, hard kick. At the same time, the front foot determines the direction of flight as the skater jumps into the air, taking all of the weight off the board. If the trick succeeds, the skater flies through the air while the momentum of the board propels it upwards at the same time, making it appear as if it were somehow stuck to the bottom of the skater's feet.

Do rocket backpacks really exist?

In the 1965 film *Thunderball*, James Bond was seen flying with the help of a rocket backpack, a device that also featured in a few other films. In reality, this technology failed to catch on because the amount of fuel it could carry only lasted for about 20 to 30 seconds.

The rocket backpack did well for a short period. The hydrogen peroxide in the jet engine turned into superheated steam, providing sufficient power to carry a human being several metres into the air. However, any slight defect could lead to a crash. Despite development work on the concept in the 1980s, the US military decided that helicopters were cheaper, more effective and safer. Rocket backpacks are still seen occasionally, but generally only in advertising stunts. However, there is a version that has proved to be useful as a tool in space. This is NASA's Manned Manoeuvring Unit, which helps astronauts change direction during spacewalks.

How does a swing work?

Children have to practise to learn how to play on the swings without help. As their body leans backwards and forwards it does not take long for a child to sense the rhythm that is required. The trick is always to supply the swing with energy at the right moment in order to maintain or increase its movement. Just as with a pendulum, a swing will always sit at the bottom of its swing unless it is pushed. If it is pushed away from its rest position, gravity will pull it back down, but it will keep overshooting the rest position and swinging up again. The energy is gradually used up by air resistance and the friction of the swing itself. A child can prevent the swing from stopping by providing it with a regular supply of energy to drive the pendulum. Weight displacement is sufficient to do this. At the highest points of the movement, you lie back and stretch your legs out in order to lower the centre of gravity. While passing the lowest point, you sit up and lift your feet in order to relocate the centre of gravity upwards.

The car of the future

The car of the future will use technology to take some of the strain off drivers and to support them in dangerous situations. Sensors will provide help when parking or warn of nearby pedestrians, especially in the dark. Navigation systems will ensure we reach our destinations with a minimum of fuss.

In science-fiction films, people board futuristic vehicles that find their way around all by themselves. There is often a chase scene, during which the hero switches off automatic drive and careers at high speed – and in complete contravention of the highway code – through traffic. According to some research and development departments, cars of the future will have these capabilities. Automated vehicles will free drivers of many tasks, ensuring greater safety and improved highway systems. However, it will also be possible to drive them manually – just in case. The engine, fuel and material used to make cars will all be subject to innovation.

Is it possible to build plastic cars?

Plastic bodywork is familiar from the former East Germany's Trabant cars and a few others. But under the plastic bonnet there was a steel skeleton for structural strength. Nowadays, there are many more materials to choose from to achieve a lightweight yet robust vehicle. Even parts that run hot can now be made from heatproof synthetic materials, ceramics and light alloys. New types of steel are malleable, strong and extremely lightweight. Light and strong aluminium foam strengthens crumple zones, while window glass can be replaced by scratch-proof polymer panes. The car of the future will have to become lighter in order to reduce fuel consumption. The only heavy cars on the roads might then be armoured limousines used to protect statesmen or celebrities – although bullet-proof materials are also becoming increasingly lightweight.

Isn't the price of petrol making cars pointless anyway?

Petrol and diesel vehicles must eventually be phased out because the Earth's crude oil reserves are finite. For cars, a few alternative options are already available. In Brazil, almost half of the fuel – alcohol – is produced from renewable sugar cane, and biodiesel is increasingly being used in Europe. It is also possible to use agricultural waste to make an environmentally neutral fuel, and methane from sewerage and landfill sites can be used to run engines. The period of adjustment is slow because car manufacturers need to be certain that there are enough fuel stations to supply any new models they produce.

Many people are placing their hopes on natural gas, which is easy to store and supply. Gas deposits are also finite, but the infrastructure could be reused for biogas. Electric engines are also promising, since they can be charged at any socket outlet and are now powerful enough to drive racing cars. However, the engine of the future is most likely to be fuelled by hydrogen, with a fuel cell at its heart. This is because hydrogen turns into water when combusted – which is environmentally friendly – and more hydrogen can always be produced from the water simply by reversing the process. The technology for this already exists, although it is expensive. Once again, there is the problem of distribution and to be really environmentally friendly, the hydrogen would need to be

A large proportion of the fuel used in Brazil already comes from sugar cane.

generated using alternative energy sources, in the same way as the electricity for electric cars. Until these problems are sorted out, the car industry is supplying hybrid vehicles, with engines that can be run interchangeably on electricity and petrol, petrol and natural gas, alcohol and petrol or other fuel combinations.

How can cars help to make drivers' lives easier?

Technological driving aids are now increasingly widely used and help to ensure safety. They include the anti-lock braking system, which maintains steering control during braking far better than the average driver can manage; power-steering, which takes the hard work out of everyday driving; and, the airbag, which provides protection in the vital fractions of a second before an impact. Modern cars are full of microchips that control everything from the engine and automatic gearbox to the central locking and navigation systems. While it is still the driver who must adapt to the car, this is something that will change in the future. Driver assistance systems will adapt to suit a driver profile, adjusting the engine performance to the preferred style of driving, and steering the vehicle with a great deal of autonomy. Headlights will automatically follow the course of the road; radar and

heat sensors will recognise humans, animals and objects on the roadway and assist in parking. In an association that will involve the engine, gears, steering and brakes, cars will soon be able to take over emergency stops and parking manoeuvres completely.

But this is not where the vision ends. If vehicles were able to communicate with each other, rear-end collisions and accidents caused by bad weather would become a thing of the past. Thanks to car-to-car communications, a vehicle would know the instant that the car in front jammed on the brakes and would therefore apply its brakes automatically. It could be informed of traffic jams ahead, and tyre sensors could warn of black ice, conditions that are high risk for aquaplaning and the presence of potholes. Before setting off on holiday, we will download the maps we need into the system, as well as the engine-control software that optimises driving in the conditions we expect to encounter – anything from steep mountain roads to high speed motorways. Perhaps the ultimate goal might be a drive-by-wire system, similar to the

fly-by-wire system used in aircraft. This would enable vehicles to plug into a steering-control system built into the roads electronically , and to be driven safely to their pre-programmed destination.

Is the new technology safe?

Some people will see the new systems as a great help, while others will feel patronised. In the end, it is consumers who will decide what could be improved and what is just superfluous gadgetry. One of the most serious problem is what to do about system errors and computer viruses. Even today, most new vehicle breakdowns are caused by electronic defects, and without a computer to carry out a fault analysis, repairs would be next to impossible. Vehicles that have external communication systems will be vulnerable to even more undesirable influences. Every new development must be thoroughly examined to work out the safety implications – from simple steps like providing a firewall for the computer system to creating an emergency action plan for dealing with a cable break.

How has electricity changed our lives?

To answer this you just have to look back a few hundred years, to a time when sunrise and sunset determined the length of the working day – if there was no oil left to light a lamp; when it took weeks for news to get about – if the messenger survived the journey; when passing on even a brief message could involve a lengthy hike; when data was stored in the brain, rarely on paper. In all these areas, the harnessing of electricity has helped us overcome the limitations of mind and body.

How does a wind-up radio work?

Clockwork radios work on the same principle as a car's electrical system – a generator transforms mechanical energy into electrical energy, which is used to charge a battery. The transformation involves electrical induction – using a wire coil rotating inside a magnetic field, which causes electrons in the wire coil to move and thus provide an electric current. The generator inside a wind-up radio – also known as a dynamo – supplies a battery that stores the energy so that the radio can continue playing.

Energy of motion – kinetic energy – is used in a similar way to supply some torches and watches with electric power. It is enough to just shake the object to move the coil in the magnetic field. The more energy hungry an appliance is, the larger the generator required. However, watch mechanisms are small so their generator can be kept tiny.

In future, the same principle may be used to turn clothing into energy generators. A pair of tracksuit trousers could be used to power an MP3 player, while a pair of walking boots could power mobile telephones other gadgets.

Mechanical energy is transformed into electrical energy inside a wind-up radio.

Do energy-saving light bulbs really consume less power?

Energy-saving lights need no more than one-fifth of the energy consumed by an incandescent light bulb. This is due to the different methods by which the light is produced. An incandescent bulb heats up its wire coil to give off light. Four per cent of this electricity is transformed into light, the remainder is released as heat resulting in a huge waste of energy. Energy-saving lights, regardless of their shape, are fluorescent lamps in which gas molecules emit ultraviolet light, which causes the inner coating of the lamps to glow. The bulb stays cool because less energy is required. However, the future is likely to belong to light-emitting diodes. They are extremely bright but consume only one-tenth of the energy used by incandescent light bulbs.

Many old batteries need a bit of help to get the car started, particularly on cold mornings.

Why do batteries go on strike when it's cold?

The power supplied by a battery is the result of a chemical reaction in which electrons flow between two different metals, such as zinc and copper. One of these (zinc) binds electrons less tightly to its atoms than the other (copper). If the metals are immersed in an electrically conductive liquid, which provides a connection between them, the electrons are able to travel and form a surplus on the copper. This surplus flows as a current when an external circuit connected to the battery provides an opportunity.

Cold slows down every kind of chemical reaction, which is why fewer electrons are released in batteries when the temperature drops. Warming a battery up often returns it to near full power. This is why many photographers carry a spare camera battery in an inside pocket during winter. Although special additives have made car batteries less sensitive to the cold, they can stop working in winter – especially if they are old.

Why aren't all batteries rechargeable?

Ordinary batteries run out and have to be replaced. As with the first galvanic elements developed by Alessandro Volta in the 18th century, a chemical reaction leads to the decomposition of one of the two metals used in the cell. The current keeps on flowing until all the chemical energy that is stored in the battery has been transformed into electrical energy.

Accumulators, on the other hand, are rechargeable because the electro-chemical reaction that occurs in them is reversible. A typical car battery contains lead and lead oxide, which are connected by dilute sulphuric acid. They react to produce lead sulphate and water, but the reaction can be reversed by supplying electricity.

Why is some electrical equipment earthed?

In order to create an electric circuit when connecting lights or electrical equipment, an electric cable has to consist of at least two wires. The current comes in through one wire and flows out through the other. Electricity becomes dangerous when it flows where it isn't expected.

When a fault in a piece of equipment causes the electricity to flow into its external metal parts, a potential difference (voltage) is created between the equipment and the outside world (ground). The current wants to flow from the device to the ground, and this becomes possible when someone touches the outer surface. Then the electricity flows through the conductive human being and into the ground. This can be fatal, especially if the equipment is still connected to the power supply so that the potential difference is maintained.

This is why many power cables have a third wire – the green-and-yellow earth wire. This is connected to the equipment's metal housing or other metal parts, and because it is a better conductor than a person, it allows the current to flow safely to the ground if there is a problem. This extra wire is not usually present in devices that have non-conductive cases.

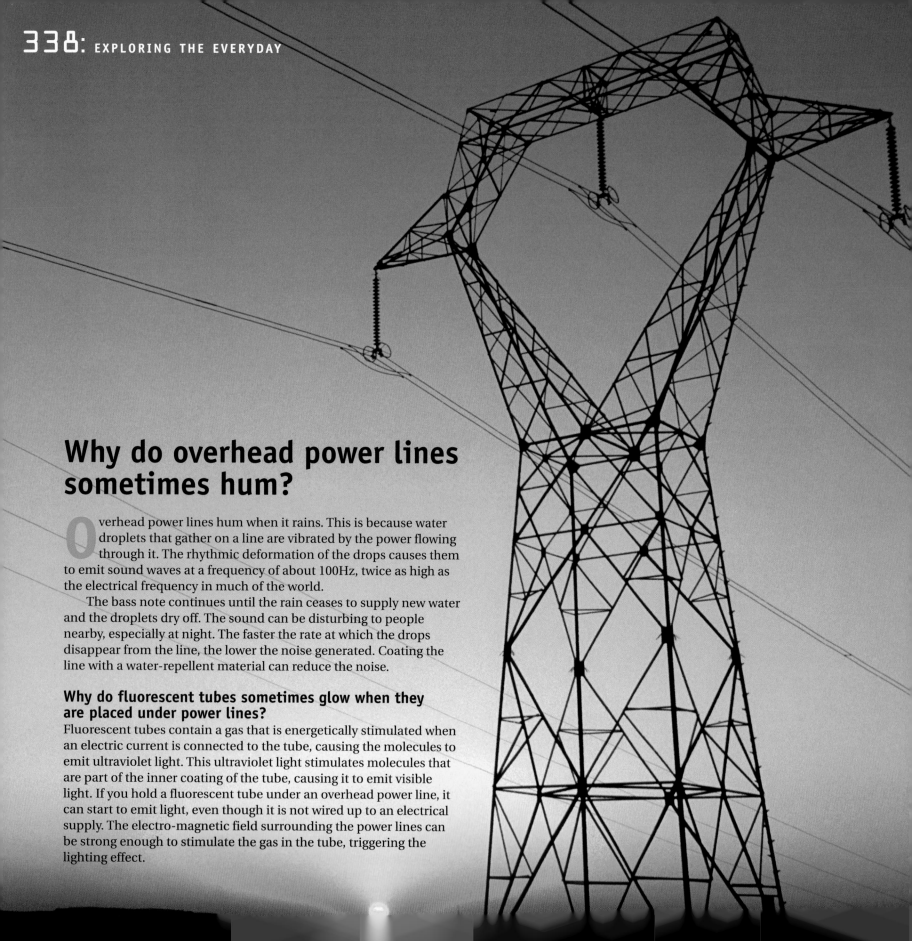

Why do overhead power lines sometimes hum?

Overhead power lines hum when it rains. This is because water droplets that gather on a line are vibrated by the power flowing through it. The rhythmic deformation of the drops causes them to emit sound waves at a frequency of about 100Hz, twice as high as the electrical frequency in much of the world.

The bass note continues until the rain ceases to supply new water and the droplets dry off. The sound can be disturbing to people nearby, especially at night. The faster the rate at which the drops disappear from the line, the lower the noise generated. Coating the line with a water-repellent material can reduce the noise.

Why do fluorescent tubes sometimes glow when they are placed under power lines?

Fluorescent tubes contain a gas that is energetically stimulated when an electric current is connected to the tube, causing the molecules to emit ultraviolet light. This ultraviolet light stimulates molecules that are part of the inner coating of the tube, causing it to emit visible light. If you hold a fluorescent tube under an overhead power line, it can start to emit light, even though it is not wired up to an electrical supply. The electro-magnetic field surrounding the power lines can be strong enough to stimulate the gas in the tube, triggering the lighting effect.

How do e-mails find their destination?

Just like any ordinary letter, every e-mail has a sender address and recipient address. For e-mail, these take the form 'myname@mydomain.domain.mycountry', where 'mycountry' stands for actual countries (.au, .jp, .fr, .uk, etc.), 'domain' stands for a class of organisations (.org), networks (.net), individuals (.name), government bodies (.gov) or educational institutions (.edu), among others. The 'mydomain' part may be the name of your company. When an e-mail is sent, it is divided into several data packages, which are numbered and dispatched through the widely branching network. This ensures that the message arrives, even when lines or servers are defective. Every computer involved in the process leaves a kind of stamp on the e-mail, so it is possible to reconstruct the route. The target computer does not announce an e-mail until all its parts have arrived and been put together again.

How are the pages on the Internet counted?

Some terms need to be clarified before answering this question. First, a web page is an individual page on a screen. This is often confused with a website, or web presence, which means a group of pages that belong together. The important thing when it comes to counting is therefore correct identification of exactly what is being counted. Each site can be found via its domain address (such as, www.mydomain.com).

Sites are counted using special software that interrogates all known sites that operate as servers or hosts for one or more sites. Domains can also be counted through the various national registers to which applications have to be made. However, this number is different to that of the sites, because sometimes several domain names refer to the same content. It would be possible to find out how many individual web pages there are on the Internet, but as they change so rapidly any total would be out of date immediately.

Where does the @ sign come from?

The 'at' sign probably originated in the Middle Ages, either as an abbreviation of the Latin 'ad' (at, towards, by) or as a mercantile abbreviation for 'amphora'. The symbol has endured to this day in Spain, Portugal and France, as the unit of weight 'arroba', which is equal to about 15 litres or 10kg. It also occurs in old German legal texts, while in English-speaking countries it served as an assignment of price (five eggs @ 20p).

The symbol made its way onto typewriter keyboards, where it waited until 1971 when Ray Tomlinson, author of the world's first e-mail, used it to provide an unambiguous separator between the two parts of an e-mail address.

The @ sign around the world

The @ sign is quite unusual in appearance – which has caused it to be given all sorts of imaginative names around the world

Monkey's tail	South Africa
Cat's tail	Finland
Pig's tail	Denmark
Dog, little dog	Russia, Ukraine
Monkey, little monkey	Poland
Maggot, worm	Hungary
Snail	Italy, France, Israel, Korea, Ukraine, Esperanto
Cinnamon twirl	Sweden
Strudel	Israel
Ear muff	Iceland
Ear	Turkey

How does a navy communicate with its submerged submarines?

Sound and radio are the two communication methods used underwater. Sound waves carry further and faster underwater than they do in air, although they are distorted. Ultrasound transmits signals underwater, where they can travel for up to 40km. The acoustic waves of human speech are transposed into the ultrasonic range and transmitted from submarines by means of underwater loudspeakers. The recipients have to turn the signals back into human speech. Short radio waves are blocked by water, which damps anything in the VHF and shortwave ranges. Long radio waves – with a wavelength over 100m – can penetrate several dozen metres deep into water, but require large aerials, which would give away a submarine's location. If a submarine cannot get to the surface to use ordinary radio for communication, a message can instead be recorded and attached to a buoy that is allowed to surface independently of the submarine so that the signal can be transmitted.

Submerged submarines are almost sealed off from the outside world.

Can mobile phones be used anywhere?

Mobile telephones function as long as they are in the vicinity of a mobile phone tower. Once the telephone is switched on, it will search for the strongest signal from the nearest towers. The telephone then identifies and registers itself to this base station. When the identity of the caller has been confirmed on a central database, a telephone call can be placed. Once the call starts, the speech is digitised, filtered and then transmitted by radio waves, first into the network and then to the recipient of the call. If a participant moves out of a tower's range – to the edge of the radio cell – the signal is generally passed on automatically to an adjacent cell. Whether there are 'dead zones' depends on the mobile phone provider's network coverage. In the sparsely populated areas of some countries, mobile telephones cannot be relied on. The only means of mobile communication in these areas is via satellites in low orbits, such as the Iridium satellite network, which will work even at the poles.

How do the police trace telephone calls?
Old crime movies will need to be rescripted, as there is no need to keep dangerous criminals talking on the line nowadays in order to trace their whereabouts. Since the telephone networks were digitised in the 1980s, the caller's number is transmitted with the call, which means that the recipient's handset can be made to display the number prior to the call being accepted.

Individuals can apply to override this feature and have their number withheld, but this facility does not work for calls made to the fire, police or the ambulance services. Malicious Call Identification (MCID) enables a recipient to mark a phone call as malicious and then the phone system will automatically trace its source. More detective work is required only if a call was made from an analogue telephone or through international networks, used by companies specialising in providing cheap calls.

Is there a 100 per cent secure encryption code?

Credit cards, cash machines, passwords, the Internet, e-mail and online banking put huge amounts of personal and private data into circulation, all of which has to be encrypted to keep it secure. Fast, modern computers are able to test massive numbers of potential codes in a short space of time, and this makes them the ideal tool for hackers, spies and password crackers. The Data Encryption Standard (DES) of 1977 – with its 56-bit key – provided around 72 million possibilities, and this was supposed to make the task of cracking a code by persistent trial and error almost impossible. However, computers were able to overcome this obstacle as early as 1999. Today, there is the improved, fast Advanced Encryption Standard (AES) – with its 128-bit key – which is aimed at providing security for the 21st century. A computer capable of cracking the DES in one second would require 149 trillion years to crack the AES. Given the rapid development of computers, it may not take quite that long.

Real security will probably only be provided by quantum cryptography, which is still at the experimental stage. In this system, the key cannot be stolen because it changes on access and becomes useless.

How do we know...

...that messages can be concealed in images?

The most secure hidden message must be one that no one suspects is being transmitted. While an encrypted e-mail may arouse suspicion, an e-mail with an apparently harmless picture as an attachment can easily slip past the most vigilant watcher. Pictures are wonderful places in which to hide information, especially with the help of a computer. If the sender changes individual pixels in a photograph – merely by a slight change of colour – a recipient who knows the original order will be able to detect the difference and decode the hidden message. The art of hiding messages is called steganography.

A semagram is even harder to detect and interpret. One might involve an image that contains a pre-determined element. Hence, a message could be concealed in the number of people appearing in a group photograph, or the colour of a baby's dummy. For outsiders, it is impossible to know where to start.

In steganography, secret information is hidden in such a way that even the presence of the information cannot be deduced.

Milestones in encoded communication

For thousands of years, people have been trying to send each other messages that nobody else could read

from 1000 BC	Primitive people develop the 'bush telegraph' to communicate across distances
c. 50 BC	Julius Caesar uses a code that shifts the letters of the alphabet
c. 1550	Blaise de Vigenère improves Caesar's code with the help of key words
from 1800	Invention of the first cipher machines
1837	Samuel B Morse develops a means of transmitting telegraphic information with an alphabet made up of dots and dashes
1936	German military deploys the Enigma cipher machine, in which cylinders shift the letters of the alphabet
1943	The Colossus calculator, built in wartime Britain, is used to decode German ciphers
1977	Data Encryption Standard (DES) based on a 56-bit key
1999	Advanced Encryption Standard (AES) based on a 128-bit key
2002	Successful quantum cryptographic transmission experiment

Electricity in action

What is electricity?

Electricity is the movement of electrical charges, especially electrons. These charges may flow as current or collect in a location as static charge. If the charge is sufficient, it can jump across a gap as a spark. This is what happens when you get an electric shock from a door handle or when there is lightning during a storm.

What is a big voltage?

Voltage is the commonly used term for 'potential difference' and is related to the energy associated with an accumulation or flow of charges. The voltage across the walls of cells in your heart might be about 0.01 volts. The voltage between the ground and the clouds before a lightning strike might be ten billion volts!

What is the difference between direct current and alternating current?

Batteries, which were invented in 1776, send electrons in a single direction and so produce 'direct current' (DC). In 1831, the theory arose that turning a cable loop in a magnetic field could also cause electrons to move – in this case the stream changes direction with every half turn of the loop and we get 'alternating current' (AC).

Why is alternating current the one used for household power?

When Edison constructed the first electricity power station in 1881, he provided his customers with direct current. However, as the number of customers grew the cables had to become thicker and thicker, which had obvious disadvantages. Alternating current was superior because it could be 'transformed' to high voltages for efficient transmission over long distances and then back to lower voltages for domestic use. In 1893, Edison's competitor, Westinghouse, conquered the US market – and thus the world market – when he introduced his new alternating current generator.

Why are there two-pin and three-pin power plugs?

Every power plug in the world has at least two pins – whether direct or alternating current is in use. The two wires in the lead form part of the circuit, which is completed when a lamp, toaster or computer is switched on, so a two-pin plug is sufficient for these jobs. The third wire connected in a three-pin plug is the earth wire, which takes currents safely to earth if a fault occurs in a device (see p. 337).

Which energy source generates the most electricity?

The efficiency of a source of energy is determined by how much of the chemical, mechanical or thermal energy that has 'gone in' is rendered usable as electrical energy – depending on the process and machine used to effect the transformation. Typical efficiencies are as follows: hydro-electric power 80–90 per cent, wind power up to 85 per cent, natural gas power 55–60 per cent, coal-fired power 25–45 per cent, nuclear power 33 per cent, fuel cell 20–70 per cent, solar power cell 5–29 per cent.

Do we really consume electricity?

No, electrical charges and even electrical energy cannot be 'consumed'. In an alternating current, the electrons simply move back and forth in the cable. This electric energy is not 'consumed', but transformed into other forms of energy, including thermal energy (in a light bulb or toaster), mechanical energy (in an electric motor) or chemical energy (in a battery).

What is meant by the terms ampere and volt?

The volt (V) is the unit of potential difference (voltage) output by a power supply. Voltage that is too high can destroy equipment. The ampere (A) is the unit measuring current – how many electrons flow in a particular time. If the current is too low, a lamp won't shine or motor run.

Why does the electricity meter measure in kilowatt hours?

The consumer has to pay for the work performed by the electric current while it is being transformed into other forms of energy. The kilowatt hour (kWh) is a unit measuring that energy. It is the product of the power in kW multiplied by the time over which that power is delivered, measured in hours.

Experiments with electrostatically charged hair provide entertainment in many science centres.

Why can't I use my hair dryer in New York?

In Britain, Australia and many other countries, electrical power is delivered to homes and businesses at a voltage of 230 or 240V. In North America, it is only 110 or 120V. Because a British hair dryer is designed for a higher voltage, it would only get a quarter of the power it required if plugged into a socket in the USA. It would either work slowly or not at all. The reverse is also the case. An American hair dryer in Europe or Australia would receive four times the voltage it expects, and would most likely break down immediately. However, some modern appliances, such as laptop computers, have power supplies that can cope with the difference and deliver the correct voltage to the device.

Why does hair stand up on end after it has been combed?

A plastic comb collects negatively charged electrons when combing hair, while the hair gives off electrons and is therefore given a positive electric charge. Things that have the same charge repel one another, so the individual hairs seek to get as far away from each other as possible – which makes them stand up, giving you a rather wild appearance.

Why does current flow in water?

In electrolytes – water and other liquids containing dissolved salts – it is charged atoms or molecules called ions, rather than electrons, that provide electric conductivity. These charged molecules carry too few or too many electrons, which is why they are drawn towards electrodes in the water connected to an external circuit.

Is it possible to run a computer with an exercise bike?

The efficiency factor of human muscles generating electricity through a dynamo is about 37 per cent. It would take five hours of cycling to produce 0.5kWh of energy. An average computer and monitor consume energy at a rate of up to 500W (0.5kW) per hour. At that rate, an hour of operation consumes 0.5kW of energy – just the amount produced by five hours of cycling.

Why does getting an electric shock throw you backwards?

A strong current flowing through a human body causes the muscles to contract convulsively, and can push the body with great force off a floor or wall. The current usually flows in through the arms and out through the legs, where the largest muscles are found. It is these that can catapult a body for several metres through the air.

How can electricity kill?

A weak electric current is something that is found naturally in a human body. Cells in the brain and muscles use it to transmit signals. However, quite moderate currents – such as those from a domestic electrical socket – can flow deeply through the body and disrupt the rhythmic pulsations of the heart. This can be especially deadly if the arm muscles tense up and contract so that the victim cannot let go of the 'live' source of the electricity. In addition, larger currents from high voltages or lightning strikes cause body tissue to burn.

Which were the first inventions to use electricity?

The first device for storing electric charge – the Leyden jar – was invented in 1745. In 1752, Benjamin Franklin erected the first lightning conductor, which was followed by the first battery, invented by Volta in 1776, the first simple arc lamp by Davy in 1810 and, in 1820, the electromagnet by Ampère. In 1833, Gauss and Weber conducted tests on the first telegraph, while Siemens did trials on the first dynamo in 1866 and Edison and others developed the first light bulb in about 1870. The first solar cell is said to have been made in 1883, a year after the first power plant went into operation.

It's hard to think what life in a modern city would be like without electricity.

Marvels of
science &
technology

Ever since the discovery of fire and the invention of the wheel, human innovations have made our lives easier. We now understand the causes of many diseases, and can search for ways of curing them by looking at individual cells and genes. After the steam engine and the Internet, nanotechnologies are poised to usher in another technological revolution. More smart ideas will be needed if we are to feed the planet's huge population – approaching eight billion – and provide environmentally friendly energy.

Were all the important things invented back in the Stone Age?

Many important developments did begin during the Stone Age, but not all. Religion, art, tools, hunting weapons, clothing, jewellery and farming are, however, important innovations that did emerge during this period.

What were the milestones of Stone Age innovation?

The use of flint to make tools was a vital first step. The oldest shaped flints – found in Tanzania – are about 1.8 million years old. Early humans used tools of this kind for a few hundred thousand years. It was not until about 800 000 years ago that people learnt to tame fire, and it probably took another 500 000 years or so before humans began to turn their tools into weapons, fashioning spears to hunt large animals. About 10 000 years ago, human beings began to form settlements, and this new lifestyle made it possible for them to farm and keep stores of food. It was at this point that large clay containers were developed.

Early stilt structures like these reconstructed buildings at Unteruhldingen on Lake Constance, on the Swiss, German and Austrian border, emerged in the early part of the Stone Age.

Was Stone Age technology revolutionised by the addition of a handle?

It took not one but two handles. This is because there were two 'moments' in the early history of humans when a handle fitted to an object had great significance for the development of civilisation.

The first was when someone came up with the idea of attaching a pointed flint to a stick. This produced a flint with a handle, or a spear. This happened about 300 000 years ago – the age of the oldest spear found so far. The simple step of attaching a handle to a flint opened up new sources of food for humans, since it had previously not been possible to kill mammoths and other large animals using only stones as weapons.

The second use of handles on objects was when they were attached to containers. Jars and pots were not developed until humans began to form settlements, about 10 000 years ago. Only then did keeping food stores become important. Jugs with handles did not appear until about the 5th century BC. This invention coincided with the beginning of a new era – although it did not bring it about. In about 4200 BC the climate became warmer. New types of settlements such as stilt houses and cave communities were developed. Being able to grasp a container by its handles to transport grain from place to place made good sense.

True or False?

Writing was invented to help trade

For a long time it was assumed that writing emerged to record language, but it seems that the desire to document calculations was more pressing. Egg-shaped clay vessels containing small stones have been discovered in Mesopotamia with an imprint of each stone on the containers' external clay surface. Some of these vessels and stones date back to the 9th millennium BC. The sealed pebbles are seen as serving as a record of quantities. Next, someone came up with the idea of scratching simple lines into the clay, rather than imprinting the pebbles prior to firing. This marked the birth of the use of arbitrary symbols to identify objects – and the first step towards a writing system.

Won't all humans come up with the same inventions because they are so obvious?

The Incas had a hierarchically structured society within which each individual had his or her allotted place. They originally lived in the Cuzco highland region of southern Peru, but their empire spread to take in a much larger area. Naturally goods needed to be transported across this vast empire, and although roads

Llamas, which can carry up to 35kg – even in the mountains – were more useful than wheeled vehicles would have been.

existed, many routes led up such steep inclines that carts would have been useless. The Incas had only one pack animal, the llama, a beast that is sure-footed on steep terrain and can carry heavy loads. As a result, the Incas never invented the wheel. Although it's a fundamentally simple idea, wheels would not have been much use to them. It seems to be the case that people only come up with inventions for which they actually have a use – or at least this is what happens in societies where people do not have the spare time to be inventors.

When did human beings become civilised?

Civilised doesn't mean the ability to use a knife and fork. The things that constitute a human being's degree of civilisation are the signs of culture and reason – characteristics that distinguish humans from the rest of the animal kingdom. These signs became apparent early on, and even Neanderthals had religious concepts. It is thought that by the late Stone Age people were already capable of bartering desirable goods rather than simply snatching them away from each other. Furthermore, the development of the key cultural technologies of writing and arithmetic took place in the final 5000 to 10 000 years BC.

True or False?

Some cultures have no religion

Religion is part of the history of human evolution. As soon as human beings recognised their own mortality they developed religion – the idea of an afterlife providing comfort and solace. Palaeoanthropologists and students of pre- and early history consider the use of burial rites in the Middle Palaeolithic period – which began about 200 000 years ago – to be the first indication of a religious sensibility and early cultural activity. It is probably correct to say that there are no cultures without religion.

It is thought that the Neanderthals were the first humans to make clothes, an important cultural development.

How long ago was money invented?

Money – in the sense of a means of exchange – emerged early in the history of humanity. Evidence for this is found in the term 'pecuniary', which is derived from the Latin *pecus*, which means cattle – an accepted means of exchange in ancient times.

Scientists have found the earliest traces of money as a means of exchange in the Upper Palaeolithic period, starting about 30 000 years ago. Prior to that, objects found in graves gave no indication that they were being used as a means of exchange rather than as adornments. Coins were a relatively late invention, and had to be preceded by the realisation that cattle and other goods were an unwieldy means of exchange and posed a transport problem. The world's oldest coins are thought to have originated in Greece during the 6th century BC.

One of the oldest coins ever discovered – a tetradrachmon from Athens.

Did cities exist in the Stone Age?

This depends on how you define a city as opposed to a settlement. Fundamentally, a city displays a greater range of characteristics than a mere settlement does. Some experts believe that a city must contain inhabitants who are not all related or related by marriage. Others see the characteristics of a city in the existence of norms to regulate coexistence. A wall to separate residential areas from the surrounding country is also considered to be a feature of early cities.

Competing for the title of oldest city in the world are Jericho, in what is now Palestine, and Çatalhöyük, in what is now Turkey. In Jericho, traces of human settlement have been found that date back to about 8500 BC. Çatalhöyük is probably 1000 years younger, although that wouldn't necessarily mean that Jericho is the oldest city. However, the fact that Jericho was surrounded by a city wall from an early stage seems to count as evidence in its favour. Çatalhöyük's claim has been a matter of dispute, because it is possible that its inhabitants were still hunter-gatherers.

If we stick to a strict definition of a city, the earliest period in which we can safely speak of the existence of cities is the Bronze Age and not the Stone Age.

The walls of Jericho may be the oldest city walls in the world.

What makes humans unique?

Stonehenge in Wiltshire, England is considered one of the earliest manifestations of human culture. This Neolithic site probably had religious and astronomical significance.

As far as we know, no animals other than humans have a concept of time, are aware that they must die one day and have an appreciation of beauty. We can only speculate about why and when these ideas became so significant for humans.

Why were graves made of stone for thousands of years?

Burying the dead in wooden boxes was inconceivable to people of the Stone Age. The dead had to be laid to rest in a stone chamber or graves protected by stones. Early European and Egyptian tombs provide strong evidence that stone was thought to preserve the afterlife of the dead.

Why do calendars sometimes differ from country to country?

It is possible that the recognition of their mortality gave human beings the desire to give time a structure, which they did with the help of calendars. Based on observations of nature and the sky, the significance of different phenomena varied between cultures. The Moon was the principal god of the Sumerians, who ruled southern Mesopotamia from about 2900 BC as the reliability of the lunar phases commanded their respect. A thousand years later, the Sun became the chief god of the Babylonians and the solar year replaced the lunar year. In Egypt, the Nile was responsible for the structure of the calendar, as its floods were punctual and divided the year.

When did time begin?

In addition to marking various stages in the course of the year, human beings have recorded the passage of time by counting the years. In the widely used Gregorian calendar, the current year is counted from the birth of Christ. In Islam and Judaism this event is not considered as important. Most Muslim countries use the Islamic calendar, which began with Mohammed's flight from Mecca to Medina. In the Gregorian calendar this event took place on 16 July 622 AD. In Judaism, the years are counted from the creation of the world. According to the Jewish faith this occurred on 7 October 3761 BC (according to the Gregorian calendar). As a result, in Israel the year 2008 is regarded as being 5768 or 5769.

How do we know when events occurred in the ancient world?

Since people at the time couldn't have known of the birth of Christ prior to the event, they were also unable to call a year 500 BC. Dates in texts are converted with the help of ancient calendar systems. The Romans named the years after their consuls' terms of office. Mesopotamia used periods of office to calculate the year, and the Greeks began counting with the first Olympic Games.

In order to calculate dates accurately, historians need to know when in antiquity a dating system became established. The Greek system that was based on the Olympics did not start until the 4th century BC, although there had been Olympic Games since the 8th century BC.

Have there been cultures that were dominated by women?

Government has not always been the exclusive preserve of men and, in antiquity, it was not unknown for women to become rulers – the most famous being Cleopatra. A woman coming to power was often because she had been the wife or sister of a deceased male ruler.

Many historians now believe that matriarchies – societies in which women dominated men – are the stuff of legend. Nevertheless some matrilineal agricultural societies do exist today, such as the Jaintia in India and the Mosuo in China. The principal characteristic of these matriarchies is that descent is traced through the mother. Men live in their mothers' houses and may move in with a sister later on. It is men who look after the children and perform the more arduous work. Reproduction is organised through marriages in which men are mere visitors – they are permitted to have sexual relations with women at night, but must leave at dawn.

How old is fashion?

Humans are the only animals that wear clothes, and clothing existed as early as the Palaeolithic period. No remains of clothes from that period have been found, but tools that are thought to have been used for working on furs have been discovered. Fashion sense may have developed early in human history. Perhaps someone in a tribe wore a loincloth that was more attractive than the others, and so his design was copied by other members of the tribe.

When did people first begin to wear jewellery?

It is thought that human beings have been adorning themselves for over 100 000 years. Shells, ivory, bones and animal teeth

Her word counts: the social order of the Mosuo people in south-west China is based on a matrilineal system.

Part of the Aztec calendar is recorded on this carved stone from the 15th century.

Amber was used to make jewellery and ornaments as early as 4000 BC.

have all been used to make jewellery. Beauty may not have been the only motive, since jewellery may also have had magical significance. Some historians suggest that humans had jewellery before they began to wear clothes. This is based on observations of primitive peoples, some of whom do not have clothing, although all wear jewellery. Jewellery has also been buried with people from early times, presumably in the hope that the deceased would not find him- or herself short of anything in the hereafter.

Will we all soon live for more than 100 years?

There are already more than 135 000 people over 100 years old, and in the year 2050 their number will have swelled to two million, according to the World Health Organization. People who live to a great age have always existed, but there has been a rise in life expectancy because of improvements in hygiene and nutrition, accident prevention and the diagnostic and healing tools offered by modern medicine.

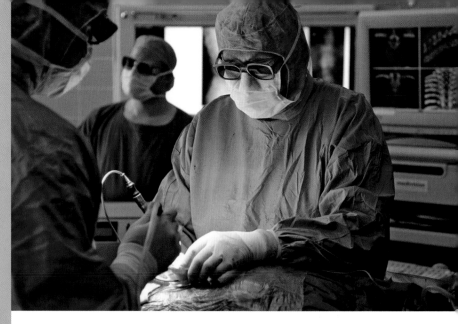

Surgeons still use ordinary scalpels for most operations – the new, high-tech scalpels are used only in special cases.

Do surgeons still use scalpels for cutting?

Scalpels are still the best tools for making simple surface cuts. Their super-sharp metal or ceramic blades cut through tissue as cleanly as possible – the sharper the cut, the fewer cells are damaged and the fewer nerve cells transmit pain signals.

Greater sharpness and precision are possible. Laser scalpels emit an intense beam of light that can be used to make surgical cuts. Depending on the setting and wavelength of the laser, the beam can also connect tissue or seal vessels that are bleeding. A scalpel that uses concentrated sunlight was developed recently. This emits a sunbeam that has been amplified 10 000 times by means of a system of mirrors – an innovation that provides a low-cost solution for developing nations with plenty of sunshine.

Plasma knives are even more precise and avoid heating cut surfaces. High electric field strengths in the instrument's filigree tip produce plasma – collections of free electrons and ions – in the form of minute bubbles. Within fractions of seconds these collapse at extremely high speeds, producing a shock wave as they do so that destroys adjoining tissue. The cut is even easier to close than the one that results from laser surgery, which is why plasma knives are considered suitable for complicated eye and brain procedures.

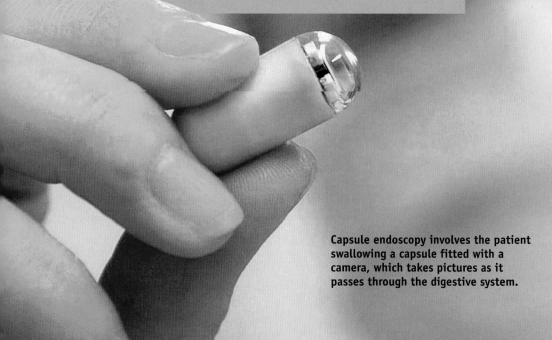

Capsule endoscopy involves the patient swallowing a capsule fitted with a camera, which takes pictures as it passes through the digestive system.

Is it possible to operate with water?

If water is funnelled into a concentrated jet that strikes a surface under great pressure, it can cut even the hardest material. The application of water jets for cutting in medicine is limited because of the large quantities of water involved. However, since the body tissue into which a surgeon will make a cut is often surrounded by liquid anyway, it is sometimes better to make incisions underwater rather than having to pump away the body fluids. It is in those circumstances that a liquid-jet or micro-jet scalpel comes into its own. It uses the surrounding fluid and briefly heats it in the jet to a temperature of about 10 000°C, producing a rapidly expanding vapour bubble that pushes a tiny spurt of water out of the tip of the jet at about 144km/h. Rapid pulsation produces a precise stream which is suitable for cutting through tissue as well as bone.

What is an 'intelligent scalpel'?

Whenever surgeons cut out a tumour there is always a risk that a few malignant cells will be left behind. Because of this, they often err on the safe side by cutting generously. A more precise way of performing such operations is provided by an innovative scalpel from the Netherlands, which is able to distinguish between different types of cells, informing the surgeon during an operation if more tissue needs to be removed. This trick is made possible by a glass fibre probe on the instrument that guides a laser beam onto the tissue. A tiny optical sensor then captures the scattered light reflected from the cells and analyses their optical 'fingerprint', with the pattern generated by cancerous cells differing from that of healthy cells. The optical scalpel may one day even eliminate the need for preliminary tissue sample investigations in suspected tumour cases.

How does 'keyhole' surgery work?

In the 1990s, minimally invasive surgery became the established approach to many operations, especially those performed in the abdominal area. Instead of opening a large area of the patient's body, a surgeon makes small incisions through which surgical instruments can be guided on long, flexible rods. One of these is fitted with a light source and video camera, which enables events inside the body to be checked on a screen. The major benefits of keyhole surgery are that surgical wounds heal rapidly, there is less pain and the patient recovers more quickly. Experienced surgeons are just as able to keep track of progress with this technique as they would be with the larger incisions required in traditional surgery.

Do gastroscopies and colonoscopies always have to be unpleasant procedures?

The traditional method of looking inside a patient's gastrointestinal tract is by inserting a long, flexible endoscope fitted at its tip with a camera and light. This enables the surgeon to conduct a detailed examination of the interior surface of a patient's oesophagus, stomach or large intestine. However, this procedure can be fairly unpleasant for the patient, who is therefore sometimes sedated. Much more agreeable is a capsule endoscopy. For this, a 1.5cm-long capsule containing a tiny camera is swallowed by the patient and then allowed to pass naturally through the digestive system, where it compiles images over a period of about eight hours. Behind the camera's wide-angle lens a light source emits two flashes per second. The data that is collected is either transmitted to a computer or stored for subsequent access. The capsule cannot be guided, however, and so its successor is already in development. This is a worm-like robot with a video head, fitted with flippers to allow its movements to be controlled, even inside the slippery intestines.

Robots have become particularly useful during delicate operations that call for an extremely steady hand.

True or False?

Robots can now replace surgeons

Robot systems have been in use for some time in operating theatres, providing support for surgeons. They have a steady 'hand' and can be guided with extreme precision. With the da Vinci surgical system, the surgical arm unit executes movements precisely as directed by the surgeon, who controls events from a screen. The robot's steady 'hand' is particularly valuable during brain operations. Another area is hip replacement operations, for which robots are capable of drilling a hole into bone to within five micrometers of the size required.

What are lasers doing in the dentist's surgery?

Having a tooth drilled by a dentist hurts – especially when the tip of the drill hits a nerve ending. Thanks to recent developments, the laser has become a pain-free instrument suitable for use inside a human mouth. Although conventional surgical lasers have been used to vaporise soft tissue for some time, hard enamel requires considerably more laser power. Intense infrared light with ultra-short pulsations – less than a microsecond – can remove tooth enamel only where required, without heating the adjacent tooth material or causing it to crack. Removing healthy tooth material with this type of pulsed laser technology takes about 100 times longer than it does with a conventional drill. Damaged tooth enamel is softer, so removing that only takes ten times as long.

How can harmful bacteria be removed from your mouth?

Regular brushing and flossing and the use of mouthwash are a good start, although this won't get rid of some harmful bacteria. Dentists now have a new weapon in their arsenal: blue light. Just two minutes of blue light treatment inside the mouth of a patient can destroy harmful oral bacteria, leaving useful bacteria untouched. This interesting discovery was made during tooth bleaching treatments using blue light, which also relieved inflamed gums. Subsequent experiments showed that blue light killed some types of bacteria that were known to cause and aggravate gum inflammation. Iron molecules inside the bacteria are responsible for the effect. They absorb the blue light, triggering a sequence of metabolic reactions fatal to bacteria.

Why do dental implants take so long to make?

Inserting synthetic replacement teeth into a jaw requires great precision, as well as a considerable amount of time taken up with drilling, measuring, fitting and securing. In the future, robotic assistance should mean the procedure will take no more than a few hours. Robotic equipment will help with both analysis and drilling, and its software will use data obtained from physical impressions and X-rays to calculate the optimum drill direction and drill depth for the synthetic teeth, while avoiding sensitive nerves.

Depending on the material used, it may be possible to accelerate the production of implants and bridges by making use of techniques similar to those used in rapid prototyping. This is widely used in industry to produce accurate, three-dimensional objects, either by laying down successive layers of finely powdered materials to build up a solid form, or using light beams to trigger the hardening of selected areas in a block of gel to create a solid object.

None of this amazing new technology will have much of an impact on one of the most important factors involved in synthetic tooth replacement – an individual's capacity to heal.

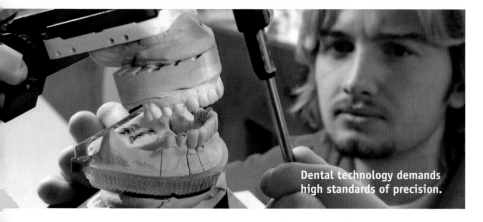

Dental technology demands high standards of precision.

What are drug carriers?

The active agents in tablets, injections or the drugs used in chemotherapy are distributed evenly through the body, although they are more effective if they only enter those areas of the body where they are actually needed. This is why scientists are testing all kinds of molecules in which active agents can remain enclosed until they reach their target. Minute particles are ideal for the task, such as hollow spheres made from carbon (fullerenes) or detergents (micelles), which dissolve easily in water and can carry water-insoluble substances within them. Scientists in the US conducted experiments in which they loaded tiny transport vehicles with an anti-cancer drug and minute amounts of iron oxide, and by using a magnetic field were able to direct the carriers to a tumour. British scientists have been able to turn bacterial spores into resiliant stores for sensitive vaccines – ideal for developing countries. The scientists gave a harmless bacterium a gene for a substance that kills the bacteria responsible for tetanus and then starved it until it used a typical survival mechanism and formed spores. Inhalation of these spores provided sufficient vaccine to guard against tetanus.

Do injections have to hurt?

Before injecting a patient, the doctor may slap the surface of the skin. This distracts the pain sensors and makes the injection less painful. This old method is becoming superfluous as the needles of syringes become smaller and more sophisticated. Measuring a few thousands of a millimetre in diameter, the latest glass, metal or plastic needles are too thin to stimulate the ends of pain nerves significantly, although they are still thick enough to permit the drug molecules to pass through them.

Is it possible to place active agents under the skin using patches?

The familiar nicotine patches work because their active agent is passively diffused through the skin. Some substances have to be brought into the body actively – as with an injection – and for this there are new kinds of patches that are not only painless but can also be applied by a lay person.

One type of patch features a large number of micro-needles, while another has tiny blades that imperceptibly make small incisions in the skin. A battery-operated mini-pump or electrode then ensures that the active agent gets under the skin. Large molecules, such as those of insulin or proteins, can be administered as easily as viruses for inocculations or drugs packed into nanoparticles for transportation through the bloodstream.

Will diabetics have to prick their fingers for very much longer?

A prick on the finger is still the standard way of measuring blood sugar, and an injection is still the standard way for administering insulin, although new methods are being developed. These include a variety of sensors that can be placed beneath the skin, which react to fluctuations in blood sugar levels. One is a kind of tattoo that will work with fluorescent molecules on plastic beads to produce light, the intensity of which is measured by machines that then recommend the dose of insulin required. Another is a plastic sensor, not unlike a familiar security tag, that will change size, triggering colour changes that signal the blood sugar level to the wearer.

The most useful method is one that eradicates the need for insulin injections. This involves a patch of micro-needles that take tiny blood samples and pass them to a device that analyses the blood, determines the required dose of insulin and then pumps the drug directly into the bloodstream.

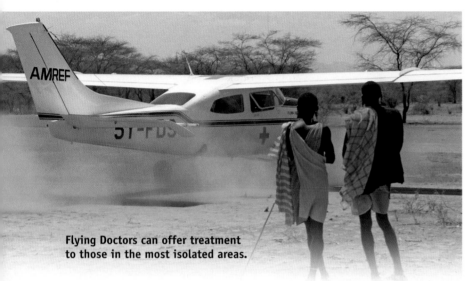

Flying Doctors can offer treatment to those in the most isolated areas.

What is electronic underwear?

Patients with cardio-vascular complaints are sometimes compelled to spend days in hospital just being monitored. With the help of 'intelligent underwear' this could become unnecessary. Sensors woven into underpants or bras are in direct contact with a patient's skin and can monitor heartbeat. If the sensors detect dangerous irregularities, the system automatically calls for help.

Additional sensors to monitor blood pressure or body temperature can also be integrated into the clothes. A cable takes all the data collected to a microchip that can either store them for some months or send them to a physician. The little power that is required by this system comes from a battery. The garments can be machine washed and ironed just like ordinary underwear.

How can lone elderly people who have had a fall at home get help?

To be lying on the floor helpless and alone after a fall, is one of the greatest fears that elderly people have. There are now several systems that can tell if someone has had a fall and trigger an alarm. These include a video system fitted to the ceiling, which analyses images and reports back if it sees someone motionless. A more mobile alternative is a device that fits into a watch or ring. This sets off an alarm if it is turned away from the body axis at an angle above 45° for a lengthy period of time. It can also be fitted with sensors that monitor blood pressure and pulse rate, as well as the blood's oxygen saturation, with the ability to send this data to a physician. A positioning system can be added to help the elderly should they collapse or get lost.

What happens when there are no doctors nearby?

The faster a doctor can get to a patient, the better are his or her chances of a full recovery. However, there are many places around the world that do not have a regular health service. In the age of computers it is possible for a digital image or a webcam to allow a remote doctor to make an initial diagnosis. This is not unlike the way in which telemedicine works in developing countries. In an emergency, a consultant at the other end of the line makes use of the eyes and hands of less experienced personnel on the spot. The first ever transatlantic operation took place a few years ago when surgeons in New York performed an operation on a woman in France with the help of a remote-controlled robot. This method is of interest to the armed forces, as soldiers are often injured in remote areas.

Do astronauts have first-aid kits?

There is first-aid equipment aboard every rocket and spacecraft, as well as the International Space Station. Individual crew members are given special first-aid training and teams often include qualified medical personnel. As a preventative measure, all the astronauts' flight suits are fitted with a variety of sensors that monitor a range of health data that is sent to receiving stations on Earth, where specialists are always on call in case of acute illness or injury. Using the data, specialists can diagnose the condition and direct treatment as necessary. Also being tested are lipstick-sized surgical robots – controlled from Earth – designed to operate on patients in outer space.

Despite a severe lack of room, every space flight has sophisticated first aid.

Can every part of a human being be replaced?

There is little prospect of building a Frankenstein's Monster in the near future, but devices that perform the same function as some human body parts can already be manufactured. It will take a while before an artificial heart looks like a real one, but a mechanical heart can already take on all of the heart's functions until such a time as a suitable replacement is found for a transplant. Cochlear implants for those with defective hearing offer a good replacement, as do nerve-controlled hand and foot prosthetics. Artificial kidneys and livers still operate outside the body, a blood substitute is still being developed, and replacement retinas for the blind can only provide shadowy sight.

Many scientists are hoping that biology rather than technology will hold the key to future developments. Using tissue-engineering techniques, they hope to build three-dimensional organs from real cells. Already available in real life are bone and cartilage parts, as well as tendons, heart valves and replacement skin grown in the laboratory from the body's own cells, which is used to help heal large wounds.

How a cochlear implant works: 1 An external speech processor transforms sound waves into digital signals; 2 The signals reach the implant; 3 The implant transforms them into electrical energy, which is transmitted to a group of electrodes in the cochlea; 4 The electrodes stimulate the auditory nerves.

How are new tissue and organs made to grow?

The lizard's tail 'knows' how to grow back, but in the laboratory the cells that are to become artificial organs don't have the information they need to do this. To tackle this dilemma, scientists face several problems. Cells grow into unstructured heaps if not provided with a three-dimensional framework. Furthermore, organs are almost always made up of four basic tissue types – muscles, nerves, connective tissue and epithelial tissue – which are traversed by blood vessels that supply the structure with nutrients. A biological version of rapid prototyping is an innovative approach to duplicating this kind of complicated structure. Layer by layer, a kind of three-dimensional printer sprays a mixture of cells and solvent – including nerve and blood vessels from the corresponding cell types – into a prepared form. This method is still in its infancy.

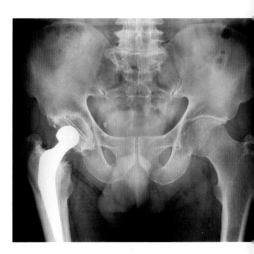

Artificial hips are the most common replacement implants. In the UK, there are more than 43 000 such operations each year and the number is rising annually.

Is it possible to grow teeth in the laboratory?

False teeth made from a person's own tooth material would be useful and attractive. It may take a few years before the idea is translated into a practical proposition, but scientists have successfully grown mouse teeth in the laboratory by implanting tooth germs from a mouse's stem cells into a gum. These grew into incisors, pre-molars and molars, complete with nerves and blood vessels, and anchored themselves into the jaw.

Another team set the stem cells onto plastic frameworks in stomach tissue and grew them into recognisable molars. Scientists have also been able to coax human cells into producing the hard tooth components like enamel and dentine. Another group of researchers has been analysing the way in which nature uses an interplay of proteins and biological minerals to form tooth enamel.

A tiny implant designed to replace a defective retina.

How do artificial electronic eyes work?

Specialists from many disciplines are working on the development of visual prosthetics. Many approaches are being tested, but success so far has been limited to enabling a blind person to make out black and white outlines.

The type of prosthesis required is determined by the cause of the blindness, whether the cornea is cloudy, the retina defective or the optic nerve damaged. The best results can be achieved in the first case, where a camera on spectacles can be used to transmit images into the eye where they are projected onto the intact retina. If the optic nerve is damaged, electrodes can be placed directly on the visual cortex, although the resulting vision is pixelated. Most current research is looking at artificial retinas, using light-sensitive ceramics or infrared photodiodes to replace the natural cells.

Is it possible to 'see' with any of the other senses?

Until there are fully functioning eye implants, technology will at least be able to help people to put together a mental image of their surroundings using the other senses. A walking cane for the blind developed in France analyses the reflections of a laser beam from objects in the environment, transforming the information it receives into vibrations or sounds.

One innovative project in the USA is working on a so-called seeing tongue. This uses images from a video camera that are transformed into electric signals and transferred onto a plate with 144 electrodes that rests on the tongue, making use of its high nerve cell density. The electrodes emit weak electric pulses that are equivalent to the stimulation provided by a nine-volt battery.

Could artificial limbs give a runner an unfair advantage?

Runner Oscar Pistorius was banned from competing at the 2008 Olympic Games in Beijing – because the International Association of Athletics Federations (IAAF) decided his prosthetic limbs would give him an unfair advantage.

South African-born Pistorius, had both his feet amputated as a child, and runs using carbon-fibre transtibial artificial limbs. He was the first disabled athlete to compete officially against able-bodied runners when he took part in the 400m at the Golden Gala in Rome in July 2007, finishing second. But after studying an independent scientific investigation the IAAF ruled that the 21-year-old's 'Cheetah' blades constituted technical aids and were therefore in contravention of the rules. Tests carried out at the Institute of Biomechanics and Orthopaedics at the German Sport University in Cologne compared Pistorius's performance with that of five able-bodied sprinters of a similar standard. The results showed that the blades returned 30 per cent more energy than a healthy ankle joint.

Pistorius, who started running when he was 17, is now Paralympic champion and world-record holder at 100m, 200m and 400m. His J-shaped artificial limbs are made by a specialist firm in Iceland, and cost £15,000 a pair.

How are pacemaker batteries replaced?

Pacemakers need to be fitted with a new battery once every eight to 12 years, and this involves an operation every time. It is hoped that this may be avoidable in future. One Japanese invention is a device with a solar cell that is inserted under the skin. When an infrared laser beam is directed at it, the battery is recharged. In America, scientists are working on implants that are automatically provided with a lasting supply of energy from tiny, thermoelectric elements that derive their energy from the body's own heat. Other researchers are working on a biofuel cell – a mini-battery that takes its energy from the blood. It imitates natural metabolic reactions, involving blood sugar and oxygen, transforming chemical energy into electrical energy.

How is it possible to take a person's temperature without touching them?

Parents are grateful for the fact that they no longer have to insert a thermometer into a child's mouth or bottom in order to take their temperature. Modern thermometers can provide a quick, accurate reading from either the forehead or an ear. However, heat cameras eliminate the need for touch at all. When the SARS virus was spreading in 2003, British engineers modified an infrared camera to take precise temperature readings without any physical contact. With an ability to measure temperature to within a fraction of a degree, the camera was first used at Singapore airport, where it took up to 30 images per second of the faces of passing passengers. The temperature detector helped to filter out SARS suspects – those passengers with noticeably higher temperatures – so that they could be closely examined.

Milestones of modern medical engineering

Major developments over the past 150 years.

1846	First operation under anaesthetic	W Morton, dentist, USA
1895	X-rays are discovered	W C Röntgen, physicist, Germany
1902	First electrocardiogram (ECG)	W Einthoven, physiologist, Netherlands
1906	First successful transplant	K E Zirm, surgeon, Germany (cornea)
1916	First artificial limbs to provide useable prosthesis	F Sauerbruch, surgeon, Germany
1924	First haemodialysis using an artificial kidney	G Haas, internist, Germany
1953	First use of heart-lung machine during open heart surgery	J H Gibbon, surgeon, USA
1959	First ultrasound image of an embryo in the womb	I Donald, gynaecologist, UK
1967	first heart transplant	C Barnard, surgeon, S.Africa
1972	First computer-assisted tomography	G N Hounsfield, electrical engineer, UK
1978	Birth of first 'test-tube' baby	P Steptoe, gynaecologist, & R Edwards, physiologist, UK
1973	Development of magnetic resonance imaging	P C Lauterbur, radiologist, USA P Mansfield, physicist, UK
1996	Birth of first cloned mammal	I Wilmut, embryologist, UK ('Dolly' the sheep)
2003	Decoding of the human genome	Human Genome Project and Celera Genomics Corp., USA

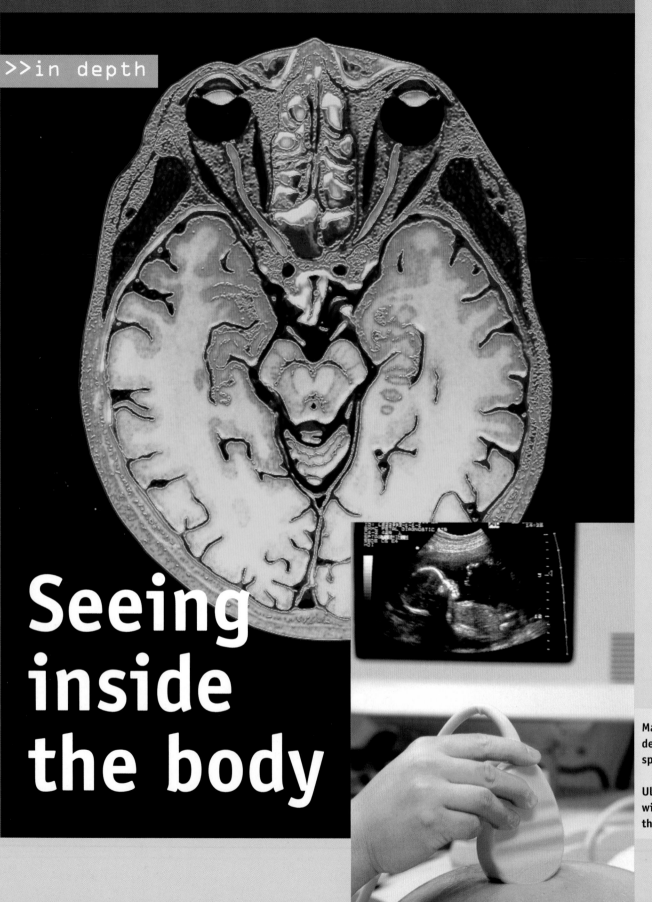

Seeing inside the body

The revolution in medical imaging began shortly before the turn of the century. Towards the end of 1895, newspapers reported that X-rays were making it possible to see through bodies. As proof, photographs were shown of a shadowy hand with clearly visible bones. So revolutionary was the discovery of this previously unknown radiation that Wilhelm Conrad Röntgen was awarded the first Nobel Prize for Physics six years later. For the first time in the history of humankind it was possible to look deep inside a human being without having to open him or her up first. Nowadays, doctors have a range of modern tools and techniques, from ultrasound and ECG to magnetic resonance imaging and the measurement of brainwaves. But X-rays are still the standard tool as they are simple and effective to use.

How did Röntgen know that invisible radiation existed?

X-rays are electromagnetic waves just like light rays or radio waves. They differ only in that they have extremely short wavelengths, or very high frequencies. They are formed when electrons are excited and

Magnetic resonance imaging was developed in the 1970s and allows spectacular views into the body.

Ultrasound examinations are the most widely used imaging procedures in the field of medicine.

diverted by energy pulses as they orbit atomic nuclei and then fall back into their old orbit. Röntgen – who became a university professor in Würzburg at the age of 30 – was just one of many scientists who were studying electric currents in evacuated glass tubes at the time. Inside these cathode ray tubes, electrons travel between two electrodes when a high current is passed between them. Today, most television receivers are based on this principle. Röntgen discovered more or less by accident that the unknown radiation escaped during this process. Some sources suggest a fluoroscope that was some distance away lit up, even though electrons can only travel a few centimetres through air. Other accounts report that photographic paper in a drawer was accidentally exposed. Whatever occurred, Röntgen recognised the importance of the phenomena and verified his assumption through a series of tests. Eventually he captured the effect on photographic paper – an X-ray of his wife's hand.

What modern methods allow us to see inside the body?

Physicians now have three ways of looking inside a body: they can guide long endoscopes fitted with cameras deep into the body; place measuring instruments such as an ultrasound head onto the skin; or, they can take readings without making any physical contact at all by using electromagnetic radiation or fields. These processes can also

be distinguished by the method they use. Some pick up electrical signals from certain types of cells, such as the heartbeat or brainwaves. Others are imaging procedures, supplying an image directly through a camera or providing an indirect image of a tissue's make-up. X-rays show bones, but barely any muscle tissue because bone tissue is considerably denser and more difficult for the radiation to penetrate. Ultrasound investigation (or sonography) uses a transducer to send out sound waves and then picks up their echo. Gas or dense tissue inside the body returns a more powerful reflection than liquids such as blood.

What are the procedures that examine patients inside a tube?

The best known of these procedures – computer tomography (CT) – uses X-rays. An emitter is rotated around a patient who is lying inside a tube, producing a three-dimensional X-ray. Magnetic resonance imaging (MRI), or nuclear spin tomography, also involves a patient lying inside a tube, although this method uses magnetic fields rather than radiation. It exploits the fact that the atomic nuclei of hydrogen in the body rotate slightly and behave like a kind of magnetic spinning top. The atoms are repeatedly lined up in a magnetic field and then disturbed. During this process the nuclei give off weak radio signals that vary according to the type of tissue being examined. Giving the patient a

With magneto-encephalography it is possible to conduct highly detailed examinations of activity in particular areas of the brain and – most importantly – of their sequence in time. At Tübingen University in Germany, this process is also being used in pain research.

contrast agent – such as barium meal in the case of an X-ray – can improve the images produced by these processes.

Nuclear medicine imaging techniques use radioactive tracer isotopes. Positron emission tomography (PET) is based on them, using short-lived radioactive tracer isotopes that the body metabolises and absorbs into its cells. The isotopes emit positrons as they decay, and these can then be measured. PET scans are most often used to detect tumours.

How is it possible to measure brainwaves externally?

The brain processes information chemically and electrically. The electric potential and the impulses that run along nerve cells can be measured from outside the body. As early as 1924, Professor Hans Berger from Jena, Germany, discovered that the electrical activity of the

cerebral cortex can be measured externally if electrodes are placed on the head in a particular arrangement. An electroencephalogram (EEG) shows rhythmic oscillations from which it is possible to determine whether a person is in a particular phase of sleep, or whether he or she is relaxed, excited or in a state of deep concentration. More specific brain activity can be investigated with the help of imaging procedures. It is even possible to observe the brain as it thinks, because functional MRI and PET reveal which areas of the brain are active at any particular time.

The greatest detail is obtained from the most recently developed method – magneto-encephalography (MEG). This can even reveal the activity of individual groups of nerve cells with accuracies in the order of a few millimetres and milliseconds.

Will computers just keep on improving?

Since the invention of the first programmable calculators in the 1930s, computers have continued to develop at an incredible speed. According to Moore's Law, computing power is projected to continue doubling every 12 to 24 months. With today's silicon technology this is likely to continue for the next decade, after which many expect nanotechnology to take over.

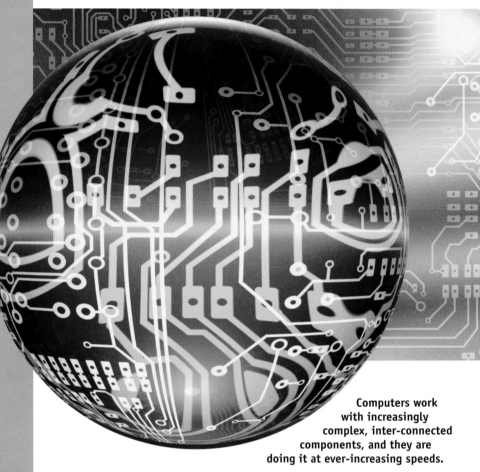

Computers work with increasingly complex, inter-connected components, and they are doing it at ever-increasing speeds.

How much more powerful can microchips become?

The power of a computer chip is directly dependent on the number of transistors it can accommodate. The individual features etched into a silicon chip now measure only 65 millionths of a millimetre (65 nanometres) in width. By 2006, chip manufacturers could already integrate about 200 million transistors onto a single chip. During 2008, this number could rise to one billion transistors, and with even smaller structures below 30 nanometres coming onto the market, this number could multiply in the next few years. A chip's computing power will increase in parallel.

True or False?

All computer chips are made of silicon

The semiconductor silicon is the key element of today's computer chip industry. The industry's growth has been extraordinary, with usable silicon transistors first appearing as late as the 1950s. The first ever transistor – developed in 1947 by the Nobel laureates (physics, 1956) William B Shockley, John Bardeen and Walter Brattain – was made not of silicon but from the semiconductor germanium. In recent years this and other compounds have again attracted the interest of scientists seeking solutions that will enable them to develop even faster and more powerful chips. More transistors are produced than any other component. In 2002, more than 1000 billion were manufactured.

Are actors about to become obsolete?

It is now possible to create computer-generated characters whose appearance is amazingly lifelike. Since the first fully computer-generated movie, *Toy Story*, hit the screens in 1995, computer animation has made incredible progress. However, the stars of computer-animated movies continue to be almost exclusively cartoon characters. *Final Fantasy* (2001) represented an effort to make computer-generated characters as realistic as possible, demonstrating in the process that no programmer can replace the skill and talent of real actors. It is unlikely that movie stars will be unemployed in the forseeable future – although this cannot be said of stuntmen and women. Computer-generated doubles can be used to shoot dangerous scenes, making the most perilous stunts risk-free.

Even when real-life actors such as Tom Hanks (in *The Polar Express*) lend their faces to computer-generated characters, the figures still have a cartoonish feel about them.

Are silicon chips always used in computers?

All computers currently on the market have silicon chips as their key components. But the silicon age is nearing its end – at least as regards the most powerful chips. The future could see the computing function of silicon being taken on by the individual atoms and molecules of materials such as carbon and genome strings. Which material will emerge as most promising, we don't yet know.

How do quantum computers compute?

Quantum computing is still in its infancy, although the theoretical foundations have been developed and early experiments are giving cause for optimism. Researchers foresee quantum computers with almost unimaginable computing power. It is possible to compute using the special physical properties of individual atoms and ions. Quantum scientists make use of the quantum state of, for example, metal atoms from the elements rubidium or potassium. Short laser pulses can affect these quantum states. The most important way in which they differ from electronic circuits is in the versatility of quantum states. They cannot be set to digital zero or one because they can hold these, and many other states in between, simultaneously. In principle, they can perform several computations at the same time, whereas today's transistors are limited to performing tasks in sequence.

How do DNA computers work?

A DNA computer is just like an ordinary computer in that it uses zeros and ones to make calculations. These basic digital units are not manipulated using electronic circuits and the flow of electrons. Instead, computations are carried out via complicated chemical reactions using the building blocks of a single DNA strand. DNA computers will never provide a replacement for the classic computer, although they could be used to rapidly calculate complex problems with multiple solutions.

Almost 100 per cent pure silicon is used in the production of microchips.

Will machines learn to think and feel?

When it comes to the game of chess, human beings are no longer any match for machines. Computers understand language (at least in a limited sense), are able to network and arrive quickly at a solution for complex problems. In the 1980s, many experts thought that computers would develop artificial intelligence and learn to think and feel independently. Scientists have since rejected this as a possible scenario.

Can computers prevent wars?

Computers have become indispensable in the development of combat strategies, but the thought that they could prevent wars is just a hope. Going to war is a decision that only humans make, and it is unlikely this will change. Computers can provide unemotional projections of the realistic consequences of a war, and it is to be hoped that outcomes involving senseless loss of life may convince warmongers to avoid rash and dangerous strategies.

Robodogs – computer systems that look like dogs – are more than technological gimmicks. They provide the foundations for much more complicated robots in the future.

...that automotive electronics are now more advanced than those of the first space shuttle?

Columbia, NASA's first space shuttle, took off on 12 April 1981. For its maiden voyage it had five computers on board – four IBMs and one Honeywell – with two computers there as back-ups in case of failure. The combined computing power of those computers does not come anywhere near the complexity of the electronics in a modern motor car. Numerous microprocessors control the engine as well as the braking and security systems. It also seems that the trend is for even more electronic gadgets to be installed in cars, with many components duplicated to overcome any problems with reliability.

Can the human brain be replaced by a computer?

Theoretically, sometime in the future there may be a computer that is capable of duplicating the memory feats and computing abilities of a human brain. But with about 100 billion nerve cells, each of which is connected to another 10 000 or so neurons, the natural neural networks inside our heads still make any computers look slow. In particular, the ability to process many stimuli and signals in parallel is still underdeveloped in computers. The world's fastest supercomputer – BlueGene/L at the Lawrence Livermore National Laboratory in the USA – performs 280.6 trillion calculations per second (in computing jargon, 280.6 teraflops). It has been calculated that even a rough replication of the human brain would require a machine capable of at least 10 000 teraflops.

Can computers be creative?

If a computer is fed sufficient data, it can combine these in unexpected ways. Solutions to problems that have arisen may appear to be the result of creative processes, however their origins still lie in the natural creativity of the programmers. The human brain is more complex and appears to be capable of more creativity. If a sufficiently sophisticated computer becomes available, will we recognise a difference or will the real differences between humans and machines then become apparent?

Is there a standard for testing a computer's intelligence?

For work on artificial intelligence (AI) to be taken seriously, there had to be a way of proving that 'intelligence' was actually present. The Turing test was developed in 1950 by British mathematician Alan Turing and is still considered the most important hurdle for so-called intelligent machines to overcome.

Turing stipulated that a computer could be described as intelligent when a human judge, addressing an unknown source at the other end of an electronic link – a phone line or terminal – could not tell whether the source was human or machine.

During the annual contest for the Loebner Prize for Artificial Intelligence, which is based on the Turing test, the aim is for three out of ten judges to be fooled for at least five minutes by the answers provided by a computer. So far no machine has achieved this.

What will computers of the future look like?

In coming years computers are going to become even smaller and more versatile. There is also a trend towards hiding these powerful machines when they are used for certain applications. With electronic chips becoming ever smaller, they can be hidden in the walls of buildings, furniture or even clothing. The only parts that would remain visible are those required for entering and viewing data. However, even those components will change. Keyboards could be superceded by tiny microphones once speech-recognition software becomes sufficiently reliable. Concealed cameras could also be used to pick up particular gestures as commands. These might then be transmitted to a central computer housed in a building's basement that would initiate the action or event called for. Computer screens will not only become flatter, but they will also be flexible enough to roll up, and could even be integrated into windows, wallpaper and jacket sleeves.

Will the car of the future be speech-driven?

Modern speech-recognition systems are already sufficiently well developed to enable cars to be controlled by voice alone. The problem is that the route a road takes cannot be described in sufficient detail from moment to moment to ensure that the vehicle arrives safely at its destination. Compared to the steering wheel, this solution is just too complicated. This is why systems that assist drivers to control cars with the help of cameras, or ultrasound transmitters that recognise their surroundings, are much more promising. Systems of this type have been tested in some countries, but so far they are too bulky to fit into an ordinary vehicle.

Speech-recognition software is being used extensively with telephone hotlines, and it is also now much easier to use spoken commands with mobile phones. Before speech-recognition programmes can become more widespread – especially for use with tasks such as the routine entry of text into a computer – the error rate will have to be reduced. At present, the best speech-recognition software can achieve accuracy rates in excess of 90 per cent, which is too low for most practical applications.

James Bond's wonder car in *Die Another Day* (2002) even had the electronics to make it invisible – an option not yet rated as standard by most car makers.

Cyborgs already exist

Half-man, half-machine creatures, such as the Terminator, or the Borg from the television series *Star Trek*, are still the stuff of science fiction. However, advances in modern prosthetics are pointing the way forward, with mechanical limbs controlled by brainwaves already boosting the mobility of amputees. Cyborgs of a different type have already seen the light of day in a German laboratory. Peter Fromherz of the Max Planck Institute for Biochemistry combined living nerve cells from a snail with a silicon chip. It is possible for electric impulses to be transmitted between these two worlds, and work such as this is laying the foundations for great improvements in prostheses.

For the time being cyborgs, like the Terminator, remain in the realm of the imagination.

How does electronic paper work?

Electronic paper is based on a novel screen principle. Early prototypes can be rolled up and placed in a coat pocket just like any newspaper. The chief advantage of this novel 'paper' lies in its low energy consumption, compared to that of current flat screens. Stimulated by small pulsations in current, tiny black and white capsules move backwards and forwards within a thin sheet. When the power is turned off, the last image to be displayed remains on view. The distribution of the capsules forms the text and images, which are as clear and easy to read as a printed page. Although some products are already on the market, they are not yet flexible. Coloured 'e-paper' already exists in the laboratory, and either entails the movement of red, green and blue spheres, or the placing of a controllable colour filter in front of a black-and-white screen.

Can a cursor be controlled by the power of thought?

It is possible to control a computer cursor with the power of thought. To do this, brainwaves are recorded by means of electrodes attached to a person's head. Every thought produces electrical impulses that filter through to the scalp, where they can be measured. Experiments show that these brainwaves follow characteristic patterns when commands such as 'up', 'down', 'left' and 'right' are thought. If signals from the 'mind reader' are synchronised with the cursor controls, it is possible to move the cursor across the screen by using thought alone.

Other researchers are looking at implants that measure electrical impulses in the body that correspond to thoughts about a movement. The hope is that this kind of technology will one day help disabled people to control their mechanical prostheses.

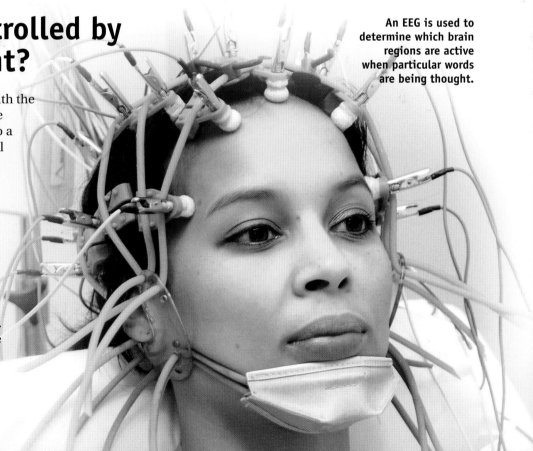

An EEG is used to determine which brain regions are active when particular words are being thought.

Do physics and chemistry still hold surprises?

Even the simplest of real world phenomena are complex in their details. Physics and chemistry continue to reveal different ways of doing everyday tasks. New technologies and materials are required to support our modern way of life, and even traditional ways of doing things benefit from being made environmentally friendly and less energy-hungry.

Is aluminium a better material than steel for cars?

Steel is harder but aluminium is lighter. The choice is a difficult one for car manufacturers, who would like to develop vehicles that are robust yet light and economical. New classes of lightweight steel being developed may solve this problem by offering a high degree of protection with reduced fuel consumption. They are super-light, extremely strong and yet flexible. These alloys combine iron with manganese, aluminium and silicon to create a specific crystal structure. Depending on its precise composition, this kind of steel can be stretched enormously without tearing, is almost twice as stress-resistant as conventional car body steel, absorbs impact energy quickly and then dissipates it, and can be rolled into strong yet thin and therefore lightweight sheets. Manufacturers expect to be using this kind of steel in cars within a few years.

What is blacker than black?

Things appear black if they reflect hardly any light, instead absorbing the energy in the light waves. The blackest things of all are thought to be the all-engulfing black holes in outer space (p. 27). Scientists in the US have discovered a method for creating true black metal surfaces using lasers. Ultra-brief, ultra-intense laser bursts form tiny ridges and furrows on the surface of metal, and these absorb almost 100 per cent of the light. This provides a durable black and will result in improved solar cells and light sensors. The process also increases the metal's surface area, which is likely to lead to more effective catalytic converters and fuel cells, as well as making paintwork on cars redundant. This process has worked on all of the metals that have been tested, including aluminium and gold.

A car body should be robust but provide a generous crumple zone in case of accidents.

Are all flame-proofing agents potentially harmful?

Flame-proofing of furniture, curtains or even heat insulation material has generally involved the use of chemicals that are bad for our health and the environment. Chemists are now trying to develop less harmful methods of providing protection. The addition of tiny amounts of clay or loam to plastic makes it fire-resistant and also stronger. Ceramic foam is an effective and safe insulation material, and wood can be treated with a protective film that foams up in the event of fire, keeping the heat at bay. It is also possible to add tiny spheres of melamine resin to plastics. These micro-capsules do not release a flame retardant until they melt, which means that although they do contain conventional chemicals these are safely contained as long as there isn't a fire.

Is it possible to melt wood?

Wood cannot be melted, nor can it be dissolved in common solvents. However, a 'liquid wood' is now being made that can be used to produce everything from loudspeaker casings and pencils to rifle stocks. Unlike other synthetic thermoplastic materials, liquid wood is not derived from oil but from renewable raw materials. Its principal component is lignin, a polymer found in wood, making the material almost identical to the real thing. Lignin, which provides plant cells with their stability, is a waste product of the paper industry because its presence causes paper to yellow. Once the discarded lignin has been enriched with other natural fibres, additives and pigments it turns into a resin that, when heated, can be formed and jet-moulded. As an added bonus, discarded liquid wood products and waste will be compostable.

New methods are being developed to flame-proof furniture without the use of harmful chemicals.

Can glass be softened without using heat?

Light can cause some special types of glass to melt. American scientists accidentally discovered the cold-melting properties of the special glass compound when carrying out research. They found that weak laser light at room temperature was enough to make the material soft and pliable. When the light was turned off, the glass became hard again. The more intense the laser beam, the softer the glass became, although its internal structure seemed to remain undamaged. This special compound – known as germanium-selenium glass – forms part of a group of semi-conductive glasses used in the electronics industry.

True or False?

Water always freezes at 0°C

Korean scientists have discovered that fresh water can be turned into ice even at room temperature. Water molecules inside a strong electric field of above 100 volts per metre arrange themselves into ice crystals at temperatures well above 0°C. It is thought that the electric field promotes the formation of the hydrogen bonds that are required for crystallisation.

Similar conditions to those produced in the laboratory can occur naturally in storm clouds and narrow clefts in rock.

Is it possible for anything to be harder than a diamond?

Diamond is the hardest natural material, but it is possible to make harder synthetic materials from compressed C_{60} molecules, a material known as buckyballs. As with diamond, buckyballs are pure carbon, except that the atoms are arranged in the shape of a football. Due to their symmetry, the molecules can be packed densely. Although the material is as soft as a lead pencil under standard pressure, it becomes ultra-hard when tightly compressed.

Scientists have been able to press the buckyballs into tiny rods at very high temperatures and pressures of about 200 000 times that of atmospheric pressure. At 492 gigapascals, the pressures are the highest ever achieved. Diamonds are produced at pressures of about 442 gigapascals.

How do you get a frog to levitate?

The answer is: place it in a magnetic field, as was done for a striking photograph that appeared in the press a couple of years ago. Frogs are not the only things that levitate in magnetic fields – strawberries, nuts and drops of water do so as well. Water, and all objects containing water, are diamagnetic. Diamagnetism is a property of every atom. When the electrons that encircle an atom's nucleus are disturbed by a magnetic field, they modify their paths in such a way that a reverse magnetic field builds up. This effect is so slight that it is usually masked by the much stronger para- and ferro-magnetisms. With water it takes effect on its own. If the magnetic field is sufficiently strong, the counterforce in the droplet – or frog – can become strong enough to overcome the downward pull of gravity. The forces on the frog's atoms balance and so the frog floats. This also works with human beings, but only with an extremely strong magnetic field and a very light person.

Can air be magnetic?

Scientists have developed electrically conductive plastic, shatterproof ceramics and floating metal. Among their latest creations is a transparent magnet. To make this possible, physicists used an extremely porous material made of silicon dioxide, known as an aerogel. Made almost exclusively of air, aerogel is extremely lightweight and transparent. About 99 per cent of the material is made up of pores, which are regularly distributed within a branching structure. Scientists deposited the tiniest particles of a neodymium-iron-boron alloy in these pores. The particles are so fine that they barely change the aerogel's appearance but, as a result, it becomes magnetic. This could be used for new kinds of flat display screens and magnetic storage media on which data can be recorded using light impulses that pass right through the material.

The diamagnetic force affects all atoms simultaneously, so the frog used in the levitation experiment was not harmed.

Are any new superglues being developed?

A patch of glue no larger than a postage stamp could support a load of five tonnes, provided scientists are able to reproduce the enormous adhesive strength of chemicals produced by some marine bacteria. Scientists in the USA have tested the strength of adhesives produced by the micro-organisms and are trying to understand the chemistry involved. The aim is to develop a strong yet environmentally friendly glue. This was achieved to some extent when researchers developed a medical superglue by duplicating the adhesive material that mussels use to stick onto rocks or ships' hulls. The shellfish extract iron from sea water and mix it with proteins they manufacture to make a strong, sticky tissue.

Do bombs exist that only destroy electrical installations?

It seems likely that e-bombs were deployed in the 1999 Kosovo conflict. They exploded with a bright flash and a thud and paralysed major power plants. E-bombs develop highly intense electromagnetic fields in short periods of time. They barely affect people, but inflict a heavy toll on electronic circuits. The effect of the explosion is so sudden that surge protectors are unable to react quickly enough. The electromagnetic field overloads cables, circuits and junctions, which then fail or are destroyed. Data stored magnetically is lost when the field rearranges the magnetic particles. The effects of the bomb are indiscriminate, however, paralysing hospitals as well as weapons.

Are there any effective non-lethal weapons?

Law enforcement agencies need to be able to free hostages or defuse crisis situations without resorting to bloodshed. Tear gas and water cannons are the traditional non-lethal defence weapons, but newer, more effective weapons are now being tested. In the USA and elsewhere, Tasers are already in use. These devices shoot two wires into an attacker through which 50kV to 750kV are discharged, causing the victim's muscles to fail temporarily. The next step involves using liquids instead of wires. The liquid Taser shoots two jets of electricity-conducting liquid and is able to bridge distances of up to 10m. It could be used to prevent potential suicides from making that fatal jump.

Other developments include infrapulse generators, which produce shock waves that stun, as well as unpleasant sounds and irritating gases. High defensive barriers, such as road blocks, can be erected in seconds with the aid of pyrotechnic motors.

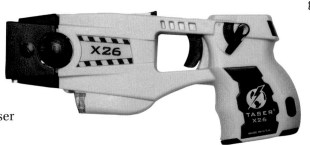

Tasers may look futuristic, but the principle behind them was patented back in 1974.

Why do dunes sing?

When he returned from his travels, Marco Polo described a low-pitched rumble that could be heard when a cool wind passed over desert dunes. This natural phenomenon can be as loud as a circular saw, and its sound can carry more than 10km. The cause was thought to be friction on the surface of the sand dunes. French researchers analysed samples of sand and believe the noise is a result of the synchronised motion of huge numbers of grains of sand inside the dunes during a slumping or avalanche motion. The wind triggers the motion of the dry, round grains – which have gel-like surfaces – and this in turn makes the air vibrate and produces audible sound waves.

Is it really possible to drown in quicksand?

Movie-makers have rather exaggerated the risk. As long as a person trapped in quicksand keeps his or her head up, there is not a lot that can happen. This was proved recently by Dutch physicists during laboratory experiments. Quicksand is sand that is supersaturated with water, with the result that friction between individual grains is reduced and they are able to glide past each other just like the molecules of a liquid. Even a small object can shift them out of their position and will sink beneath the surface. The denser a body is in comparison to the quicksand, the deeper it will sink. However, a human body is not dense enough to sink completely. The only risk to life for someone stuck in quicksand is from the tide if it happens to be coming in. This is because it is practically impossible to get out of the viscous mixture of sand and water, and doing so slowly requires a great deal of strength.

How do trees resist powerful winds?

The more powerfully a wind blows, the more a tree will bend. It is only this flexibility – which reduces the flow resistance – that allows the tree to avoid the wind's destructive power. The flow physics underlying this effect are highly complex, which is why US scientists have only recently been able to describe in mathematical terms the behaviour of trees in wind, or of jellyfish in water.

What colours did Gutenberg use to print his bibles?

Only 180 or so copies of the Gutenberg Bible, considered among the most beautiful books ever printed, were ever issued. The secret behind the brilliant colours used for the illustrations was investigated by a team of Anglo-American researchers in 2005. The team analysed the pigments and identified nine principal colours. They were able to determine the chemical composition of seven of these. The radiant red proved to be vermilion, yellow was lead-tin yellow and black turned out to be coal. Azurite was used to make blue, and malachite for green, both of these being alkaline copper carbonates. Calcium carbonate provided white and verdigris the darker shade of green. Gold and dark red are still a mystery. The gold is likely to be painter's gold and the red may have been derived from plants or insects.

Only 48 copies of the beautifully illustrated Gutenberg Bible still remain in existence.

What colour is the universe?

The intuitive answer would be a deep black or dark blue, but the actual colour – as seen by human beings – is a pale beige. This is what American astronomers have suggested after examining and combining the light of 200 000 galaxies within a radius of up to three billion light-years. Their initial results indicated that the universe was turquoise tending towards green, but they had not taken into consideration the adaptability of the human eye, which can automatically correct for slight colour casts. If it were possible to look out of a neon-lit room onto the universe, it would appear turquoise. If seen from a dark room, the human eye will perceive it as beige – and this was the conclusion that the astronomers came to.

By measuring the colour of the light from galaxies, astronomers are able to draw conclusions about the sequence in which the stars were formed.

What makes lobsters blush when they are cooked?

It wasn't until quantum theory was developed that it became possible to explain why indigo-coloured lobsters turn bright red in hot water. The red pigment astaxanthin is a component in the crustacyanin protein molecule that is found in the carapace of live lobsters. Cooking destroys the proteins, so that the underlying red colour becomes apparent. However, there was no explanation for the lobsters' original indigo colour, which absorbs almost the full spectrum of light. Because computer programmes have recently become available to perform complex quantum mechanical calculations, Dutch scientists were able to explain this. Astaxanthin pigments always occur in pairs and change their energy state as they interact with each other. Because of this they are able to absorb most of the light and reflect only the wavelengths that we perceive as indigo.

Why do biscuits crumble?

This is the fault of the humidity in the air, according to British scientists. Assuming that they weren't damaged during packaging or transport, biscuits tend to crumble when the climate in the bakery wasn't right. Physicists observed freshly baked biscuits as they cooled. As a biscuit cools, the moisture slowly moves from its centre to its edges, and the inside contracts while the edges expand slightly. Depending on the type of biscuit, this leads to hairline cracks and, in the worst cases, the biscuit crumbles. This effect is aggravated when humidity in the bakery is high. The higher the sugar content in the dough, the more likely it is that a biscuit will crumble when the air is damp; the greater the fat content, the tougher the cookie.

What is the most delicious way to eat biscuits?

Biscuits taste best when accompanied by cold chocolate milk. This is because milky, slightly fatty drinks keep the biscuit flavour in the mouth for longer, and release up to 11 times as much flavour as a dry biscuit eaten on its own. Adding cocoa powder to the milk amplifies the effect, or so one British scientist claims.

Tea and coffee wash the bisuit flavour away too quickly, preventing the consumer's mouth and nose from having sufficient time to enjoy it. The worst drinks to have with biscuits are soft drinks like cola and lemonade. They reduce the aromatic enjoyment of a biscuit by a factor of ten.

Why does dry spaghetti break into so many pieces?

Anyone who's ever tried to shorten spaghetti before cooking will know that the strands shatter rather than break cleanly. French physicists discovered that this was the result of elastic waves generated in the spaghetti. Experiments conducted in front of a high-speed camera confirmed what physicists had predicted. Bending initially causes a clean break at the point of highest curvature, but the break triggers elastic waves that make the pasta rod vibrate. This causes excessive bending in other areas, which creates further breakpoints. This is probably also true of glass fibre and metal rods.

How can sealed wine bottles be checked?

There was a time when a cork had to be taken out of a bottle before it was possible to tell whether or not a precious vintage had turned to vinegar. Thanks to nuclear magnetic resonance, wine bottles can be quality-controlled before they are opened. Californian chemists placed a wine bottle into a magnetic resonance imaging scanner and within five minutes were able to determine whether the alcohol had oxidised into acetic acid or acetaldehyde. In both of these cases protons in the molecules within the wine shift to new positions, and this shift is sufficiently clear for the machine to recognise and even quantify the extent of any change. Wine is considered spoilt when its acetic acid content exceeds 1.4g/litre or when the acetaldehyde content is greater than 300mg/litre. However, this method cannot reveal whether or not the wine is corked.

Is laundry detergent about to become a thing of the past?

Australian chemists have discovered a way of getting fat stains out of laundry without the need for detergent. The only precondition is that the water has to be pure; in particular, it must be free of the tiny gas bubbles that it normally contains. Filled with oxygen and nitrogen, these bubbles envelop water-repellent surfaces ensuring that drops of oil and fat clump together instead of being finely distributed. If all gases are removed from the water – with the help of a vacuum pump – the task of removing fat stains with water alone becomes easy. This will help to reduce the amount of laundry detergent used – which would be good for the environment and household budgets – but new washing machines will be called for to exploit the discovery.

True or False?

Dry air is best for drying things

It is easier to get things dry with hot water vapour than it is with hot air. Using superheated water vapour in drying processes can make energy savings of 50 per cent and time savings of up to 80 per cent. This applies to a range of products, from dried fruit and crisps to surface coatings and ceramic materials. The secret lies in a two-chamber system and relies on the fact that superheated water vapour at temperatures of up to 180°C can transfer its heat to the items to be dried at twice the speed of dry air at the same temperature. During this process, the vapour loses a large part of its humidity through condensation. In the second chamber the remaining dry air completes the task.

Why are so-called clap skates faster across ice?

The blade of a clap skate is attached to the front part of the boot only. It can come free at the heel end and a spring makes it slap back to the boot. This allows the blade to remain in contact with the ice for longer, even when the foot is tilted forward. The longer the blade is in contact with the ice, the longer the skater is applying a force to the ice and the faster the skater will go. With conventional skates, only the tip of the blade is in contact with the ice when the foot is tilted forward. This can be hard on the skater's calf muscles, especially for beginners.

Why do curling crosses so often end up in the goal?

Don't blame the goalkeeper if the ball inexorably heads into the goal after a curling cross – he or she had little chance of stopping it. Shots with a great deal of spin can follow trajectories that are almost impossible to calculate.

In one study, professional footballers were asked to look at records of shots and estimate whether the ball would land just inside or just outside the goal. Most of the players got it wrong. Balls that spin up to 600 times a minute around their own axis are not something that occurs in the natural environment. They outwit the goalkeeper because human visual processing is simply not built to deal with a curling cross.

When does a Mexican wave form in the stadium?

The obvious answer is, when the mood is right. When hundreds of people suddenly throw their arms into the air and cause a Mexican wave to move around the stadium it is possible to see their behaviour as being similar to that of molecules in chemical reactions that bump into one another. Scientists in Dresden conducted complex computer simulations and discovered that an understanding of the dynamics of Mexican waves doesn't belong in the field of chemistry at all, but in that of chaos theory. According to their model, it is not sufficient for a few individuals to want to start a wave, there have to be at least 25. They also discovered that random influences play such a large part in the behaviour of human crowds that it's impossible to predict what might happen – unlike processes at work in chemical reactions. Why does the wave move in only one direction? Why aren't all the spectators 'infected'? Scientists hope their analysis of human crowds as multiple particle systems will help direct traffic in railway stations, defuse panic situations in stadiums and avoid traffic jams on motorways.

Can sports shirts still be improved?

Today's sports shirts are highly effective at drawing moisture away from the skin so that the wearer remains comfortable. However, they could be a lot more informative, as is the case with innovative basketball shirts from Australia that provide spectators with the kind of information usually only available to television viewers. This includes the score and remaining playing time as well as the individual player's point score and foul record. This is accomplished using a display on the fabric and a small computer – attached by means of a chest strap – which is wirelessly connected to a central computer. A light-emitting diode information display strip indicates the player's personal goal quota, while areas on the front and back provide other data.

It is often futile for a goalkeeper to stretch or leap in an attempt to save a curling cross.

Leonardo da Vinci – a genius without parallel

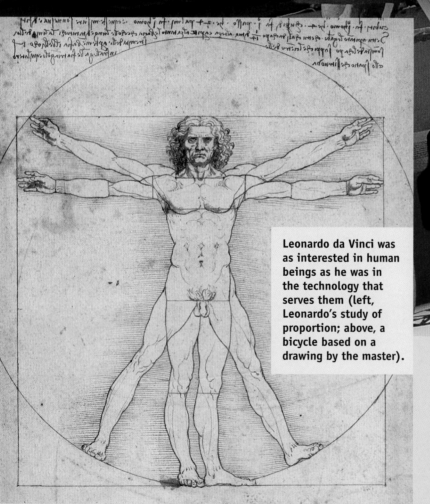

Leonardo da Vinci was as interested in human beings as he was in the technology that serves them (left, Leonardo's study of proportion; above, a bicycle based on a drawing by the master).

It is only every couple of centuries that humanity produces a genius who makes important inventions and discoveries in several fields at once. Leonardo da Vinci, the polymath Italian was a supreme example. Born the son of a notary and a peasant girl on 15 April 1452, he was denied a university education because of his illegitimate birth. Instead, he was apprenticed to the Florentine artist Andrea del Verocchio in 1465. Leonardo never seems to have forgotten that he was excluded from university and for the rest of his life despised academia and valued the individual mind above all else.

What was so special about Leonardo?

Leonardo's interests and talents were unusually wide-ranging and he was a pioneer in many fields. Although anatomy formed part of every artist's basic training during the Renaissance, it was Leonardo who most frequently applied the insights he gained from anatomy. He worked constantly to deepen his understanding of the subject, producing thousands of anatomical drawings of humans and animals. Although most famous as an artist, he promoted himself as an engineer, most famously with his numerous ingenious 'machines of war'.

Was Leonardo an engineer?

At the time, it was far from evident that Leonardo's interest in engineering would meet with a wider response. It was only during the Renaissance, that the previously held hostility to experimentation began to subside. Plato – one of the great philosophers of antiquity, who had been revered since the heyday of scholasticism at the beginning of the 13th century – eschewed experiments and sought understanding through reason alone. For him, experimentation was a 'base, mechanical art'.

Manuscripts by other thinkers of Greek antiquity did emerge during the Renaissance, concerning mining, agriculture, architecture and astronomy. But as few people could read Greek, it was a long time before these works were brought to the attention of the broader public.

When Leonardo went to Milan in 1482, it was not because the court was expecting him to work as an engineer. To begin with, Leonardo had to make a living from his painting, and in this period he created the famous *Virgin of the Rocks* and *Lady with an Ermine*.

His great opportunity to demonstrate his engineering prowess came in another artistic endeavour, when he was given the task of designing a sumptuous stage set for the court in Milan. On the occasion of the Paradise Festival on 13 January 1490, a huge model of the planets orbiting through their astrological signs was presented. Although none of Leonardo's records for this project remain, the name 'Leonardo' was on everyone's lips after the festival.

Between 1490 and 1500 Leonardo was at the peak of his fame. He was also in demand as an engineer, and received commissions to produce works to improve the waterways of Lombardy and to develop weapons and fortification technology. He also created one of his major works of art, *The Last Supper*, which took four years to complete.

What were later scientists able to learn from Leonardo?

Leonardo evidently had time to spare, even when involved in his most elaborate commissions. From the middle of the 1480s he kept notebooks in which he wrote his thoughts on scientific and engineering matters. These included physiology, anatomy, optics, acoustics, the design of musical instruments and, as always, weapons technology and automation. He was obsessed by automata, which is surprising given that in his day practically nothing was automated. Leonardo developed a tank and a car that drove themselves, and also pondered the problems of flight, studying the anatomy of birds and their ability to fly. Subsequent to these studies he developed – on paper only – a helicopter and a parachute. Some of these ideas, including Leonardo's parachute and a self-propelled car, have been realised by modern-day researchers. The parachute actually worked and the car is remarkably similar to modern motorised vehicles.

Towards the end of the 15th century, Ludovico Sforza, Leonardo's chief employer, became involved in armed conflicts with France. He was captured, and so Leonardo's time in Milan came to an end. For 20 years he spent periods of time in Florence, Rome, back in Milan and, towards the end of his life, he went to France, where he died on 2 May 1519. Over the centuries, his work was scattered and lost so that many of his groundbreaking ideas on optics and aeronautics had to be developed again centuries later.

The idea for this flying machine, found among Leonardo's papers, was constructed for an exhibition in London.

Where will our energy come from in the future?

Energy is becoming an increasingly sought-after commodity – especially in the booming economies of China and India, where the demand is rising all the time. There is now a worldwide search for new energy sources, preferably ones that makes a minimal contribution to global warming. Meanwhile, engineers are increasing the amount of power produced by conventional power plants.

Are we about to run out of energy sources?

Most of the power generated by the industrial nations is based on the fossil fuels gas and coal, and on nuclear power. Although stocks will last for several more generations, the need to reduce the environmental impact is making it more important that new energy sources are found. Hydroelectric power has been generated for decades and, in many countries, there is general agreement that its potential has now been fully exploited. Wind power, heat from within the Earth and solar energy can now be developed with modern methods and will soon be available at economical prices. Some experts predict that it should be possible to supply Germany entirely from renewable sources by 2025.

How long will fossil fuel stocks last?

The International Energy Agency (IEA) regularly assesses the world's energy reserves. At current consumption rates, oil will run out in about 45 years' time, and all the known natural gas deposits will be exhausted within approximately 60 years. Many experts believe that there are still more reserves to be tapped, and that these will provide sufficient gas for an additional 50 years. The greatest stocks of energy are those locked up in coal reserves. The largest reserves are in the USA, Russia, China, Australia and South Africa. If consumption remains at current levels, all the reserves that can now be mined commercially would provide sufficient coal to last another several hundred years. However, global warming demands new approaches to burning coal if this resource is to be exploited.

A large proportion of the electricity generated in Europe still comes from nuclear power stations.

By 2025, 25 per cent of the German electricity supply should be generated by wind power.

What alternatives are there to fossil fuels and nuclear power stations?

Nature offers many alternatives to burning coal and splitting atoms. The most important renewable energy source today is wind power. Generating power from sunlight also works well, but is expensive. So-called 'energy plants', such as rapeseed, are suitable for making fuels, while terrestrial heat, ocean tides and waves also have great potential. Electricity can be generated from all of these energy sources without further polluting the atmosphere with the greenhouse gas carbon dioxide.

Are wind turbines with two, three or four blades most effective?

No wind turbine can transform more than 59 per cent of the wind's energy into electricity. The number of blades makes little difference. Of greater importance is the area covered by each blade as it rotates. Doubling the blade diameter quadruples the power yield. Most wind turbines today have three blades, as this offers greater stability. The vibration stress would be reduced even further with five blades, but production would be more expensive. Three blades are therefore an ideal compromise between construction cost and durability.

What is a solar updraft tower?

A solar updraft tower is a type of power plant that uses upward-flowing air masses to produce electricity, instead of the familiar horizontal winds we experience every day. The blade's rotation axis is vertical rather than horizontal. A solar updraft tower is a special type of solar power plant, because it is the Sun's rays that heat the air sufficiently for it to flow upwards. The towers, which are tall and hollow, are situated in the middle of huge glasshouses that are open at the sides. The Sun heats the air in the glasshouse, which flows inwards to the base of the tower at the centre and then travels rapidly upwards, as if through a chimney, reaching speeds of up to 50km/h. The force of the air drives a turbine in the tower, which rotates to generate electricity. A 100m-tall solar updraft tower is in the planning stage in Australia, where it could be used to supply power to a town with 200 000 inhabitants.

How can we save energy without sacrificing comfort?

To ensure that our future energy needs are met in an eco-friendly way, we need to do more than simply expand the use of renewable energy sources. Energy-consciousness and energy-saving measures are at least as important. According to current studies, we could reduce our electricity consumption by at least ten per cent without any loss of comfort. More savings could be made on the roads by increasing the use of public transport and the number of economical cars that consume less than five litres of fuel per 100km. Intelligent building services and modern insulation could reduce heating and cooling costs without anyone having to freeze in winter or bake in summer.

Where can the greatest energy savings be made?

Industrial and commercial businesses are already focusing more on their energy costs than private households. In much of Europe, the greatest potential for energy savings lies in the generation and use of hot water for residential properties. Most existing buildings require about three times as much energy for heating as new buildings with well-designed energy strategies. Moving into a new energy-saving home fitted with the best insulation and heat pumps, and solar panels for heating water, is not the only option. Professional renovation measures and modern building services engineering can help save up to 80 per cent of heating costs. On average, only about one-third of possible saving measures are implemented during a renovation.

Are there any natural sources of energy that have not yet been exploited?

Energy sources are to be found everywhere in nature – from bacteria to the oceans' tides. Scientists are examining chemical, biological and physical processes to see whether there are any that could provide electricity or fuel. These include spectacular and highly controversial ideas. Bursting bubbles in a jar has been claimed to make the fusion of atomic nuclei possible. This would solve our energy problems once and for all. Some well-regarded scientists are enthusiastic about 'bubble fusion', but most experts regard it with scepticism and there is no proof that the effect exists.

Can bacteria produce electricity?

Microbes from the stomachs of cattle can generate electricity out of crop waste. American scientists have built a prototype of a living battery. A test model filled with cellulose as fuel and the micro-organisms from a cow's stomach provided a potential difference of 600 millivolts for several days. The micro-organisms are highly concentrated in the cattle's rumen – the first and largest part of their stomach – where they help to digest the vegetarian fare. In the digestive system that was replicated in the laboratory, protons moved through a separating membrane, and electrons moved via connected cables from one side of the microbiological battery to the other. Despite its low efficiency, scientists are hoping to find a useful application for it. Perhaps straw and crop waste in large tanks could be transformed into electricity in rural areas and developing regions. If cattle were kept in these areas, the microbes needed would also be available.

Is there a potential source of energy hidden on the ocean floor?

In 1971, mysterious ice structures were found at the bottom of the Black Sea. These methane hydrates – a solid compound of water and the flammable gas methane – are stable at a temperature of 2°C under the high pressure that exists underwater at depths exceeding 500m. The deposits are readily combustible in air since one litre of solid methane hydrate provides 164 litres of methane gas. Since the initial discovery, these substances have been found on many continental slopes in deep coastal waters. There is some doubt about the advisability of trying to bring this source of energy to the surface. Scientists have warned that doing so could lead to dangerous landslides, perhaps resulting in tsunamis.

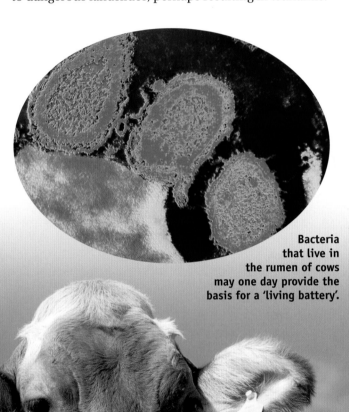

Bacteria that live in the rumen of cows may one day provide the basis for a 'living battery'.

Hydrogen – seen here in its liquid form – is considered to be a promising future source of energy.

How can energy be stored?

When it comes to oil, coal and natural gas, nature shows us how best to store the energy made by the sunlight that warmed the Earth many aeons ago. Human beings have yet to invent a comparatively efficient and easily transportable means of energy storage. Developing one is among the greatest challenges facing today's power industry, particularly because wind turbines and solar power plants only produce energy in areas that have a lot of wind or Sun, while the energy itself is generally needed elsewhere. Batteries work well in motor vehicles and mobile phones, but they would be far too bulky and expensive for storing larger amounts of power. Some experts believe that hydrogen will provide the power source of the future. One way of exploiting it is to use surplus power to split water into its component parts of hydrogen and oxygen. Using these two gases – which can be stored and transported – could allow a fuel cell to generate electricity anywhere. One problem still to be solved is how to save the large amounts of energy that are lost in the conversion.

Is it possible to store energy in plastic?

Modern, rechargeable batteries are mostly made of metals and their compounds. Plastics would make light, economical storage to replace lithium, lead, nickel or cadmium. There are several electro-conductive synthetic materials that are already beginning to be used in electronic components and flexible display prototypes. Scientists hope it may be possible to store energy in conductive plastic only, but so far this has not proved possible. Nevertheless, a combination of metal and plastic based on lithium-ion polymers has led to what are now the best available batteries for mobile phones.

Can energy be stored with the help of compressed air?

Compressed air is being considered as a method for storing energy, which would make use of a modern form of energy conversion. To exploit it, electricity would be used to power machinery to compress air into containers. The gas stored under pressure could be released later so that it flows through turbines to produce electricity. Some energy would be lost in the conversion between the various forms of energy, but losses could be kept to a minimum. Researchers are now exploring the possibility of using empty natural gas reservoirs to store the compressed air. If this proves to be practicable, surplus electricity – generated by wind farms that operate at night, for example – could be used to compress air into the reservoirs. If more power was required than could be provided by the wind farms during peak periods, the compressed air stores could be used instead. This would be a solution to fluctuating demands for power.

Will there be new kinds of engines for the cars, aeroplanes and ships of the future?

The internal combustion engine, whether it uses diesel or petrol, powers most forms of transport. Global warming and decreasing oil reserves have made it necessary for engineers to think of alternatives. The first hybrid cars that feature an additional electric engine are now being mass-produced, and electric buses and fuel cell-driven submarines have passed their first performance tests. Some cargo ships are being fitted with sails, and aeroplanes may one day be powered by clean electric engines.

A computer model of the HyFish fuel cell aircraft, showing the location of hydrogen and oxygen tanks.

Will cargo ships make use of sails again?

Concepts for using wind to help move cargo ships are being tested. The German company SkySails has had initial successes with experiments using towing kites. Ships could be pulled along by huge kites measuring $5000m^2$ flying in front of them at altitudes of 100m to 300m. With a good wind, the kite could halve fuel consumption. Overall, SkySails claims, a fuel saving of 10 to 30 per cent seems realistic. Another German firm, the wind turbine company Eis, in collaboration with others, is pursuing a concept that originated in the 1920s. A cargo ship, fitted with cylindrical rotors that behave like efficient sails, is due for launch in 2008.

How do we know...

... that rockets have been around for 800 years?

The American engineer Robert Goddard is a pioneer of the modern rocket. After the Second World War, with the help of scientists and engineers from the German wartime rocket industry, the Americans and the Soviets developed their rocket programmes. But rockets were first used as weapons of war in China in 1232. The Chinese used gunpowder-powered missiles to frighten the horses belonging to their Mongol opponents. In 1955, more than 300 years later, Europe's first rocket took off in Sibu, Romania. There are also unconfirmed accounts of bamboo sticks being shot into the sky over Byzantium shortly after the birth of Christ, using a mixture of naphtha, saltpetre and sulphur.

Is it possible for an aeroplane to take off using an electric motor?

Once it is in the sky, a glider can use warm updraughts to glide for hours without using any power. But in order to take off it has to be towed into the air by a cable winch or motorised aircraft. HyFish is a prototype fuel cell-powered aeroplane without a combustion engine that should be able to take off without assistance. Researchers at the German Aerospace Centre (DLR) are testing the new engine on small, unmanned prototypes. Energy is produced from the hydrogen in the fuel cells and is used to drive an electric motor. The first prototype was battery-powered, but it is hoped the hydrogen engine will power forthcoming versions of the futuristic aircraft and eventually allow it to fly at altitudes of 7000m and achieve speeds of between 200km/h and 300km/h.

Will we one day fill our cars from the water tap?

The idea for a water-fuelled engine should not be confused with fuel cell-engines, where the power to run the electric motors comes from hydrogen and oxygen. In the late 1990s, there were reports that an engineer in the Philippines had developed a water-powered car. It was said that five litres of tap water were sufficient to drive the car for 500km at a maximum speed of 200km/h. At the heart of the water engine was said to be a coil, which used 'ether energy' and water to power the vehicle. Alas, the story turned out to be an urban myth.

Nuclear fusion, power for the future

S cientists around the world are looking for new sources of energy. For the past 50 years, physicists have been putting their hopes in nuclear fusion. Fusing hydrogen atoms produces the inert gas helium and, at the same time, huge amounts of energy are released. This is the theory, at least, although so far no working machine has been produced. Nevertheless, nuclear fusion has great potential. Just 1kg of hydrogen fused into helium could produce as much energy as burning about 11 000kg of coal.

Do fusion reactors re-create the Sun?

The Sun is the model for terrestrial nuclear fusion reactors. For billions of years, light hydrogen nuclei in the Sun have been fusing into helium. The energy radiated by the Sun in a single second is 100 million times as much as the energy used by the entire population of the Earth in a year. The fusion process works so well there because of the immense pressure and a temperature of 15 million degrees Centigrade in the Sun's core.

Before the atomic nuclei fuse, they separate from their surrounding electrons and form a plasma made up of charged particles. Every second, the fusion reactor that is the Sun fuses around 564 million tonnes of hydrogen into about 560 million tonnes of helium. The difference in mass is converted into energy.

The best current fusion reaction experiments are also based on hot plasma. Although the largest existing test reactor consists of a plasma chamber measuring 80m³, less than 1g of hydrogen is actually involved. Temperatures of 100 million

Preparations for the next experimental fusion reactor are underway across Europe.

Hydrogen plasma being shaped and contained with the help of a magnetic field.

degrees Centigrade are needed to initiate fusion, and microwaves are used to heat the gas made up of hydrogen atoms. Despite the heat, the only time nuclear fusion was successfully initiated for a few seconds was in 1997 at the European Jet (Joint European Torus) fusion experiment in Culham, UK – and the scientists celebrated this as a great success. The process used up 25MW of energy and only yielded 16MW in return.

Nevertheless the construction of the next reactor – ITER (International Thermonuclear Experimental Reactor) – which will be about ten times larger, was approved in 2006 and will be built in Cadarache, France. In about ten years' time, it is hoped that this machine will for the first time release more energy than it uses.

In what way do fusion reactors differ from today's nuclear power stations?

In a conventional nuclear power plant, heavy atoms are split in a process called nuclear fission, whereas in a fusion reactor light atoms are joined. For a nuclear power station to have a controlled chain reaction, the radioactive fuel uranium – a very heavy metal – is required. Nuclear fusion uses hydrogen, the lightest of all elements. Both processes convert mass into energy – as described by Einstein's famous formula $E = mc^2$ (E = energy, m = mass, c = the speed of light). Since nuclear fusion converts relatively more mass, the energy yield should be greater.

Why are such powerful magnetic fields needed?

One of the greatest challenges facing nuclear fission is the ability to control the chain reactions. Once a reaction has started and there is sufficient fuel available, it keeps going. The atomic bomb is an example of this kind of reaction. Nuclear fusion, with its hot hydrogen plasma, is much harder to ignite. Furthermore, slight temperature and pressure fluctuations are sufficient to extinguish the solar fire. Extremely strong magnetic fields are required to control and contain the plasma, since no material can withstand the 100 million degree Centigrade temperature of the hot cloud of plasma. In order to avoid any contact with metals or ceramics, the magnetic fields hold the plasma's hydrogen nuclei in place as if with invisible threads.

Is nuclear fusion dangerous?

Extreme temperatures, strong magnetic fields and the large amounts of energy being released can seem frightening, although scientists are convinced that fusion reactors are safe. This is because an independent chain reaction – of the type that occurs in a traditional nuclear (fission) reactor – would be impossible. If the magnetic fields or the microwave heat source fail, the

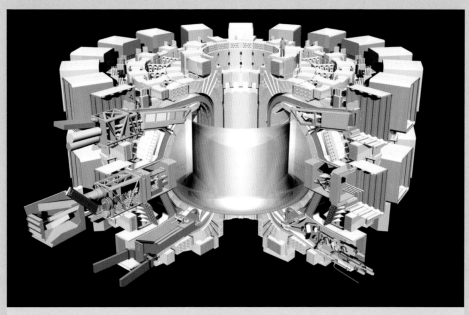

A computer image of the planned ITER research reactor shows the hot, 100 million degree Centigrade, energy-producing hydrogen plasma (pink), which is held in place in the combustion chamber by strong magnetic fields. Construction work on ITER began in 2008.

fusion process stops and meltdowns cannot happen.

However, a fusion reactor can't function without radioactivity. The fuel – the heavy hydrogen isotope tritium – and the chamber walls that are bombarded with rapidly moving particles during fusion both emit radiation, although far less is emitted than from the spent fuel rods or the reactor vessel of a fission power plant. The worst possible accident would be a leak in the water cooling system. Even if tritium were to escape from the reactor, the radioactive dose released would be around one-tenth that of the natural annual dose received from cosmic rays. Even contaminated building components that are replaced annually do not really qualify as long-term nuclear waste. They will only need to be stored for about 100 years in order to be rendered harmless.

When is the first fusion plant due to go into operation?

It will be several decades before a plant can start to provide power to the grid. JET will be succeeded by ITER, which is ten times larger. However, even if ITER were to prove successful in about ten years' time, this vastly expensive project would not produce a single kW of electricity. It will only demonstrate that commercial power generation by means of nuclear fusion is possible in principle – which is something nobody is able to say with certainty at the moment.

The next project is already being planned. DEMO will be at least twice the size of ITER, and may actually produce electricity. However, construction is not due to start for 20 years.

Are we about to see a new industrial revolution?

The invention of new, high-resolution microscopes two decades ago allowed scientists to examine the fine structure of matter in much greater detail. The tiniest particles were a surprise because they exhibited characteristics not previously observed in larger objects – nanotechnology was born and it has gone from strength to strength ever since. The term 'nano' is derived from the Greek for dwarf. Self-cleaning surfaces, quicker computer chips and possibly even new cancer treatment methods are based on nanotechnology.

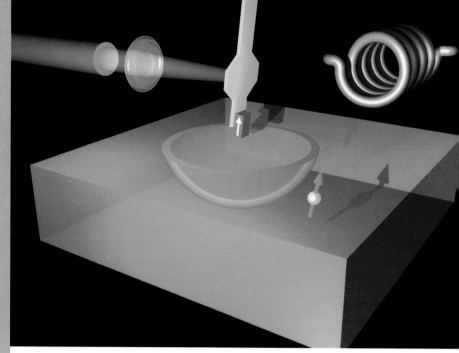

The magnetic resonance force microscope uses a magnetic tip (blue), which is a nanometre in diameter, to measure the magnetic signal of an electron.

How are scientists penetrating the world of atoms?

A hydrogen atom is approximately one-tenth of a nanometre in size – or one-tenth of a millionth of a millimetre. An individual atom is not visible through a microscope but it can be revealed in a scanning tunnelling microscope with the assistance of tiny electrical currents. Minute pulses from an equally fine metal tip can move individual atoms forwards on a surface. Special optical 'tweezers', made by focusing laser light, that can trap and move atoms and molecules are another recent development.

Could we play a game of billiards with atoms?
The ultra-fine tips of the latest microscopes can be used like billiard cues. Atomic force microscopes were developed to scan surfaces atom by atom and then use these signals to generate an image. They can also be used to move atoms. They do not simply scan the current between an atom and the tip but emit an additional, external current impulse that is capable of moving the atom precisely. The atoms and the tip of the microscopes don't touch, because their shells of negatively charged electrons cause them to recoil from one other. Atomic force microscopes are now among the most important tools of nanotechnology.

An atomic force microscope can show the location of single atoms in this sample of yttrium oxide.

How are microstructures engraved?

The latest silicon chips have structures etched onto their surfaces that are considerably smaller than 100-millionths of a millimetre. Ultraviolet light and chemical etching processes provide the basis for the photolithographic methods required to make these chips. Firstly, an image of a complex set of devices is transferred from a photomask to a blank, coated silicon chip or wafer using ultraviolet light. The image created on the light-sensitive surface is then 'developed', using chemicals that reveal the areas to be kept. Acids are then used to etch away the unprotected areas of silicon to make structures that are up to 65 nanometres in size. The remainder of the light-sensitive layer can then be washed away. The complicated structure on a chip means that this process has to be repeated several times.

Is computing with individual atoms possible?

Silicon circuits are becoming smaller and smaller, although every transistor on a chip is still made up of millions of atoms. Scientists would like to use individual atoms in what, in theory, would be the smallest and most powerful computer ever made. To achieve this, individual atoms would be locked into the electromagnetic field of a 'quadrupole ion trap'. Light pulses would allow the spin of an atom to switch between two different states, which would be used to represent the digital base values 0 and 1.

Experiments with single switchable atoms are still in their infancy. We shall probably have to wait a few more decades before there is a computer that operates on the basis of just single atoms.

The scanning tunnelling microscope is a large instrument that allows scientists to see the smallest of all structures. It was invented in 1981 by Gerd Binnig and Heinrich Rohrer.

The first image of a single atom

The physicist Gerd Binnig invented the scanning tunnelling microscope and was the first human being to see a single atom.

Gerd Binnig from Frankfurt and the Swiss Heinrich Rohrer developed the scanning tunnelling microscope (STM) at the IBM Zurich research laboratory in 1981. Other scientists had been able to see regular atomic structures with the help of electron microscopes, but Binnig and Rohrer's STM provided a much more precise view. It became possible to scan surfaces in great detail with the help of a fine tip. From fluctuations in the current (known as tunnel current) between the tip and the surface, they were able to reconstruct an image of the atomic structure of the surface. Every subsequent high-resolution scanning probe method, including the atomic force microscope, is based on the STM microscope. As a result of their work – which opened the doors to nanotechnology – Binnig and Rohrer, together with Ernst Ruska, became joint recipients of the Nobel Prize for Physics in 1986.

What do materials look like when seen up close?

Metals with smooth, polished surfaces are shiny and appear perfectly even and free from irregularities. When we stroke the surface of a piece of metal or plastic, we do not meet with much resistance. This is because the nerve cells that give us our sense of touch are not sensitive enough to allow us to feel the detailed structure of these kinds of surfaces.

How smooth is a smooth surface?

Seen through an electron microscope, a smooth metal surface is revealed as a landscape of craggy hills. Mountain ranges of atoms with micrometre-deep canyons leap into view. Beneath the surface lies a strict symmetry, in which every atom has its allotted place. Glass has a chaotic structure and, on the atomic level, there is no order at all. It forms what is known as an amorphous structure.

Extremely smooth surfaces are vital for making microchips or telescope mirrors. The mirror in the ROSAT X-ray satellite is thought to be one of the world's smoothest surfaces, with no bump on it being more than 0.35 nanometres high – equal to the diameter of an atom. Until 1999 it orbited Earth at an altitude of 580km. During this time it discovered X-rays from the comet Hyakutake.

How do crystals grow?

Snowflakes seen under an optical microscope provide an impressive and intricate spectacle, thanks to the beautiful structure of the small ice crystals, no two of which are ever the same. The salts and most metals are also composed of comparatively fine crystals. The shape of their structure is dictated by the chemical bonding behaviour of individual atoms and molecules. Starting with a condensation nucleus, they always attach themselves to previously solidified neighbours so as to form a bond that is as stable as possible. Depending on the number of connection points, a symmetrical crystal pattern is formed, usually in the shape of squares and hexagons.

An unfamiliar view of a familiar item – a single crystal of salt seen through a microscope.

Will factories build themselves in the future?

The field of nanotechnology is full of ambitious ideas, one of which is the 'nanofactory' conceived by the American scientist Eric Drexler. Drexler's vision is one in which complex objects and nanomachines would be assembled atom by atom, molecule by molecule. Hitherto unknown materials with undreamt-of characteristics could be produced in this way, he believes. However, even if the individual machine parts of such an assembler – for example nanolevers and tiny bucket wheels – are conceivable now, the idea of a fully functioning nanofactory is still science fiction.

How can nanomachines help us?

Test-tube replacement organs – a dream shared by patients and physicians – could become reality thanks to tiny nanomachines. Scientists are trying to discover how individual molecules could join together in a self-organised way to produce the structures needed for such organs. It is conceivable that this would also work for organic tissue, like a liver or heart, but so far this remains a dream. In contrast, stem cell research seems to be opening up a promising new field in which undefined cells can be turned into nerve, muscle or skin cells with the help of specific biomolecular stimuli. But even in this controversial field of research, the idea of producing complete organs is a distant hope.

How great is the danger of nanorobots getting out of control?

Nanorobots may form the basis of the nanofactories of the future. Designed to operate independently with no human intervention, they have given science-fiction writers and futurologists plenty of material. In Michael Crichton's novel *Prey*, nanorobots gang up and turn into dangerous mobs set on destroying humanity. But as nanorobots are among the least probable results of nanotechnology, we should be safe. Due to their minuscule size, there is a risk that nanoparticles could be inhaled without being noticed and affect health adversely. However, these nanorobots could have beneficial effects for the body by directly approaching individual cells and treating them. They would require a power supply and would need to be fitted with a navigation system, but they represent a development that may be feasible in the not too distant future.

True or False?

Synthetic viruses can pose a danger to humanity

Making synthetic viruses out of individual molecules might be done using nanorobots, but this is not possible yet. The real danger is designer viruses that could be developed by molecular biologists using existing micro-organisms, such as the smallpox virus. This kind of virus could be deployed as a biological weapon, and humans would be helpless to fight it since the virus could be engineered to be immune to countermeasures.

Anyone who developed this kind of virus would also need to create an effective antidote – and this would, as a result, make the rest of humanity dependent on them.

Pie in the sky? An artist's impression of a nanorobot at work injecting a cancer cell with a drug that will kill it.

What makes nanoparticles so special?

The possible benefits of nanotechnology outweigh the dangers. Because nanoparticles are so small, features that are not so important at a larger scale become significant at the nano scale. Platinum particles are more effective as catalysts than large areas of platinum, and carbon nanotubes make better electrical conductors than lumps of coal. The reason for this is the relatively large surface areas of nanoparticles. Nearly all the atoms in a nanoparticle are on the outside, while with larger items most are on the inside.

How small is the smallest engine?
The world's smallest engine was not designed by nano-engineers but is found in natural living cells. The ATPase protein measures only 10 nanometres across – 10 millionths of a millimetre – and is fuelled by high-energy molecules of adenosine triphosphate (ATP). The 'engine' works by accumulating fuel, which undergoes a chemical reaction accompanied by rotation of a 'rotor'.

Nanoscientists have isolated these and other biological motor systems and used them to move tiny nanoparticles. There have also been successful laboratory experiments in which nano-engines were driven by electrical pulses or flashes of light instead of chemicals. A model for this is provided by bacteria, which are able to travel forwards with the help of long, slender tails called flagella that are capable of moving at about 12 000 rotations per minute.

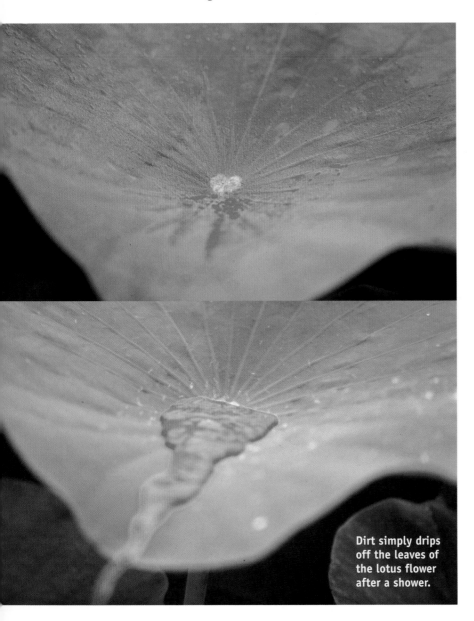

Dirt simply drips off the leaves of the lotus flower after a shower.

Is there such a thing as a dirt-repellent coating?

The first dirt-repelling paint (nicknamed 'lotus-effect' paint) came onto the market in 1999. Botanist Wilhelm Barthlott got the idea from the plant world – from nasturtiums and lotus flowers (*Nelumbo nucifera*). The surface of a leaf forms a structure with many little peaks measuring millionths of a millimetre in height. It is this structure that the dirt-repellent paint imitates when it is dry. Dirt particles adhere loosely to the peaks, while drops of rainwater are repelled. As they roll off, they take the dirt particles along with them. Lotus paint is one of the first commercial products to make use of nanotechnology. The dirt-repelling lotus effect will be made available to clothes manufacturers, but first it will be necessary to develop suitable fabrics with the special surface structures that can survive everyday wear and tear.

How can paint be made scratch-proof?
An average car has up to 8kg of paint distributed over its body. This is applied in several layers, dried and then cured, so that it is fairly scratch-proof when complete. Nanotechnology can make significant improvements in this area. There are already paints that contain microscopic ceramic particles that, during the curing process, form a dense network structure. Experiments in car washes have shown that these paints are three times as scratch-proof and 40 per cent shinier than conventional paints.

There will soon be mini-submarines able to travel through blood vessels

Computer graphics showing tiny submarines navigating through blood vessels are spectacular, but a prototype is still many years away. Optimistic nanoscientists are working on a suitably small engine to drive the nanosubmarine. Even if it were possible to construct these machines, just one or two of the vehicles could not achieve much on their own. Thousands, or even millions, of nanosubmarines would be called for, requiring many fast-working nanotools – which don't yet exist.

Tiny submarines like this could one day navigate through the arteries of the human body looking for defects. A great deal of developmental work will be needed if this scientific dream is to become a reality.

When will we see the benefits of nanotechnology?

The first nanotechnology products have been around since the beginning of the 21st century. Most are found in surface finishes that have become stronger and easier to clean thanks to the application of nano-structures and nanoparticles. Composite materials that have been strengthened by the addition of nanoparticles are also appearing on the market. Nanomaterials are used in the manufacture of longer-lasting rechargeable batteries with greater current densities. Nanomethods for use in medicine are also in development.

Are there any commercial products made with the help of nanotechnology?

A few products that use nanotechnology are on the market. The range includes dirt-repellent or scratch-proof paints and stronger tennis rackets. Small quantities of nanoparticles are incorporated into existing paints, sprays or plastics products. If we define 'nano' as being any size smaller than 100 nanometres, then every contemporary computer chip is a nanoproduct because the circuits that are being etched into the silicon blanks are only 65 millionths of a millimetre in size.

What are nanocapsules good for?

Nano-sized containers are of great interest for medical applications. They could transport small quantities of an active agent – say one that affected cancerous growths – directly to the tumour site. Once the capsules had been injected into the bloodstream, they could be directed to unload their cell-destroying drugs only at the site of the disease, avoiding any side effects to healthy tissue. This development is not advanced enough for clinical applications to proceed, although nanocontainers with lids and steerable casings have been developed. The lids or membranes can be opened from the outside with the help of focused light beams, chemicals or current impulses.The containers also have external molecules that only attach themselves to cancerous tumours.

Why are metals so versatile?

Today's world would be unimaginable without metal, its uses ranging from car bodies to nanowires for computer chips. Cast from liquid, drawn to make wire or rolled into sheets, metals can be shaped and formed as required yet still remain strong. They are better conductors of electricity and heat than other materials, and they are available in great quantities at a reasonable cost. Research on stronger types of steel and tiny nanoparticles indicates that the range and versatility of applications for metals will continue to grow.

What makes premium steel premium?

The basic raw material for steel is iron. The less carbon, sulphur and phosphorus the iron contains, the higher the grade of the resulting steel. To be classed as premium or even stainless steel, several different elements must be added. These are usually chrome, nickel and molybdenum, and they ensure that the final product will have the characteristics that are wanted. High-grade steels are in demand for mechanical engineering purposes, where corrosion-proof, wear-resistant, electrically conductive or magnetic materials are called for. It would be a mistake to assume that premium steels are always shiny and free from rust. Many pieces of rusty iron are better quality than they might appear to be.

Stainless steel comes in many grades. It is widely used in the home, but there are varieties used in the chemical industry that have to meet far more stringent standards to survive attack from highly corrosive substances.

It is now possible to clad buildings with metal. Der Neue Zollhof in Düsseldorf, designed by the architect Frank O Gehry, is a striking example.

Do liquid magnets exist?

Magnets are made of metals and their composites. If melted at high temperatures, they lose their magnetism. Despite this, it is possible to make magnetically sensitive liquids. Known as ferrofluids, they consist of extremely small particles of a ferromagnetic material such as iron oxide finely distributed in water or oil. Thanks to their size, the particles don't sink and so the the liquid becomes magnetic.

There are many areas in which magnetic liquids can be used, from new sealing methods to modern lubricants. One great advantage is that a magnetic force will keep them where they are needed. Magnetic liquids may provide a useful tool in cancer therapies where they could be embedded in the tissue of a tumour and then have their polarity constantly reversed. The rapid movement of the particles would lead to a build-up of heat, which would result in the destruction of the tumour tissue.

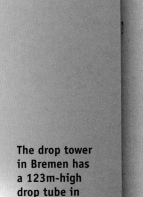

Ferrofluid drops are pulled upwards when brought into close proximity to a magnet.

The drop tower in Bremen has a 123m-high drop tube in which weightless conditions can be simulated for almost 10 seconds following the release of a capsule.

Do new materials really offer a range of novel properties?

Humans have always searched for new materials. The quest that began with a desire for stronger and sharper swords lead ultimately to the invention of stainless steel and flexible plastics, as well as a range of more exotic materials. The versatility of metals and plastics is growing steadily, with scientists examining ever smaller particles and conducting experiments with ceramics and molten metals prepared in the weightless environment of the International Space Station – all in the quest for new materials.

Can better materials be produced in weightless conditions?
Everything on Earth is subject to gravity. It even affects individual atoms in the formation of crystals or molten metals as they harden, but it is impossible to say exactly which of a material's characteristics are influenced by gravity. Gravitational effects are at a minimum aboard an orbiting spacecraft, in a drop tower or inside an aircraft during special acrobatic manoeuvres. Materials scientists make use of these environments to produce new kinds of metal alloys and plastics. In 'weightless' conditions, the process of solidification is subject only to the chemical and physical forces between the atoms. Scientists have made new discoveries about the structure of matter during their experiments. These may not have led to many technical applications so far, but this is because it would be much too costly to set up a production plant in space at this stage.

Can metal or plastic have a memory?

Some substances are known as 'shape memory' materials and do have a memory. The principle can be explained using the example of a surgical thread, which is 2cm long in its normal state but can be stretched to a length of 3cm. When subjected to heat it returns to its original state – which means that a loosely tied knot tightens up automatically when the memory shape is recalled. The thread's 'memory' is based on the way it is made. It consists of two materials: one stretches like a spring while the other holds it stretched. When the second component is 'melted', the string contracts.

Polymers – long chain molecules like those that are found in plastics – can also be programmed for particular uses by modifying their structures using heat or ultraviolet light. Some metals also have this kind of power of recall, with 'memory metals' regaining their original shape when heated.

Are there any everyday uses for memory metals?

Memory metals could prove to be useful. If they were used in the automobile industry, for example, damaged car bodies would no longer be so expensive to repair since a blow dryer might be all that is needed to restore their shape.

Greenhouses are another area where they might be used. Hinges made of memory metal could be fitted to vents and combined with temperature sensors. When the temperature in the greenhouse rose above a certain level, the hinges would stretch and the vents open.

This sand mixture will be turned into high-grade glass.

How is ordinary sand turned into glass?

Quartz sand, soda and lime are placed in a large container where they are finely ground and mixed together. This mixture then goes into a glass kiln, where it is heated to a temperature of 1400°C to 1600°C. This causes a chemical reaction that creates a soft, sticky mass full of bubbles, which is cooled down to about 1000°C for further processing.

Impurities in the raw materials cause the glass to be cloudy, so manganese dioxide is added to obtain a clear, colourless mass. This neutralises the green and brown colouration produced by the sand's iron content. Coloured glass is made by adding a range of metal oxides to the molten glass, and other colourings in microscopically fine forms can also be distributed in the glass mass. Cloudy glass can be made by using additives such as calcium phosphate, tin dioxide and cryolite. Common treatments for glass after it has been formed into a shape include polishing, sandblasting and etching.

How do we know...

... that lead pencils do not contain lead?

Lead pencils were once common, although they were made of an alloy of lead and silver, which was pressed into a pencil form. Because this material was hard, the paper that was used to write on needed to be robust. And long-term contact with the skin posed a health hazard. People switched to using graphite encased in a wooden shaft, although the word 'lead' to describe the graphite pencil remains in use to this day.

Will plastics soon replace all other materials?

Plastic is a general term that describes synthetic materials based on polymer molecules. It is possible to vary its characteristics by using different basic raw materials, modifying the manufacturing process or putting in different additives. Synthetic materials can be made to take on the characteristics of other materials that are either too expensive or too difficult to make, or too rare. In the aircraft manufacturing industry, plastics are increasingly replacing various types of steels.

How is oil turned into plastic?

Plastic is not a naturally occurring raw material. It is generally made synthetically, and the most commonly used source material is crude oil. To obtain the raw materials to make a plastic, the oil must be heated. This takes place in a distillation tower at an oil refinery at temperatures of approximately 400°C. One of the products of the distillation process is naphtha, which is reheated, this time to 800°C, and then immediately cooled to 200°C. It is during this process that the many little molecules – known as monomers – combine to make the long, powerful molecular chains known as polymers. Synthetic materials with a variety of characteristics can be produced by adding a range of chemicals.

Is it possible for plastic to be harder than steel?

Plastics can have a wide range of characteristics and can be soft, hard, transparent and even electrically conductive. They are easily shaped and are usually considerably lighter than metals or ceramics. They can be as strong as metals and some steels, although there is no synthetic material stronger than high-tensile steel. This is because of the low density of the polymer material. However, the strength of synthetic materials is being increased all the time by the addition of nanoparticles and by getting the molecules in the polymer chains to link together tightly.

The distillation of crude oil in plants such as this produces gas, diesel oil and middle distillate, which is used to make heating oil and diesel.

This aircraft, a Grob G-180 SPn, is rugged yet lightweight because it is made from carbon fibre-reinforced plastic.

Bionics – inventions inspired by nature

Bionics deals in the observation, analysis and understanding of natural processes and their use in engineering designs. This is not an attempt to copy nature, but an effort to understand its principles and use them in technological applications.

Where does the term 'bionics' come from?

The word bionics is a combination of *biology* and *electronics*, and was first suggested by American air force major Jack E Steele in 1960.

Who was the inventor of bionics?

Leonardo da Vinci, the great Italian artist and engineer, is often put forward as the first student of bionics. He studied many natural phenomena, such as bird flight, in the early 16th century. On the basis of his observations, he produced engineering designs that included flying machines and even a helicopter.

Is the parachute modelled on nature?

More than 500 years ago Leonardo da Vinci sketched the first parachute, possibly receiving his inspiration from dandelion seeds – their umbrellas of tiny hairs carry them for hundreds of metres on even the slightest breeze. Perhaps the first trial of a parachute occurred nearly 400 years ago when, in 1617, Croatian Faust Vrancic performed the first successful parachute jump and descended safely from the bell tower of St Martin's Cathedral in Bratislava.

What was Otto Lilienthal's contribution to the development of aeroplanes?

Like many other pioneers of aviation around the world – such as the Australian Lawrence Hargrave – the German Otto Lilienthal was an early proponent of bionics. He observed and studied storks, analysing the way in which the birds fly. Based on what he observed, he developed and built unpowered flying machines. Lilienthal is credited with the world's first manned unpowered flights, gliding for distances of up to 230m in the early 1890s. He was killed in 1896 when one of his gliders crashed on a windy day.

Are there any flipper-driven ships?

A system for flipper propulsion – based on the way in which penguins swim – is currently being developed. Preliminary tests have shown that it is more effective, and provides more responsive manoeuvring, than a traditional propeller. Before the idea can progress from the drawing board to a working model, some means will have to be perfected for reliably translating the circular motion of the drive shaft into the up and down movements of the flippers.

How can an earthworm contribute to an investigation of the human intestine?

German researchers have been studying the way in which earthworms move, and have used their observations for the development of a novel device for performing colonoscopies – looking inside the colon. Unlike the conventional tools for this and similar jobs, which are pushed along from the outside, this revolutionary new microrobot moves independently through the intestine, minimising the risk of injury.

What was the model for Velcro?

Velcro is another invention modelled on nature. In the mid-20th century, the Swiss scientist Georges de Mestral considered the irritating way in which burrs are able to stick to the surface of fabric, making them hard to remove. He examined the burrs under the microscope and developed a similar synthetic fastening mechanism.

Is it possible to dive underwater without getting wet?

The fishing spider manages to stay dry when it dives underwater. Fine hairs on the surface of its body trap innumerable tiny air bubbles so that its body is protected by a layer of air. Bionic engineers are trying to adapt this principle for technical applications, such as coatings for bathing suits or the hulls of ships.

What lessons can structural engineers learn from nature?

Constructing buildings that are delicate yet robust is one of the challenges facing architects today. Spiders provide nature's model for elegant roof structures, like that built over Munich's Olympia Park. Spider webs are braced by crossing gossamer threads that can withstand enormous forces. Attempts are being made to reproduce spiders' silk – which is light but very strong – in the laboratory.

Which natural example could be useful in developing robots to explore Mars?

The first robot explorers on Mars rolled around the stony surface of our neighbouring planet on six wheels, but the next generation could be equipped with six insect-like legs. This would help them to overcome even quite large obstacles. Experiments with cockroaches have demonstrated that each leg performs every step fully independently. The development of the first walking robots is founded on this principle.

Can lens manufacturers learn from the eye?

Sophisticated lenses, compound eyes and optical fibres – no other natural sense has inspired such a wide range of innovative designs as the sense of sight. Prototypes of liquid lenses are already being developed, and Japanese scientists have constructed a system made up of many tiny lenses – similar to those found in the compound eyes of flies – and have placed them on a hemisphere to provide a 360° view. An X-ray sensor on the International Space Station is able to obtain an extremely wide-angled view because it is domed like the eye of a lobster.

Are computers imitating the brain's structure?

No computer is as powerful as the human brain. The reason lies in the way that brain cells are packed into a tight network. Inspired by this structure, computer scientists have been able to build small, synthetic neural networks using electronic components. These are able to learn certain facts on the basis of training examples, without first having to be programmed.

Could the solution to the energy problem lie in synthetic photosynthesis?

No solar cell is as efficient at transforming sunlight into a different form of energy as photosynthesis is for plants. Scientists have been trying to copy this process for years. A German–Swiss team have had some initial success with a system of self-organising macromolecules and lipid membranes, which was able to store the energy from sunlight with separate electrical charges. This early development is not yet suitable for use in a rechargeable solar battery because the electric charges could only be accessed once.

Will the structure of a gecko's foot provide the model for new adhesives?

Geckos can walk up the smoothest of vertical surfaces, thanks to tiny hairs on their feet (see p.173). Scientists in the USA have copied this adhesive effect using carbon nanotubes. Early experiments demonstrated that the nanotubes had an adhesive force stronger than all traditional adhesives, and could even provide better adhesion than that achieved by geckos' feet. Although the synthetic gecko modules have tremendous adhesive properties, they won't have a real-world application until the problem of how to unstick them has been solved.

What can engineers learn from sharks?

Sharks can swim long distances with the expenditure of very little energy. One of the reasons for this is their special skin, which is not smooth but covered with numerous fine structures that help to reduce its flow resistance to water. Swimsuits are available that make use of similar microstructures, and experiments are being conducted to produce coatings for the hulls of ships that will help to reduce their fuel consumption.

A close-up view of the horny scales that make up the surface of a shark's skin, helping to reduce water resistance.

What does a gene look like?

As a rule, genes are sections of different lengths of the DNA molecule. The sequence of the gene's building blocks contains the information needed to produce proteins. Other parts of the sequence serve as a kind of 'switch' that proteins can attach themselves to in order to either block or activate a gene.

Some bacteria, such as *Mycoplasma* that have no cell walls, can survive with a small number of genes.

How alike are the genes inside humans and bacteria?

The genetic information stored in the DNA of a bacterium and a human is written in an identical language, one that uses the same four DNA building blocks as its 'letters' – the nucleotide bases adenine, thymine, guanine and cytosine (A, T, G and C). This is why it is possible to introduce human genes into bacterial cells in order to produce human proteins. There are differences, with the genes of humans typically interrupted by portions called introns that have to be removed prior to the production of proteins. The production of proteins in human cells is more complicated than it is in bacterial cells, therefore offering many more possibilities.

What is the minimum number of genes required by a living organism?

Of all the independently viable organisms that have been examined, the *Mycoplasma genitalium* bacterium has the smallest number of genes. Five hundred or so genes are all it needs, while most other bacteria have several thousand. Scientists involved in the 'Minimal Life' project have been able to further reduce the genotype of *M. genitalium* without killing the bacterium. This exercise demonstrated that a living organism needs at least 300 genes. Parasitic bacteria and many viruses have fewer genes, but they are unable to live independently.

How do mutations happen?

A mutation is a sudden change in a cell's genetic material, during which building blocks of DNA are lost, exchanged for others, chemically modified or added to the overall structure. These changes can be triggered by the presence of toxins in the environment or by faults during the duplication of genetic material. Mutations can also be generated by radiation or chemicals. Minor defects in DNA are usually recognised by a cell and rapidly corrected. However, with advancing age, the number of mutations that are not repaired gradually increases.

Discovering the rope ladder

What was it that James Watson and Francis Crick discovered?

Since 1944, scientists have known that DNA is the carrier of genetic information. But the molecular structure of nucleic acid that could explain the way DNA worked was not known. In 1953, Watson and Crick recognised that DNA is present in the form of two connected strands arranged like a rope ladder in a double helix. The strands of molecules are held together by chemical bases that always form themselves into the pairs A-T and G-C. This pairing means that the sequence of bases on one strand automatically determines the base sequence on a second strand made using it as a template. This model explains how the genetic material in the DNA molecule is reproduced identically during cell division.

Did Mendel know what a gene was?

The Augustinian monk Johann Gregor Mendel (1822–1884) conducted many experiments on the hybridisation of pea plants, which he analysed statistically.

He concluded that there had to be inherited traits that were responsible for external characteristics such as the colour of a flower, and that these traits were transferred to subsequent generations in accordance with specific rules. In 1909, the Danish biologist Wilhelm Johannsen came up with the term 'gene' to describe these inherited traits. Neither scientist knew that genes are sections of DNA molecules and that they are located in the chromosomes of the cell nucleus.

Did Mendel cheat?

Mendel used a knowledge of mathematics he had acquired as a student in his botanical research. He was one of the first natural scientists to proceed by first putting forward a hypothesis that he then sought to prove by means of experiments.

Over eight years, he analysed more than 10 000 experiments. Recent attempts to reproduce his experiments have shown that he must have enhanced his results to ensure a better match with his expectations. His published figures are too precise to be believable. Modern statistical methods prove that his figures

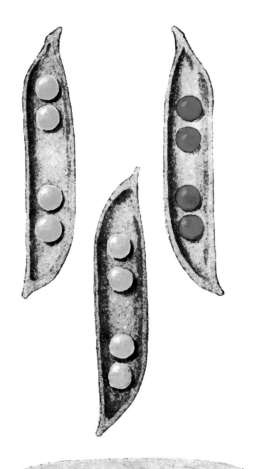

One of the discoveries Mendel made during his hybridisation experiments was that yellow is a dominant colour trait in pea plants.

do not exhibit the range of variation that would be expected as a result of biological variability. It seems likely that he disregarded less favourable tests in his analysis, perhaps assuming that errors in the experiment produced the wrong results, although this does not affect the validity of the rules of inherited traits that he discovered.

Are clones unnatural?

Bacteria, plants and some animals reproduce by forming clones – offspring with the same genetic make-up as the parent organism. In single-cell organisms this happens by means of binary division. Strawberry shoots and potatoes also develop natural clones, while cnidarians (jellyfish, sea anemones and corals) clone themselves by budding. This kind of

In 2000, Scottish scientists were the first to clone pigs successfully, producing Millie, Christa, Alexis, Carrel and Dotcom.

asexual reproduction does not occur in mammals, where the female gives birth to young whose genetic make-up is inherited in equal halves from the mother and the father. Genetically identical mammals can only occur naturally if two embryos develop from one fertilised egg cell, leading to the birth of identical twins.

How does the cloning of mammals work?

The technology of cell nuclear transfer makes it possible to produce a genetically identical clone of an adult mammal. To do this, the cell nucleus of any somatic cell (one which contains the animal's entire genetic make-up) has to be inserted into an ovum, the cell nucleus of which has previously been removed. Next, a chemical or electrical signal sets off continuous cell division and an embryo is formed. Once implanted in the womb, this embryo will develop into a clone of the cell nucleus donor. The genes are not a perfect match because the DNA undergoes minor changes during the embryo's development and after birth.

What is therapeutic cloning?

In contrast to reproductive cloning, therapeutic cloning does not involve the development of a complete living organism, but aims to produce genetically identical stem cells. The development of an embryo is stopped at an early stage, and its cells used to grow cell cultures. It may be possible to control the development of these cells to produce specific tissue for therapeutic purposes – although there is no proof that this will work.

Can dinosaurs be brought back to life?

In theory, it should be possible to grow a dinosaur by using its fully preserved or reconstructed genotype to produce a clone. However, dinosaur bones are over 65 million years old and no longer contain any usable DNA. Dinosaur blood samples found in insects preserved in amber only yielded fragments of dinosaur DNA – too little to allow the complete genotype to be reconstructed. Even if it were possible to achieve a full reconstruction of the genetic make-up, this would then have to be transferred to an ovum belonging to a close relative of the giant reptiles. It is unlikely that this would develop into a dinosaur embryo.

What are so-called bad genes good for?

Genes that cause diseases can be useful, otherwise there would be no way of explaining why 'bad' genes are found in the human genotype. The classic example of this is the gene for sickle-cell anaemia, which is only widespread in areas affected by malaria. The defective gene has few negative effects on people who have a healthy gene for haemoglobin, but it does provide them with protection against malaria. It is thought that a similar benefit is inherent in the cystic fibrosis gene, which is common in Europe. It may provide healthy carriers with protection against diarrhoea and tuberculosis.

What are 'knockout' mice?

A 'knockout' mouse has been genetically engineered so that one of its genes is inactive. This makes possible comparison of two living organisms that differ only in the absence of a single gene, which helps determine the function of that missing gene. Potential cures for human diseases are also tested on them. Knockout mice are created from embryonic stem cells in which the gene being investigated has been swapped for a defective version.

How do genetic paternity tests work?

The kinship of two people can be proved by comparing their respective genetic fingerprints – a much more precise method than the blood test used previously. The test requires DNA that has been obtained by taking cell swabs from saliva or blood. For a paternity test, the lengths of particular segments of DNA are determined with short, repeating sequences of DNA building blocks.

Every human being has a unique spectrum of these DNA segments. When placed on a gel and divided and dyed according to size, the result appears as a set of black bands similar in appearance to a barcode. Each DNA section of the child must also be present either in the father's or the mother's DNA. Paternity can therefore be confirmed or disproved with almost 100 per cent accuracy.

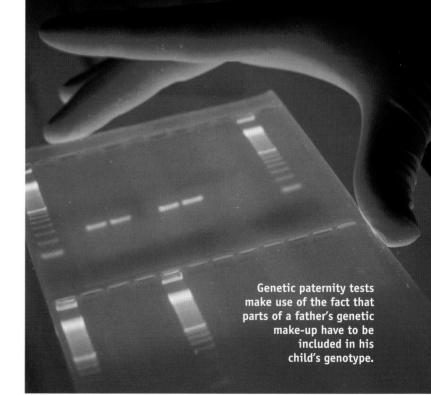

Genetic paternity tests make use of the fact that parts of a father's genetic make-up have to be included in his child's genotype.

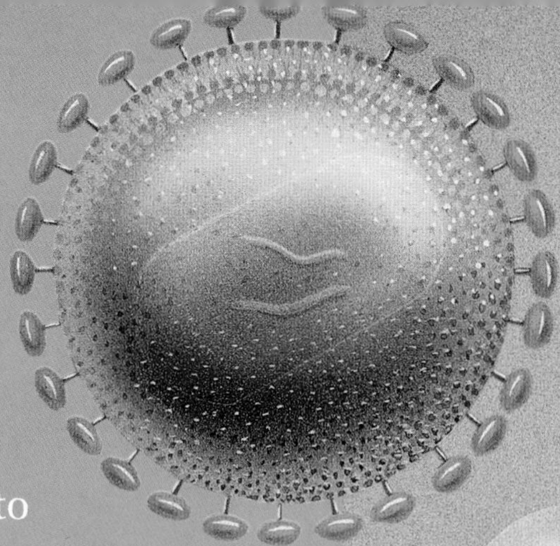

What are microbes?

Microbes are single-cell organisms that are invisible to the naked eye. They include an assortment of highly diverse life forms, including bacteria and archaeons, protozoans, single-cell algae, yeasts and mould fungi. Archaeons used to be described as archaeo-bacteria, but have been found to make up a separate group of single-cell microbes. Viruses are a special case because they lack certain attributes regarded as essential for living organisms.

How many bacteria are there?

About 7000 species of bacteria have been described and named – a fraction of all the existing species. More than 95 per cent of the bacteria found in soil and water have not yet been given names. We know of their existence only because parts of their DNA have been identified. Many bacteria cannot be cultivated in the commonly used growth mediums. It is estimated that one litre of sea water contains about 20 000 species of bacteria, so it is possible that there are several million different species worldwide. There are thought to be several hundred thousand different bacteria in the human mouth; only a fraction have been identified.

Can bacteria survive immersion in water at boiling point?

Most bacteria die in boiling water because vital proteins begin to be destroyed at temperatures above 50°C. Some soil bacteria are able to turn themselves into dry, heat-resistant spores. When favourable conditions return, the spores germinate and the bacteria once again become heat-sensitive. Although boiling reduces bacterial content, it does not result in sterility. In order to kill off the spores it is necessary to expose them to water vapour under pressure at temperatures above 100°C.

Some microbes have adapted to living in hot places, including geysers. They don't start to feel comfortable until temperatures have soared to 80°C and above. The current heat record for these bacteria and archaeons is 121°C. Organisms of this type contain proteins in which the molecular chains are tightly interlinked internally, which serves to make them extremely resistant to heat.

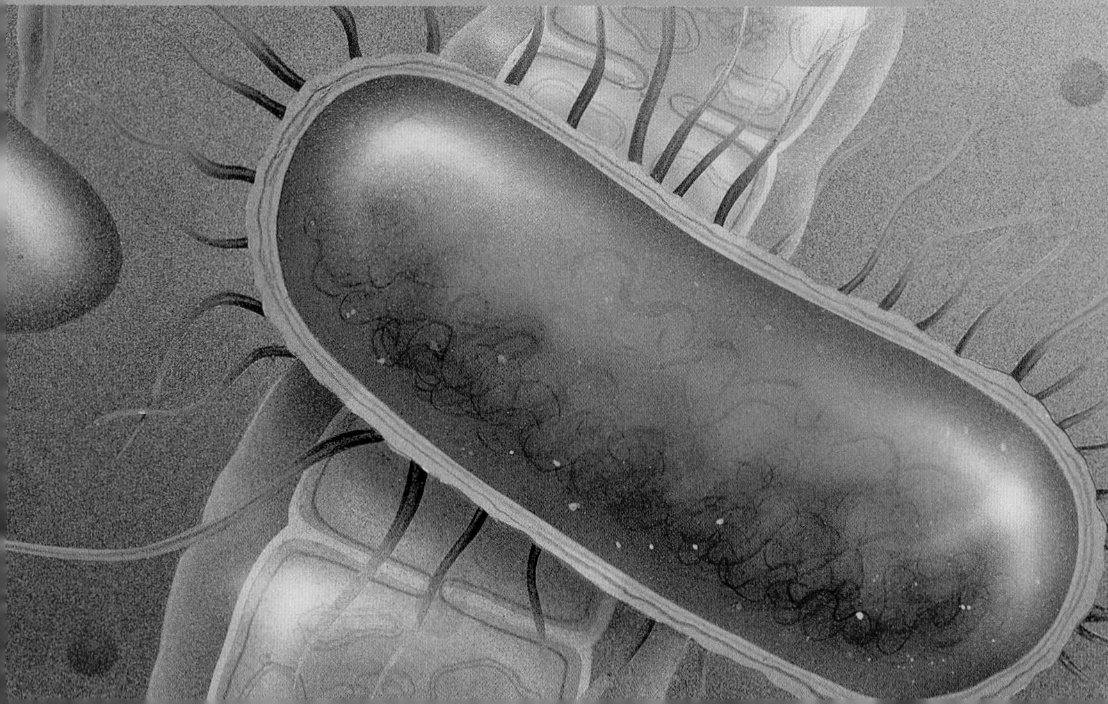

The world of microbes – seen here in an artist's impression – is only accessible through a powerful microscope.

How are bacteria and viruses different?

The most obvious difference between bacteria and viruses lies in their sizes. A typical bacterium has a diameter of one micrometre – one-thousandth of a millimetre – and measures between one and three micrometres in length. Viruses only measure between 0.02 and 0.3 micrometres in length.

The most important difference between them is that bacteria are life forms that can reproduce independently, whereas viruses are dependent for their reproduction on living host cells.

Viruses only have one kind of nucleic acid, and their genotype only exists as DNA or RNA. Bacteria are like higher life forms in that they store all their genetic information in their DNA and also contain various kinds of RNA that fulfil other functions.

Bacteria are easier to combat than viruses as they can be killed off with antibiotics, which interrupt the cell's metabolism. Viruses use the host cell's metabolism, so antibiotics don't stop them. Few drugs have been developed that are effective against viruses.

How do we know...

...that some bacteria are fitted with a compass?

Migratory birds, carrier pigeons and salmon are not alone in using the Earth's magnetic field to find their way. Magnetospirilli are waterborne bacteria that, like compass needles, align themselves with magnetic field lines. They do this with the help of tiny magnetite crystals arranged in a chain along a protein rod inside the cell, which runs parallel to its long axis. About 30 genes are required to construct this compass. The bacteria were believed to follow the Earth's magnetic field lines in a northerly direction (in the northern hemisphere). Some have now been found to move in the opposite direction, so it is not clear what decides which way they swim.

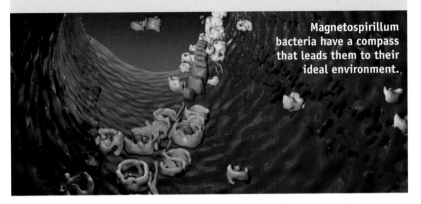

Magnetospirillum bacteria have a compass that leads them to their ideal environment.

How many viruses are lying dormant in our bodies?

Every human being has in his or her cells the genes of viruses that lodged themselves in the genotype of our ancestors millions of years ago. They make up a staggering eight per cent or so of our DNA. Most of these viruses are no longer functional, although it is possible that some play a role in triggering particular diseases.

More virus genes are being added all the time, since some of the viruses that we are infected with during the course of our lives go on to integrate their DNA into our cells. These infections do not cause any adverse effects at first, although stress or illness can reactivate these dormant viruses at a later stage and trigger a range of complaints. A good example of this is the herpes viruses, which has found a home in the nerve cells of most people. It generally only causes the familiar recurrent blisters to appear in a few of its reluctant hosts when they are stressed in some way. It is assumed that most people are infected with additional viruses that could become active when their immune systems are weak, or have been suppressed by drugs.

Where did the HIV/AIDS virus come from?

Both types of the Acquired Immune Deficiency Syndrome (AIDS) virus, HIV-1 and HIV-2, originate in the related Simian Immune Deficiency Virus (SIV), which affects monkeys. Viruses found in the blood of chimpanzees are similar to the globally spread HIV-1/AIDS virus. The chimpanzees were probably infected by other primate species without themselves succumbing to the disease. But when the virus was transferred to humans it changed and became the cause of the human immune-deficiency syndrome.

We don't know how the virus was transferred from chimpanzees to humans. Scientists suspect that bush hunters and meat vendors in central Africa were infected through contact with fresh monkey meat. Another theory is that the chimpanzee virus was transferred to humans via an SIV-contaminated polio vaccine distributed in what was then the Belgian Congo. Genotype analysis of the viruses has determined that the transfer from ape to human occurred in about 1930 – a long time before the vaccination drive in the late 1950s.

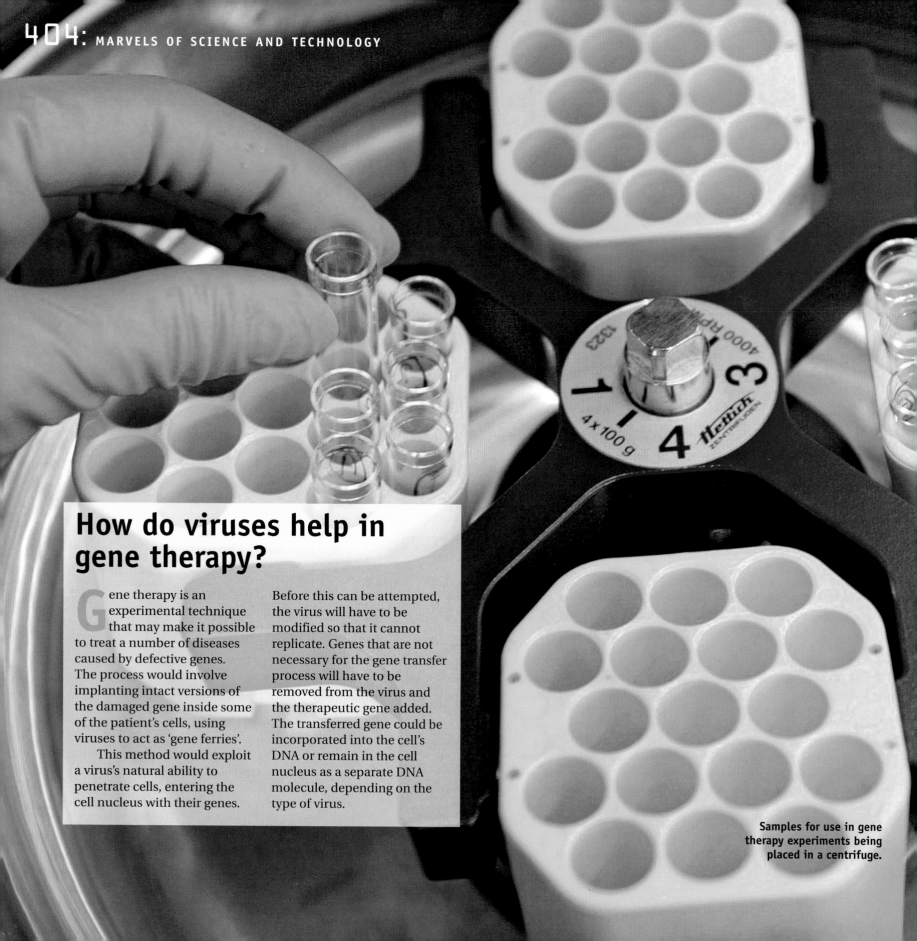

How do viruses help in gene therapy?

Gene therapy is an experimental technique that may make it possible to treat a number of diseases caused by defective genes. The process would involve implanting intact versions of the damaged gene inside some of the patient's cells, using viruses to act as 'gene ferries'.

This method would exploit a virus's natural ability to penetrate cells, entering the cell nucleus with their genes.

Before this can be attempted, the virus will have to be modified so that it cannot replicate. Genes that are not necessary for the gene transfer process will have to be removed from the virus and the therapeutic gene added. The transferred gene could be incorporated into the cell's DNA or remain in the cell nucleus as a separate DNA molecule, depending on the type of virus.

Samples for use in gene therapy experiments being placed in a centrifuge.

What different kinds of pathogens are there?

In addition to bacteria – which are known as prokaryotes because they lack a cell nucleus – infectious diseases can be caused by eukaryotes, which are organisms with a true cell nucleus. These include pathogens from the protozoa, fungus and worm groups. Unlike pro- and eukaryotes, viruses are not cell structures and are dependent on other living organisms to reproduce. Prions are not living organisms but pathologically distorted protein molecules, with the result that they are the only pathogens that do not have any nucleic acid (DNA or RNA). They multiply by attaching themselves to similar molecules and forcing them to take on their form.

In what way do pathogens differ from harmless bacteria?

Most bacteria cannot infect people or cause diseases. This requires special attributes known as virulence factors, which include the polysaccharide capsules of *Streptococcus pneumoniae* (which causes bacterial pneumonia). Other cell surface structures prevent the instant destruction of the invaders by immune cells. Pathogens also need weapons of attack in the form of released enzymes that split proteins and other molecules, thereby enabling the tissue to be penetrated. Many pathogens create toxins that damage cells and trigger inflammations and fevers.

There are many variations of the intestinal bacterium *Escherichia coli* (known as *E. coli*), including the common, harmless form as well as several that are equipped with virulence factors – some of which cause gastro-intestinal diseases and urinary tract infections.

Where do new pathogens come from?

Pathogens discovered in recent years either couldn't be found until modern detection methods were developed or the microbes previously only infected animals, making the jump to human beings relatively recently. This generally means that the pathogen has undergone modification in order to adapt to its new host. Viruses such as those responsible for haemorrhagic fever, Severe Acure Respiratory Syndrome (SARS) and AIDS, as well as the form of Creutzfeldt-Jakob disease caused by prions, have all emerged from a reservoir of animal pathogens. Random modifications to the genotype of viruses can rapidly change the characteristics of a pathogen, thereby increasing the hazard it presents to humans.

Different pathogens

When we think of pathogens, we tend to think of viruses and bacteria – but not all diseases can be blamed on these two groups of microbes

Pathogen	Examples
Worms	Roundworm, tapeworm, liver fluke
Fungi	Yeasts (e.g. *Candida*), cutaneous fungi (e.g. athlete's foot)
Protozoans	Amoebas, leishmanias, plasmodium (cause of malaria)
Bacteria	*Salmonella*, *streptococcus*, mycobacteria
Viruses	Flu virus, smallpox virus, HIV
Prions	Cause of Bovine Spongiform Encephalopathy (BSE) in cattle and Creutzfeldt-Jakob disease in humans

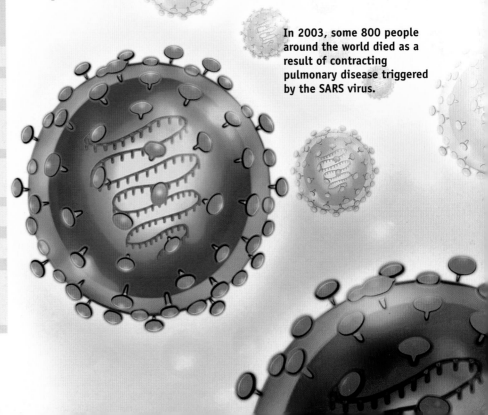

In 2003, some 800 people around the world died as a result of contracting pulmonary disease triggered by the SARS virus.

What are hospital-acquired diseases?

Hospital-acquired diseases are caused by microbes that have adapted to life in hospitals. These are generally *staphylococcus*, *enterococcus* or other bacteria that have become resistant to over-prescribed antibiotics. They are spread by doctors and nurses from one patient to the next. Inadequate disinfection of hands and insufficiently sterilised instruments help to spread these diseases, which are the cause of urinary tract, wound and pulmonary infections, especially in patients whose immune system is weak. Strict hygiene measures and the controlled use of antibiotics help to reduce the risk.

The life cycle of the predatory bacterium *Bdellovibrio bacteriovorus* (yellow) as it attacks a bacterial prey cell (blue).

Can bacteria be used as 'living antibiotics'?

The predatory bacterium *Bdellovibrio bacteriovorus* hunts and destroys other bacteria, both in the open environment and in the intestines of mammals. It recognises its prey by the structure of their cell walls and attaches itself to the largest host bacterium. Enzymes are released that create a hole in the external cell coating of the victim through which the attacker bores into the cell. Once there, it feeds on the cell plasma of its prey, which stays alive for a while. After the *bdellovibrio* has devoured the host bacterium from within, it produces up to 15 offspring. These break open the shell of the dead bacterium and set off in search of new victims.

There are various kinds of *bdellovibrio*, each of which targets a different species of bacteria as its prey. Since they never attack human cells, these predatory microbes may be used as 'living antibiotics' to fight pathogens in our bodies.

Why can't bubonic plague pathogens be exterminated like the smallpox virus?

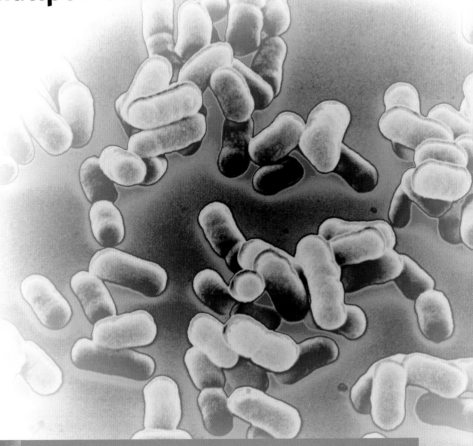

Bubonic plague is not restricted to the Dark Ages – thousands of new cases are reported each year in more than 30 countries. The plague bacteria have survived for so long because – unlike smallpox viruses – they are not restricted to infecting human beings, and can survive in animals.

Plague-infected rodents are found in remote forests, especially in Madagascar but also in the USA and elsewhere. These rodents represent a constant source of contagion, causing bubonic plague infections to recur regularly. Infection can occur by means of direct contact through a bite or scratch, or when infected rats come close to human settlements where the plague bacterium *Yersinia pestis* can be passed to humans though flea bites. To eradicate the pathogen it would be necessary to destroy all the potentially infected animals living in the wild – something that is not possible.

Infection by plague bacteria can be treated if antibiotics are prescribed at an early stage. Once bubonic plague has developed into septicaemia or pneumonic plague, the patient is usually beyond help.

In addition to the anthrax pathogen, Robert Koch discovered the pathogens responsible for tuberculosis and cholera.

Groundbreaking discovery?

How did Robert Koch prove that bacteria could cause certain diseases?

According to Robert Koch, four conditions have to be met in order to prove that a pathogen is the cause of a particular disease: 1. It must be found in all cases of the disease examined; 2. It must be possible to prepare and maintain it in a pure culture; 3. It must be capable of producing the original infection, even after several generations, in culture; 4. It must be possible to retrieve it from an inoculated animal and culture it again. In 1876, Robert Koch used the anthrax pathogen *Bacillus anthracis* to prove for the first time that a bacterium can cause a disease. However, Koch's conditions cannot be met when a pathogen requires other cells to reproduce in the laboratory, or when no susceptible animals are available for testing.

How many microbes live on and in our bodies?

Our bodies are made up of 100 trillion human cells (the number 1 followed by 14 zeros). There are around ten times as many microbes – predominantly bacteria – which have settled in our intestines, on our skin and in the mucous membranes of our mouth, throat, nose and sexual organs. These intestinal and skin microbes are normal and fulfil functions vital to our survival – such as stimulating the immune system, making vitamins and preventing pathogens from reproducing. Animals raised in germ-free environments are highly susceptible to infections and suffer from a range of disorders.

The microbial density is greatest in the large intestine, with 1g of intestinal content containing more than 100 billion bacteria. The number of microbes living on the skin depends on humidity and can vary between a few hundred and several hundred thousand bacteria per square centimetre. About ten billion bacteria live on the skin of a human being. In the human oral cavity, bacteria outnumber other single-cell organisms such as amoebas, flagellates and yeasts. Several billion bacteria of the mouth, most as yet unidentified, are present in the mucous membranes, on and between the teeth and in the periodontal pockets.

Does hand-washing kill off bacteria?

Washing your hands with soap and water is not enough to kill off bacteria – but most of them are removed with the dirt. Pathogens can find their way onto the hands as a result of going to the toilet, handling raw meat, eggs and other foodstuffs and through contact with animals or sick people. From the hands, the pathogens can get into the nose or mouth and then enter the body. Hand-washing reduces the risk of infection from bacteria, including those that cause diarrhoea. It is dangerous when bacteria from unwashed hands get into open wounds and cause infections. This is why doctors and nurses use disinfectants to kill the pathogens on their hands. Solutions that contain alcohol or other germicidal substances are used and have to be given at least 30 seconds to be effective.

A computer graphic highlights those areas on a hand (white and blue) where bacteria have settled.

What are probiotic bacteria?

Lactic acid bacteria and other intestinal microbes with health-promoting properties are called probiotics. As component parts of the intestinal flora (the microbes of the digestive tract), they suppress the growth of pathogens, regulate digestion and strengthen the body's defences by stimulating the immune system. Products containing concentrations of probiotic bacteria could, if taken daily and over an extended period of time, relieve diarrhoea and chronic intestinal inflammations. They are also helpful in restoring normal gastro-intestinal function where antibiotics have had an adverse effect. A prerequisite for achieving these positive effects is that as many of the bacteria as possible survive the journey through the stomach and its aggressive gastric acid. Often the effect is only temporary because the bacteria cannot settle permanently in the gut. There is also some controversy surrounding the use of probiotic yogurts that contain only small numbers of bacteria. It is not clear that they have any positive effect on intestinal function.

What causes body odour?

The body's sweat glands produce about one litre of liquid per day. Sweat is a water-based solution of salts, proteins, amino acids, urea and other substances mixed with fatty discharges from glands in the skin. Neither this film of moisture, nor the secretions of glands in the armpits and the pelvic area, have a strong smell of their own. Body odour is produced by bacteria on the skin that feed on these secretions. They turn fats into cheesy-smelling butyric acid, and fats and amino acids into fishy-smelling by-products of decomposition. An individual's personal odour is also affected by their hormonal balance, nutrition, state of health and moods.

Bad breath is usually the result of bacterial activity, where bacteria have multiplied excessively as a result of poor oral hygiene and are producing foul-smelling by-products derived from food fragments and dead cells.

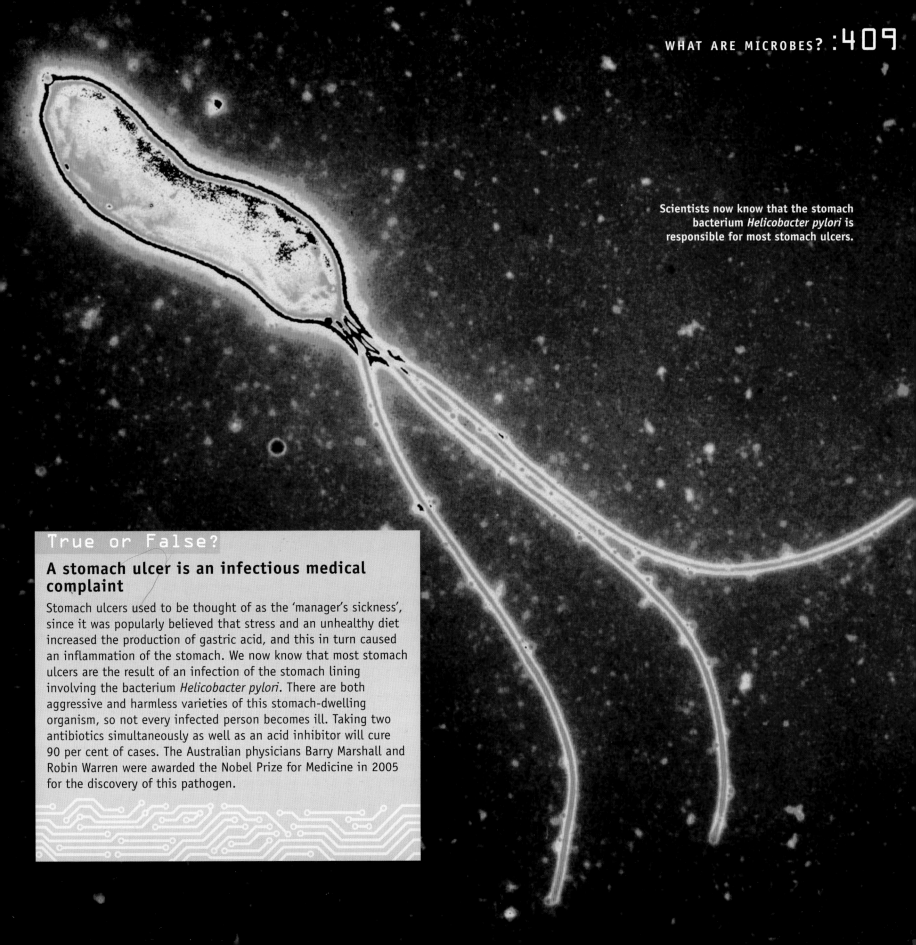

Scientists now know that the stomach bacterium *Helicobacter pylori* is responsible for most stomach ulcers.

True or False?

A stomach ulcer is an infectious medical complaint

Stomach ulcers used to be thought of as the 'manager's sickness', since it was popularly believed that stress and an unhealthy diet increased the production of gastric acid, and this in turn caused an inflammation of the stomach. We now know that most stomach ulcers are the result of an infection of the stomach lining involving the bacterium *Helicobacter pylori*. There are both aggressive and harmless varieties of this stomach-dwelling organism, so not every infected person becomes ill. Taking two antibiotics simultaneously as well as an acid inhibitor will cure 90 per cent of cases. The Australian physicians Barry Marshall and Robin Warren were awarded the Nobel Prize for Medicine in 2005 for the discovery of this pathogen.

Page numbers in **bold** print refer to main entries. Page numbers in *italics* refer to photographs and illustrations

Picture credits

t. = top; b. = bottom; c. = centre; l. = left; r. = right

Front cover (main) G. Bell/Wildlife, l picture-alliance/dpa, r Getty Images/ AFP; Back cover l NASA/ Carla Thomas, r Agentur Focus/ SPL/ David Luzzi; 1 www.istockphoto.com/ Irina Tischenko; 2-3 www.shutterstock.com/ Huebi; 4-5 Mauritius Images/ imagebroker.net, l Laif/ Kristensen, c Mauritius Images/ Steve Bloom, r EADS Astrium Satellites; 5 IFA-Bilderteam/ Jon Arnold Images; 6 l. Astrofoto/Detlev van Ravenswaay, c. mauritius images/Oxford Scientific, r. Bilderberg/Reinhard Dirscherl; 7 l. Wildlife/B. Cole, c. Ingo Wandmacher, r. Agentur Focus/SPL/Klaus Guldbrandsen; 8/9 Astrofoto/Detlev van Ravenswaay; 10 Robert Gendler; 11 W. M. Keck Observatory/Scott Kardel; 12 Astrofoto/Detlev van Ravenswaay; 13 t. picture alliance/kpa, b. NASA, ESA and Hubble Heritage Team (STScI); 14 Astrofoto/Detlev van Ravenswaay; 15 NASA/WMAP Science Team; 16/17 Wei-Hao Wang; 17 John Bahcall (Inst. for Advanced Study, Princeton), Mike Disney (Univ. of Wales) and NASA/ESA; 18 and 19 background Astrofoto/Shigemi Numazawa, 19 centre Astrofoto/Bernd Koch/Sven Kohle; 20 t. ullstein bild/ histopics, b. mauritius images/Jan Halaska; 21 l. mauritius images/ Photo Researchers, r. and 22 NASA/JPL-Caltech; 23 t. Astrofoto/ Gerald Rhemann, b. akg-images; 24 NASA, ESA and G. Bacon (STScI); 25 NASA/JPL-Caltech; 26 Astrofoto/AURA/STScI/NASA; 27 t. Astrofoto/ Carroll, b. Astrofoto/NASA; 28 t. mauritius images/IPS, b. picture alliance/dpa; 29 t. Agentur Focus/SPL/Lynette Cook, b. Kamioka Observatory, ICRR (Institute for Cosmic Ray Research), The University of Tokyo; 30 l. UCAR/NCAR/High Altitude Observatory, r. ESA; 31 Astrofoto/EIT/SOHO/NASA; 32 Astrofoto/Bernd Koch; 33 t. SOHO, b. Royal Swedish Academy of Sciences; 34 b. Bilderberg/Milan Horacek; 34/35 Astrofoto/Dorst; 35 Astrofoto/Bernd Koch; 36/37 and 37 Astrofoto/Detlev van Ravenswaay; 38 t. Astrofoto/NASA, b. NASA/ J. Bell (Cornell U.) and M. Wolff (SSI); 39 t. red.sign, b. picture alliance/kpa; 40 Astrofoto/Shigemi Numazawa; 41 Astrofoto/Bernd Koch; 42/43 ESA/DLR/FU Berlin (G. Neukum); 43 NASA/JPL; 44 mauritius images/Photo Researchers; 45Astrofoto/Keller/ Schmidbauer; 46/47 Astrofoto/Ralf Schoofs; 47 Astrofoto/NASA; 48 Astrofoto/Shigemi Numazawa; 49 NASA/JPL-Caltech; 50 NASA; 51 Astrofoto/Andreas Walker; 52 t. ullstein bild/kpa/HIP/Ann Ronan Picture, b. Astrofoto/Bernd Koch; 53 Astrofoto/NASA; 54 SOHO; 55 Agentur Focus/SPL/Mark Garlick; 56 picture alliance/ZB; 57 t. NASA/JPL-Caltech, b. ESA/AOES Medialab; 58 l. Astrofoto/Detlev van Ravenswaay, t. r. and 59 picture alliance/dpa; 60 t. ESA/P. Carril, b. NASA; 61 and 62 t. NASA, b. Astrofoto/Binnewies/Sporenberg; 63 ESA; 64 ESA/AOES Medialab; 65 t. NASA/JPL, b. NASA/JPL-Caltech; 66 picture alliance/dpa; 66/67 STS-114 Crew/ISS Expedition 11 Crew/NASA; 67 and 68 NASA; 69 ESA; 70 NHK/Japan; 71 t. Bilderberg/Klaus Bossemeyer, b. Bigelow Aerospace; 72 Liftport; 73 Astrofoto/Ralf Schoofs; 74/75 mauritius images/Oxford Scientific; 76 laif/Hoa-Qui; 76/77 mauritius images/Phototake; 77 l. IFA-Bilderteam/Reporters, r. Knut Schulz; 78 t. Agentur Focus/SPL/Dr. Gopal Murti, background photos.com, b. IFA-Bilderteam/IDS; 79 photos.com; 80 l. mauritius images/age, r. IFA-Bilderteam/PLC; 81 t. mauritius images/imagebroker.net, b. IFA-Bilderteam/FUFY; 82 mauritius images/Nonstock; 83 mauritius images/Steve Bloom Images; 84 Agentur Focus/SPL/Lauren Shear; 85 l. Theo TOP Krath, r. Bilderberg/Jean-Philippe Soule; 86 mauritius images/Photo Researchers; 87 Superbild/B.S.I.P.; 88 l. f1online/Widmann, r. and 88/89 Jupiterimages/Nonstock; 89 Agentur Focus/SPL/Arthur Toga; 90 t. Agentur Focus/SPL/Bo Veisland, b. mauritius images/Sarah Johanna Eick; 91 mauritius images/Doc Max; 92 mauritius images/ age; 93 t. mauritius images/Foodpix, b. mauritius images/Stock Image; 94 laif/Herzau; 95 mauritius images/Stock Image; 96 Agentur Focus/Eye of Science; 97 t. laif/Kristensen, b. mauritius images/ Stock Image; 98 mauritius images/Photononstop; 98/99 IFA-Bilderteam/DIAF/SDP; 99 laif/Eisermann; 100 Agentur Focus/SPL/P. M. Motta & S. Correr; 101 Agentur Focus/SPL; 102 laif/Aurora; 103 t. mauritius images/John Warburton-Lee, b. laif/Frommann; 104 mauritius images/Jo Kirchherr; 105 t. mauritius images/ Phototake, b. mauritius images/Schultze + Schultze; 106 mauritius images/Sabine Stallmann; 107 t. Agentur Focus/SPL/AJ Photo, b. Agentur Focus/SPL/Colin Cuthbert; 108 t. IFA-Bilderteam/ Photolibrary UK, b. mauritius images/Stock Image; 109 picture alliance/ZB; 110 t. Agentur Focus/SPL/Mauro Fermariello, b. picture alliance/dpa; 111 Studio X/Gamma/Franck Crusiaux; 112 laif/REA; 113 laif/Meyer; 114 t. Bilderberg/Nomi Baumgartl, b. Agentur Focus/SPL/John Bavosi; 115 IFA-Bilderteam/Int. Stock; 116 mauritius images/Pedro Perez; 117 t. mauritius images/Stock Image, b. picture alliance/dpa; 118 t. Bilderberg/Tania Reinicke, b. laif/Madame Figaro; 119 t. mauritius images/Phototake, b. picture alliance/IMAGNO; 120 and 121 picture alliance/dpa; 122 Wildlife/ P. Ryan; 123 Interfoto; 124 mauritius images/Dieter Woog; 125 t. istock, b. laif/Hemispheres; 126 laif/Westrich; 127 mauritius images/age; 128 t. mauritius images/age, b. Bridgeman Giraudon; 129 t. picture alliance/dpa, b. Interfoto; 130 TV-yesterday; 131 Agentur Focus/SPL/John Bavosi; 132 top laif/Emmler, b. laif/ Hemispheres; 133 mauritius images/Hans-Peter Merten; 134 STOCK4B; 135 laif/Hahn; 136 t. mauritius images/Haag + Kropp, b. laif/Gamma; 137 l. Das Fotoarchiv/Xinhua, r. picture alliance/dpa; 138 Agentur Focus/SPL/Friedrich Saurer; 138/139 Agentur Focus/SPL/Susumu Nishinaga; 139 Das Fotoarchiv/Xinhua; 140/141 Bilderberg/Reinhard Dirscherl; 142 IFA-Bilderteam/Kerri; 143 t. Wildlife/D. Harms, b. Wildlife/TOP Diez; 144 t. mauritius images/Frank Lukasseck, b. IFA-Bilderteam/Giel; 145 Bilderberg/Jürgen Freund & Stella Chiu-Freund; 146 l. photos.com, r. mauritius images/Botanica; 147 picture alliance/Bildagentur Huber/G. Simeone; 148 t. f1online/Horizon, b. Bayer AG/Ginter; 149 t. agenda/Jörg Böthling, b. Bayer AG/Ginter; 150 t. Agentur Focus/SPL/Andrea Balogh, b. mauritius images/ imagebroker.net; 151 t. mauritius images/age, b. picture alliance/ Klett GmbH; 152 mauritius images/Phototake; 153 tl. Bilderberg/ Jürgen Freund & Stella Chiu-Freund, t. r. photos.com; b. Agentur Focus/SPL/Andrea Balogh; 154 t. photos.com, b. mauritius images/Westend61; 155 IFA-Bilderteam/Siebert; 156 IFA-Bilderteam/ Birgit Koch, red.sign; 157 t. picture alliance/Bildagentur Huber, b. mauritius images/IPS; 158 l. mauritius images/Heinz Zak, r. Bilderberg/Peter Mathis; 159 t. IFA-Bilderteam/Willemeit, b. Agentur Focus/SPL/Eye of Science; 160 mauritius images/ Workbookstock; 161 t. IFA-Bilderteam/Jacobi, b. picture alliance/ dpa; 162 mauritius images/Westend61; 163 l. mauritius images/Fritz Rauschenbach, r. mauritius images/Nigel Dennis; 164 l. mauritius images/age, r. mauritius images/Steve Bloom; 165 IFA-Bilderteam/ Schlosser; 166 Bilderberg/Rainer Drexel; 166/167 mauritius images/John Warburton-Lee; 168 t. Wildlife/K. Bogon, b. IFA-Bilderteam/BCI; 169 mauritius images/Steve Bloom Images; 170 mauritius images/Reinhard Dirscherl; 170/171 picture alliance/ dpa; 171 mauritius images/Reinhard Dirscherl; 172 picture alliance/ dpa; 173 l. Bilderberg/Ulf Boettcher, r. IFA-Bilderteam/BCI; 174 IFA-Bilderteam/Bloch/Jung; 175 picture alliance/dpa; 176 mauritius images/dieKleinert; 177 t. picture alliance/dpa, b. mauritius images/Phototake; 178 Bilderberg/Reinhard Dirscherl; 179 l. mauritius images/Oxford Scientific, r. Interfoto; 180 l. mauritius images/Itamar Grinberg, r. Agentur Focus/SPL/Matthew Oldfield; 181–183 Bilderberg/Reinhard Dirscherl; 184 t. Bilderberg/ Allstills53N, b. mauritius images/Oxford Scientific; 185 l. IFA-Bilderteam/BCI, r. Arco Images/NPL; 186 mauritius images/Tsuneo Nakamura; 187 t. Bilderberg/Reinhard Dirscherl, b. laif/Gernot Huber; 188 mauritius images/Oxford Scientific; 188/189 picture alliance/ Bildagentur Huber; 190 mauritius images/age; 191 l. mauritius images/Oxford Scientific, r. Fotex/Ernst and Gerken; 192 mauritius images/Fritz Rauschenbach; 193 IFA-Bilderteam/R. Maier; 194 t. photos.com, b. and 195 mauritius images/imagebroker.net; 196 t. laif/ Arcticphoto, b. mauritius images/Norbert Fischer; 197 t. akg-images, b. blickwinkel/fotototo; 198 and 199 mauritius images /Steve Bloom; 200 mauritius images/Fritz Rauschenbach; 201 t. mauritius images/ Oxford Scientific, l. b. photos.com, r. b. Agentur Focus/Eye of Science; 202 l. blickwinkel/J. Hauke, tr. mauritius images/ dieKleinert, br. red.sign; 203 IFA Bilderteam/Giel; 204 l. mauritius images/Günter Rossenbach, r. picture alliance/Bildagentur Huber; 205 IFA-Bilderteam/Siebig; 206 laif/Reporters; 207 Interfoto; 208/209 Wildlife/B. Cole; 210/211 Astrofoto/Ralf Schoofs; 211 t. picture alliance/dpa, b. mauritius images/photolibrary; 212 mauritius images/Ludwig Mallaun; 213 t. IFA-Bilderteam/Walsh, b. picture alliance/kpa; 214 picture alliance/dpa, b. Bilderberg/Frieder Blickle; 215 Bilderberg/Jörg Heimann; 216 t. Aibo & Göbel, bl. Agentur Focus/SPL/Pekka Parviainen, br. Agentur Focus/SPL/George Post; 216/217 background Agentur Focus/SPL/John Howard; 217 tl. Agentur Focus/SPL/Colin Cuthbert, tc. Aibo & Göbel, tr. istock/ Bonnie Schupp, cr. Aibo & Göbel, cl. blickwinkel/Natur im Bild/R. Foerster, b. Agentur Focus/SPL/Pekka Parviainen; 218 t. photos.com, b. mauritius images/Ludwig Mallaun; 219 t. mauritius images/Photo Researchers, b. mauritius images/Michael Obert; 220 t. Bilderberg/ Keystone-CH, b. Visum/Marc Steinmetz; 221 mauritius images/age; 222 t. mauritius images/Josef Beck, b. EADS Astrium; 223 picture alliance/dpa; 224/225 laif/Gamma; 226 mauritius images/ imagebroker.net; 227 l. NASA/University of Miami/B. Evans b. P. Minnett, r. Agentur Focus/SPL/Gary Hincks; 228 and229 picture alliance/dpa; 230 mauritius images/ANP Photo;231 t. picture alliance/ZB, b. f1online/Horizon; 232 t. mauritius images/ dieKleinert, b. mauritius images/imagebroker.net; 233 picture alliance/Klett GmbH; 234 t. picture alliance/dpa, b. mauritius images/age; 235 background mauritius images/Oxford Scientific, M. ullstein bild/Granger Collection; 236 Agentur Focus/SPL/Gary Hincks; 237 IFA-Bilderteam/Int. Stock; 238 mauritius images/Bernd Wenske; 239 t. Interfoto, b. Agentur Focus/SPL/NASA; 240 l. picture alliance/dpa, r. Agentur Focus/SPL/Karsten Schneider; 241 mauritius images/age; 242 and 243 picture alliance/dpa; 244 mauritius images/Pacific Stock; 244/245 mauritius images/AVA; 245 laif/Heeb; 246/247 picture alliance/Bildagentur Huber; 247 Bilderberg/Thomas Ernsting; 248 picture alliance/Bildagentur Huber; 249 picture alliance/Godong; 250 t. Agentur Focus/SPL/Jesse Allen, b. picture alliance/dpa; 251 IFA-Bilderteam/Jon Arnold Images; 252 l. picture alliance/dpa, r. picture alliance/Bildagentur Huber; 253 picture alliance/dpa; 254 Agentur Focus/SPL/Gary Hincks; 255 l. Geo-Zentrum an der KTB, r. Agentur Focus/SPL/Gary Hincks; 256 t. Agentur Focus/SPL/Christian Darkin, b. Agentur Focus/SPL/Gary Hincks; 257 mauritius images/ Jose Fuste Raga; 258 t. mauritius images/dieKleinert, b. laif/Reporters; 259 Agentur Focus/SPL/Roger Harris; 260 mauritius images/imagebroker.net; 261 t. IFA-Bilderteam/ AP&F, b. picture alliance/dpa; 262 t. picture alliance/dpa, b. mauritius images/ SuperStock; 263 t. Agentur Focus/SPL, b. Agentur Focus/SPL/Gary Hincks; 264 Bilderberg/Frieder Blickle; 264/265 IFA-Bildertem/ Wisniewski; 265 mauritius images/Rene Mattes; 266 t. laif/Heeb, b. red.sign; 267 t. Agentur Focus/SPL/ Martin Bond, b. mauritius images/Photononstop; 268 laif/Meissner; 268/269 Agentur Focus/SPL/Gary Hincks; 269 picture alliance/dpa; 270 t. Okapia, b. blickwinkel/Hecker/Sauer; 271 Wildlife/G. Bell; 272 Agentur Focus/ SPL/B. Murton; 273 t. ullstein bild, c. Interfoto, b. mauritius images/dieKleinert; 274 Das Fotoarchiv/Thomas Mayer; 275 blickwinkel/TOP Broders; 276/277 Ingo Wandmacher, 278 mauritius images/Foodpix; 279 mauritius images/Josh Westrich; 280 StockFood/Studio Bonisolli; 281 t. mauritius images/Boris Kumicak, b. mauritius images; 282 t. mauritius images/Jörn Rynio, b. Wildlife/P. Hartmann; 283 mauritius images/Foodpix; 284 mauritius images/ imagebroker.net; 285 t. laif/Kürschner, b. laif/Volz; 286 Visum/A. Vossberg; 287 mauritius images/Arthur Cupak; 288/289 Bilderberg/ Frieder Blickle; 289 l. mauritius images/Rosenfeld, r. Agentur Focus/SPL; 290 t. mauritius images/Sean Miller, b. mauritius images/Peter Rathmann; 291 mauritius images/Andrea Marka; 292 mauritius images/photolibrary; 293 l. mauritius images/Rosenfeld, r. mauritius images/Stephanie Böhlhoff; 294 background mauritius images/Haag + Kropp, c. mauritius images/Nonstock; 295 t. picture alliance/dpa, b. picture alliance/Bildagentur Huber; 296 IFA-Bilderteam/Photolibrary UK; 297 t. mauritius images/Manfred Habel, b. IFA-ilderteam/Reporters; 298 mauritius images/pepperprint; 299 picture alliance/dpa; 300 l. Agentur Focus/SPL/Andrew Syred,

Our World Transformed was published by The Reader's Digest Association Limited, London

First edition copyright © 2008 The Reader's Digest Association Limited, 11 Westferry Circus, Canary Wharf, London E14 4HE

This book was first published as *1000 Fragen an die Wissenschaft* in 2007 by Reader's Digest, Germany

We are committed both to the quality of our products and the service we provide to our customers. We value your comments, so please do contact us on 08705 113366 or via our website at www.readersdigest.co.uk

If you have any comments or suggestions about the content of our books, email us at gbeditorial@readersdigest.co.uk

Note to readers
While the creators of this work have made every effort to be as accurate and as up to date as possible, scientific and medical knowledge is constantly changing. The writers, researchers, editors and the publishers of this book cannot be held liable for any injuries, damage or loss that may result from using the information contained within it.

CONTRIBUTORS
Dr Robert Coenraads, Dr Joachim Czichos, Cornelia Dick-Pfaff, Dr Rainer Kayser, Dr Kathy Kramer, Jan Oliver Löfken, Karen McGhee, Dr Doris Marszk, Dr John O'Byrne, Dörte Saße, Richard Whitaker, Marita Willemsen

FOR READER'S DIGEST
Project Editor Lisa Thomas
Art Editor Julie Bennett
Copy Editor Diane Cross
Designer Martin Bennett
Translator Karen Waloschek
Indexer Diane Harriman

READER'S DIGEST GENERAL BOOKS
Editorial Director Julian Browne
Art Director Anne-Marie Bulat
Managing Editor Nina Hathway
Head of Book Development Sarah Bloxham
Picture Research Manager Sarah Stewart Richardson
Pre-press Account Manager Dean Russell
Product Production Manager Claudette Bramble
Senior Production Controller Katherine Bunn

Concept code US 4374/G
Book code 400-342 UP00001
ISBN 978 0 276 44385 5
Oracle code 250001571H.00.24